NETWORKS IN CLIMATE

Over the last two decades the complex network paradigm has proven to be a fruitful tool for the investigation of complex systems in many areas of science, for example, the internet, neural networks and social networks. This book provides an overview of applications of network theory to climate variability phenomena, such as the El Niño–Southern Oscillation and the Indian Monsoon, presenting recent important results obtained with these techniques and showing their potential for further development and research. The book is aimed at researchers and graduate students in climate science. A basic background in physics and mathematics is required. Several of the methodologies presented here will also be valuable to a broader audience of those interested in network science, for example, from biomedicine, ecology and economics.

HENK A. DIJKSTRA is a professor of dynamical oceanography at the Institute for Marine and Atmospheric research Utrecht within the Department of Physics and director of the Centre for Complex Systems Studies, both at Utrecht University in the Netherlands. His research focuses mainly on the stability of the ocean circulation and on the physics of climate variability. He is author of the books *Nonlinear Physical Oceanography* (2005), *Dynamical Oceanography* (2008), and *Nonlinear Climate Dynamics* (Cambridge University Press, 2013). He is a member of the Netherlands Royal Academy of Arts and Sciences (KNAW). He was awarded the Lewis Fry Richardson Medal from the European Geosciences Union in 2005, and he was elected a Fellow of the Society for Industrial and Applied Mathematics (SIAM) in 2009.

EMILIO HERNÁNDEZ-GARCÍA is a research professor at the Institute for Cross-Disciplinary Physics and Complex Systems (IFISC), a joint research center of the Spanish Higher Research Council (CSIC) and the University of the Balearic Islands (UIB) in Mallorca, Spain. His research in complex systems, statistical physics and nonlinear dynamics includes ocean Lagrangian transport and network techniques in biology and the geosciences. He co-authored the book *Chemical and Biological Processes in Fluid Flows* (2009).

CRISTINA MASOLLER is a professor in the Physics Department of the Polytechnic University of Catalonia, Spain. Her research interests are interdisciplinary and cover a wide range of topics including neurons, lasers, complex networks, climate and biosignals. Her main research focus is on the development of novel data analysis tools for the study of complex systems (symbolic analysis, complex networks). Specific interests include novel methods for the analysis of climatological data (climate networks) and complexity measures for the classification and characterisation of complex images. In both 2009 and 2015 she received the ICREA Academia Award from the Catalan Institution for Research and

Advanced Studies (ICREA). In 2015 she was elected a Fellow of the Optical Society (OSA).

MARCELO BARREIRO is a professor of climate dynamics at the Institute of Physics of the University of the Republic, Uruguay. His research focuses on large-scale ocean-atmosphere interactions and climate predictability on seasonal to decadal time scales. He received the Edward Lorenz Award from the International Center for Theoretical Physics, Italy in 2009, where he is now an associate.

NETWORKS IN CLIMATE

HENK A. DIJKSTRA
University of Utrecht, the Netherlands

EMILIO HERNÁNDEZ-GARCÍA
Spanish Higher Research Council and University of the Balearic Islands, Spain

CRISTINA MASOLLER
Polytechnic University of Catalonia, Spain

MARCELO BARREIRO
University of the Republic, Uruguay

CAMBRIDGE
UNIVERSITY PRESS

CAMBRIDGE
UNIVERSITY PRESS

University Printing House, Cambridge CB2 8BS, United Kingdom

One Liberty Plaza, 20th Floor, New York, NY 10006, USA

477 Williamstown Road, Port Melbourne, VIC 3207, Australia

314–321, 3rd Floor, Plot 3, Splendor Forum, Jasola District Centre, New Delhi – 110025, India

79 Anson Road, #06–04/06, Singapore 079906

Cambridge University Press is part of the University of Cambridge.

It furthers the University's mission by disseminating knowledge in the pursuit of education, learning, and research at the highest international levels of excellence.

www.cambridge.org
Information on this title: www.cambridge.org/9781107111233
DOI: 10.1017/9781316275757

First published 2019

Printed in the United Kingdom by TJ International Ltd. Padstow Cornwall

A catalogue record for this publication is available from the British Library.

Library of Congress Cataloging-in-Publication Data
Names: Dijkstra, Henk A., author. | Hernandez-Garcia, Emilio, 1963– author. | Masoller, C. (Cristina), author. | Barreiro, Marcelo, 1971– author.
Title: Networks in climate / Henk A. Dijkstra (University of Utrecht, The Netherlands), Emilio Hernandez-Garcia (Universitat de les Illes Balears, Spain), Cristina Masoller (Universitat Politcnica de Catalunya, Spain), and Marcelo Barreiro (Universidad de la Republica, Montevideo, Uruguay).
Description: Cambridge ; New York, NY : Cambridge University Press, 2019. | Includes bibliographical references and index.
Identifiers: LCCN 2018039209 | ISBN 9781107111233 (hardback : alk. paper)
Subjects: LCSH: Climatology. | Climatology–Social aspects. | Atmospheric physics. | Weather forecasting. | Climatic changes.
Classification: LCC QC861.3 .N48 2019 | DDC 551.63/3–dc23 LC record available at https://lccn.loc.gov/2018039209

ISBN 978-1-107-11123-3 Hardback

Contents

Preface

Over the last two decades, the complex network paradigm has proven to be a fruitful tool for the investigation of complex systems in various areas of science, e.g., the internet, neural networks, and social networks. The application of complex network theory to climate science was the main focus of the LINC project (Learning about Interacting Networks in Climate)[1] funded by the European Commission under the FP7 program (FP7-289447). The LINC project was a *Marie-Curie Initial Training Network* (ITN) aimed at improving our understanding of the Earth's complex climate phenomena, such as the El Niño–Southern Oscillation (ENSO), by approaching the problem from a complex systems interdisciplinary perspective, bringing together experts from different fields such as physics, dynamical systems theory, computer science, and earth sciences.

In the network approach to the Earth's climate, the vertices of the network are identified with the spatial grid points of an underlying global climate data set. Edges are added between pairs of vertices depending on the degree of statistical interdependence between the corresponding pairs of time series taken from the climate data sets (pressure, temperatures, precipitations, etc.).

A crucial step for understanding the characteristic behavior of such systems consists in inferring the connection topology. When studying the connectivity structure, a main challenge is due to multiple spatial scales and temporal scales in the climate system. A basic tool for identifying connectivity is causality. There are various techniques to quantify causality, ranging from classic cross-correlation, to information theory measures, such as coarse-grained entropy or transfer entropy, to Granger causality, to recurrence-based measures. However, there is a need to develop appropriate statistical tests for several of these measures. For climate network identification, the Lagrangian description of air and water transport among

[1] https://cordis.europa.eu/result/rcn/187859_en.html and https://cordis.europa.eu/result/rcn/156590_en.html

Earth locations from observations or from the output of general circulation models (GCMs) can also be used.

This book summarizes the main results of the LINC project, which addressed these challenges from an interdisciplinary perspective. It begins with an introduction to the most relevant climate phenomena and models (Chapters 1 and 2), followed by a review of methods of data analysis (Chapter 3) and network construction (Chapter 4). Then, applications of climate networks to various climatological fields (pressure, temperatures, wind velocities, etc.) are presented. As the analysis requires the management of large data sets, high-performance computer algorithms need to be used for storing, processing, and visualizing the data. An overview of such tools is presented in Chapter 5. Then, Chapters 6 and 7 present new insights on atmospheric variability and ocean dynamics, which were gained through the use of the network approach.

As is well known, one of the most striking characteristics of the climate system, and probably the one with the greatest potential impact on humanity, is that it can operate in a variety of distinct regimes, with the possibility of sharp transitions among them. Methodologies to detect signatures of abrupt change in past climate and, most importantly, to identify warning signals of the closeness of future tipping points, are discussed in Chapter 8. Climate predictability is another challenge that has a huge economic and social impact for present and future generations, and can underpin advances in areas as diverse as energy, environment, agriculture, and marine sciences. The final chapter 9 is devoted to an overview of the progress made on this topic by using the complex network approach.

While this book is aimed at the general reader interested in climate science with a basic background in physics and math, several of the methodologies presented here will be of interest to a broader audience. For example, in the economy, banks, insurance companies, and firms interact and can be represented by interdependent networks. As another example, the network diagnostic tools developed to detect signatures of "climate shifts" can find applications for the analysis of complex biomedical signals, or in ecology.

Acknowledgments

Writing this book has been a challenge, and also a pleasure due to the interaction with many present and former students, friends, and colleagues. In particular, we are indebted to all the fellows of the LINC project (Fernando Arizmendi, Miguel Bermejo, Juan Ignacio Deza, Hisham Ihshaish, Qingyi Feng, Marc Segond, Enrico Ser Giacomi, Veronica Martin-Gomez, Victor Rodriguez, Veronika Stolbova, Alexis Tantet, Giulio Tirabassi, Liubov Tupikina, Ruggero Vasile, Yang Wang, and Dong Zhou) who put in a lot of energy, enthusiasm, and dedication during the four years of the project. Of course, the success of their work would not have been possible without the active support of the senior researchers that participated and contributed to the LINC research (Markus Abel, Karsten Ahnert, Johan Dijkzeul, Avi Gozolchiani, Shlomo Havlin, Jurgen Kürths, Cristobal Lopez, Norbert Marwan, and Manfred Mudelsee). We in particular thank Veronika Stolbova and René van Westen for providing useful comments on the near-final manuscript and also René for making the CESM figures in Chapter 1. We also sincerely thank Kendra Ng for her work during the early stages of the writing of this book.

Matt Lloyd, Zoë Pruce, and Mariela Valdez-Cordero at Cambridge University Press have been very supportive and patient in waiting for the final version of the manuscript. We thank Gary Smith for the superb copy-editing of the manuscript.

All authors acknowledge financial support from the European Commission (ITN LINC, FP7-289447). C.M. also acknowledges support from the Spanish MINECO (FIS2015-66503-C3-2-P) and the program ICREA ACADEMIA of Generalitat de Catalunya. E. H.-G. acknowledges support from AEI and FEDER (CTM2015-66407-P and FIS2015-63628-C2-1-R). H.D. acknowledges support from the Netherlands Earth System Science Centre (NESSC), financially supported by the Dutch Ministry of Education, Culture and Science.

1

The Climate System

In this chapter we will introduce the concept of the climate system. Section 1.1 starts by discussing the different components of this system and their interactions. Next, in Section 1.2., the forcing of the climate system is presented. Climate models form the main topic of Section 1.3 and, using results from a simulation of a state-of-the-art climate model, the mean atmospheric and oceanic general circulation is described in Section 1.4.

1.1 System Components

Although individual scientists' views on the climate system probably greatly differ, anyone would admit that it is a system displaying complicated spatial-temporal variability in many of its subsystems, such as the atmosphere, the oceans, the cryosphere, and the biosphere.

In a report to the NASA Advisory Council, Bretherton (1988) presented a sketch of the Earth system components and their interactions. The original figure (Bretherton, 1988) is often referred to as the *horrendogram* of the climate system. The simplification shown in Fig. 1.1 is useful for recognizing many of the subcomponents of the climate system and identifying their connections. The figure also provides a basis for understanding the transfer of properties (such as energy and mass) that are exchanged between these different subsystems. Examples of such interactions and associated fluxes are usually referred to as the energy cycle, the hydrological cycle, and several biogeochemical cycles (for example, the carbon, sulfur, and nitrogen cycles).

Multi-scale interactions between processes within and between the different components create the complicated behavior of observables (such as temperature) in the climate system. Large-scale motions induce smaller-scale ones through

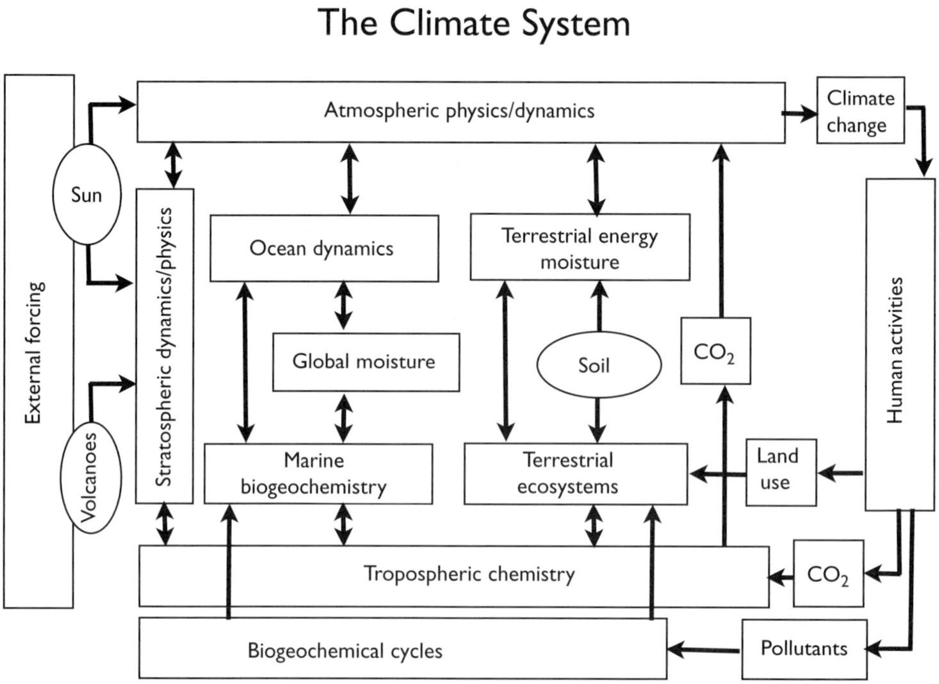

Figure 1.1 A schematic of the climate system showing the different components and their connections (simplified from Bretherton, 1988). The arrows provide an impression of the interactions between the components (one-way or two-way).

instabilities in a hierarchical cascade. In turn, the collective interaction of the small-scale processes affects the development of the large-scale motions.

Cloud formation is one of those processes in which these features are clearly evident. A cloud is formed through the aggregation of water, by means of atmospheric aerosol particles that act as initiation centers. The dimensions of these particles are usually smaller than those of a typical grain of dust. The type and extent of the resulting clouds will depend on the nature, concentration, and distribution of such agents, together with the thermodynamical state of the external environment. This mesoscopic characteristic will then impact, for example through rain, the longwave absorption and shortwave reflection of radiation, and subsequently affect the large-scale atmospheric motions.

Many feedback processes exist within and between the various components of the climate system. An example is sea-ice albedo feedback, in which a reduction in sea-ice extent will lead to less reflection of shortwave radiation and hence to a further reduction in the sea-ice extent. The complex feedback web that is created in this way, together with the forcing of the climate system (Section 1.2) gives rise to complicated climate variability and makes the understanding of phenomena in the climate system and their prediction of future development a difficult task.

1.2 Forcing

There are several external forcing mechanisms of the climate system, the most important being the radiation from the Sun. Insolation is the amount of energy received by the Earth's surface in the form of shortwave radiation. The energy difference between the incoming solar radiation and the outgoing longwave radiation at the top of the atmosphere is the main source of energy in the climate system.

The Earth can be seen as a closed system to which heat is added, so that the first law of thermodynamics applies, i.e.,

$$dQ = dE - dW, \tag{1.1}$$

where dQ is the amount of heat added, dE is the change in internal energy of the system, and dW is the work extracted from the climate system (Hartmann, 1994). Since the work done by the Earth on its environment is negligible and the incoming energy from the Sun and the outgoing energy from the Earth are mainly radiative, the dominant energy balance of the Earth is a radiative balance. If the Earth is in thermodynamic equilibrium (i.e., its internal energy is constant), the radiative flux received by the Earth from the Sun must be balanced by the outgoing radiation emitted by the Earth.

The solar radiation flux incident upon the Earth at the top of the atmosphere, often referred to as the solar constant S, has a mean value of 1370 Wm^{-2}. It is a function of the luminosity of the Sun and is inversely proportional to the square of the Sun to Earth distance (James, 1994). However, the solar constant is not the flux received by the surface of the Earth. Indeed, the Earth's area normal to the Sun's beam is πr_0^2 while its total surface is $4\pi r_0^2$, where r_0 is the radius of the Earth. Furthermore, a fraction α of the incoming solar radiation is directly reflected to space. The quantity α is called the planetary albedo and has an average value of about 0.3; the local albedo varies strongly locally due to the presence of clouds, snow, and ocean water. Thus, a total incoming flux of

$$R_{in} = \frac{(1-\alpha)S}{4} \tag{1.2}$$

is received by the Earth's surface and must be balanced by outgoing radiation for the Earth to be in equilibrium.

The Stefan–Boltzmann law states that the radiation emitted by a black body in thermodynamic equilibrium (R_{out}) is uniquely related to its surface temperature T through

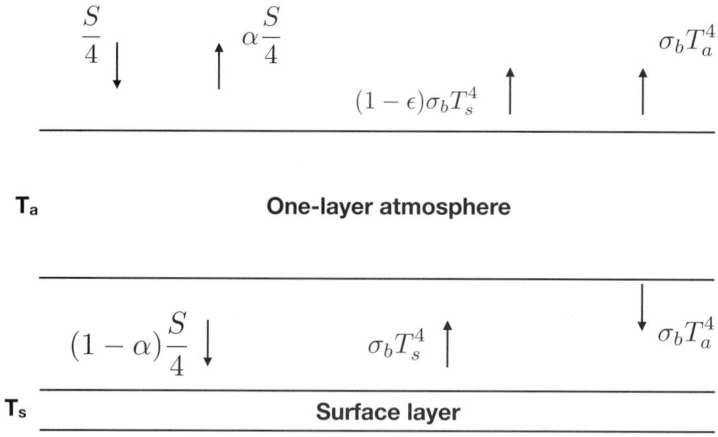

Figure 1.2 Sketch of an idealized model of the atmosphere to illustrate the greenhouse effect (after James, 1994).

$$R_{out} = \sigma_b T^4 \, , \tag{1.3}$$

where $\sigma_b = 5.67 \times 10^{-8}$ Wm^{-2} K^{-4} is the Stefan–Boltzmann constant. From the balance $R_{in} = R_{out}$, the equilibrium surface temperature T_e of the Earth is found from (1.2) and (1.3) as

$$T_e = \left(\frac{(1-\alpha)S}{4\sigma_b} \right)^{\frac{1}{4}} , \tag{1.4}$$

giving a value of $T_e = 255$ K for the Earth substituting the values for α, σ_b, and S given above (Hartmann, 1994).

The observed average temperature at the surface of the Earth is close to 288 K, which is considerably warmer than T_e. This temperature difference is explained by the Earth's atmosphere being relatively transparent to the incoming solar radiation but acting as a black body for outgoing terrestrial radiation (Hartmann, 1994). This mechanism is called the "greenhouse effect" and can be illustrated with a simple slab model of the atmosphere (Fig. 1.2).

The atmosphere is assumed completely transparent to incoming solar radiation R_{in} but to absorb a fraction ϵ of the infrared radiation R_{out} emitted by the Earth's surface (Fig. 1.2) and emit this in all directions. The radiative balance at the surface thus gives

$$R_{in} = R_{out} - R_a, \tag{1.5}$$

with $R_{out} = \sigma_b T_s^4$ and $R_a = \sigma_b T_a^4$ where T_s and T_a are the temperatures of the surface and of the atmosphere, respectively. Similarly, the radiative balance in the atmospheric layer is

$$\epsilon R_{out} = 2R_a. \tag{1.6}$$

From (1.4)–(1.6), we find

$$T_s = \left(\frac{2}{2-\epsilon}\right)^{\frac{1}{4}} T_e, \tag{1.7a}$$

$$T_a = \left(\frac{\epsilon}{2-\epsilon}\right)^{\frac{1}{4}} T_e. \tag{1.7b}$$

We can see that for an atmosphere totally opaque to infrared radiation, $\epsilon \to 1$, the temperature of the atmosphere T_a is the emission temperature T_e and the surface temperature is warmer than the overlying atmosphere. For a typical value of $\epsilon = 0.771$, the temperature $T_s = 288$ K, close to what is observed, while the temperature of the atmosphere is colder, $T_a = 227$ K. It is worthwhile noting that the value of ϵ is controlled by the concentrations of greenhouse gases in the atmosphere which absorb radiation in the infrared band. The main contributors are carbon dioxide, methane, and water vapor, providing about 80% of the current greenhouse effect (Hartmann, 1994).

Radiative balance models explain to first order the global annual mean temperature of the Earth. However, the incoming solar radiation is not uniformly distributed over the planet as it is large in the tropics and small in polar regions. At the same time, outgoing longwave radiation is relatively uniform with latitude. This leads to an energy imbalance in which the tropics are continuously gaining heat while the polar regions lose heat, hence creating a meridional temperature gradient. This gradient drives the atmospheric and oceanic circulations that redistribute energy on the planet and are important for understanding the observed climate variability.

1.3 Climate Models

Observations are crucial to the study of climate variability, but the records are very limited (see Chapter 3). As we also cannot investigate climate variability phenomena in the laboratory, climate models are a central tool in climate research. A wide range of models is in use, from conceptual climate models to state-of-the-art global climate models (GCMs). It would be impossible (and rather useless) to try to provide an overview of all the models that are being used at the moment in the climate research community.

Scales and processes are important properties of climate variability, and this motivates a classification of climate models using these two traits (Fig. 1.3). Here the trait "scales" refers to both spatial and temporal scales as there exists a relation between both: on smaller spatial scales, usually faster processes take place. "Processes" refer to either physical, chemical, or biological processes taking place in

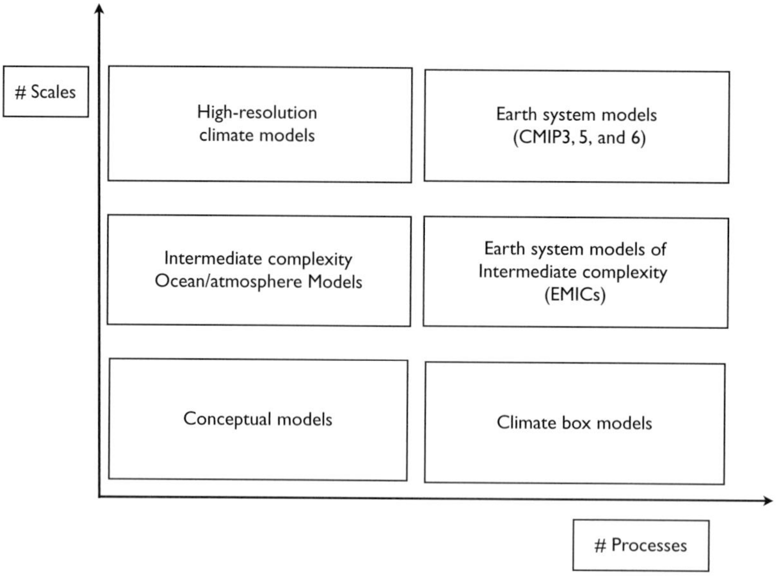

Figure 1.3 Classification of climate models according to the two model traits number of processes and number of scales. There is of course overlap between the different model types, but for simplicity they are sketched here as nonoverlapping.

the different climate subsystems (atmosphere, ocean, cryosphere, biosphere, lithosphere).

Models with a limited number of processes and scales are usually referred to as conceptual climate models. In these models only very specific interactions in the climate system are described. An example is the models of glacial–interglacial cycles (Saltzmann, 2001) formulated by small-dimensional systems of ordinary differential equations. For example, only the interactions of the ice sheet volume, atmospheric CO_2 concentration, and global mean ocean temperature are included.

Limiting the number of processes, scales can be added by discretizing the governing partial differential equations spatially up to three dimensions. A higher spatial resolution and inclusion of more processes will give models located in the right upper part of the diagram Fig. 1.3. In a GCM (Fig. 1.4), we divide the atmosphere, ocean, ice, and land components into grid boxes. Over such a grid box we consider the budgets of momentum, mass, and, for example, heat. The difference between what goes into a box minus what comes out of that box (e.g., heat) leads to an increase/decrease of a particular quantity (e.g., temperature). Once the distribution of a quantity is known at a certain time, then these budgets provide an evolution equation to determine the quantity some time later.

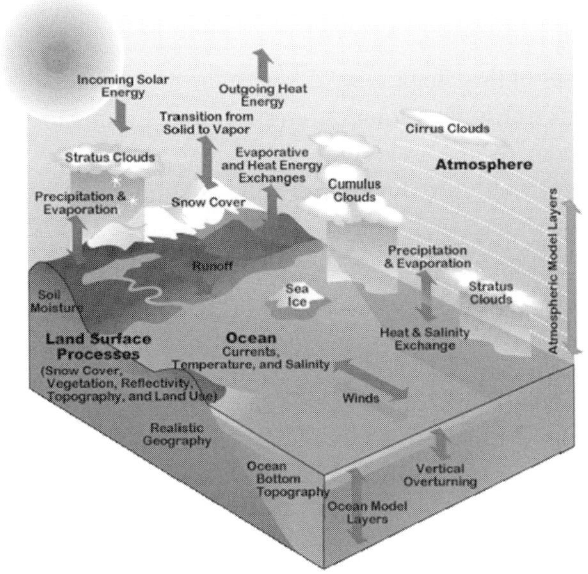

Figure 1.4 An overview of the components of the Community Earth System Model (CESM, see www.cesm.ucar.edu).

The advantage of more boxes is that we resolve the temperature more accurately (more points in a certain area). With an increasing number of grid boxes, however, the time evolution of an increasing number of quantities (at each grid box) has to be calculated. The same holds for the number of processes included in a GCM: more processes simply means more calculations. Also, the longer the time period over which we want to compute the evolution of each quantity, the longer it takes to do the calculation on a computer.

The state-of-the-art GCMs are located above the Earth System Models of Intermediate Complexity (EMICs; Claussen et al., 2002). Compared to GCMs, the ocean and atmosphere models in EMICs are strongly reduced in the number of scales. For example, the atmospheric model may consist of a quasi-geostrophic or shallow-water model and the ocean component may be a zonally averaged model. The advantage of EMICs is therefore that they are computationally less demanding than GCMs and hence many more long-timescale processes, such as land-ice and carbon cycle processes, can be included. Each of the individual component models of EMICs may also be used to study the interaction of a limited number of processes. Such models are usually referred to as *Intermediate Complexity Models* (ICMs). A prominent example is the Zebiak–Cane model of the El Niño/Southern Oscillation phenomenon (Zebiak and Cane, 1987). In time, the GCMs of today will be the

EMICs of the future and the state-of-the-art GCMs will shift toward the upper-right corner in Fig. 1.3.

In the remainder of this chapter, we describe the mean state of the climate system as determined from a high-resolution climate-model simulation. Although these mean fields deviate from observations in several aspects, the advantage of this approach is that the fields are dynamically consistent in that the basic conservation laws are satisfied.

1.4 Mean State

Model output from a recent about 250-year Community Earth System Model (CESM) simulation is used, where the ocean component (the Parallel Ocean Program, POP), as well as the sea-ice model have a 0.1° resolution (van Westen and Dijkstra, 2017). There are 42 (non-equidistant) vertical levels in the POP model, with highest resolution near the ocean surface. The atmosphere and land surface model have a horizontal resolution of 0.5°, and 30 non-equidistant pressure levels are used in the atmosphere model. The forcing conditions (e.g., CO_2, solar, aerosols) are those observed over the year 2000 (repeated for every model year). The model output analyzed has a monthly resolution.

Two time series are shown in Fig. 1.5 to illustrate the equilibration of this simulation. The global mean surface temperature and radiative imbalance at the top of the atmosphere start equilibrating after about 150 years, with only a small positive value of the global mean radiative imbalance over the last 100 years of the simulation (Fig. 1.5). In the analysis of the CESM results that follow, we will take the mean over the model years 170–199 (also often referred to as the climatology).

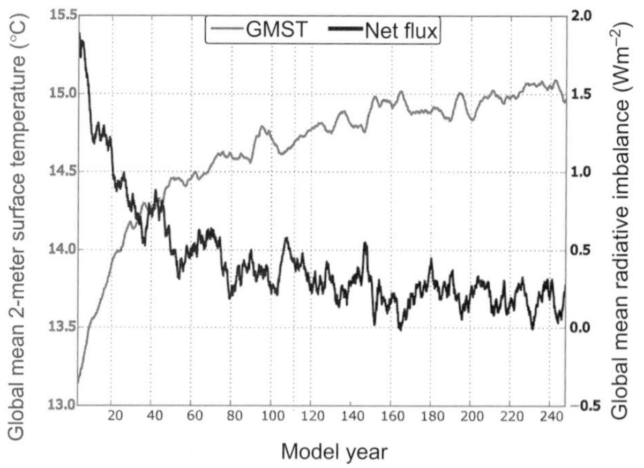

Figure 1.5 Equilibration of the CESM simulation, shown through the global mean surface (2 m) temperature and global radiative imbalance at the top of the atmosphere.

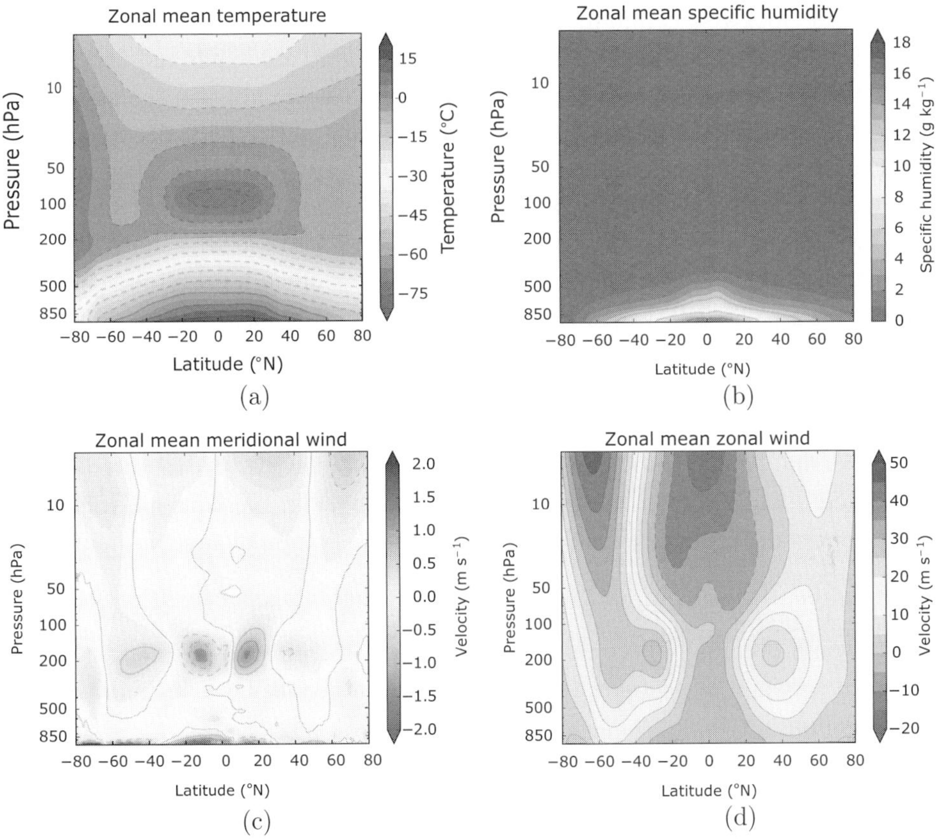

Figure 1.6 (a) Zonal mean temperature, (b) zonal mean moisture, (c) zonal mean meridional velocity, and (d) zonal mean zonal velocity. All fields are obtained by averaging over the CESM model years 170–199. (A black and white version of this figure appears in some formats. For the color version, please refer to the plate section.)

1.4.1 Atmosphere

The zonal mean temperature distribution (Fig. 1.6a) corresponds to that expected from the radiation balance, with relatively high surface temperatures in the tropics and low temperatures at the poles. As warm air can hold more moisture than cold air, also the specific humidity field (Fig. 1.6b) has a similar spatial structure as the temperature field, with high humidity in the tropics. The temperature decreases vertically up to a height of 100 hPa (according to mechanisms discussed in the previous section) but then increases again with height due to the presence of ozone.

Important characteristics of the mean circulation of the atmosphere can be analyzed from vertical-meridional cross-sections of the zonal mean zonal and meridional wind velocities (Fig. 1.6c, d). Over the latitudes 30°S–30°N, the dominant feature is a clockwise cell in the Northern Hemisphere and an anti-clockwise cell in the Southern Hemisphere. There are surface equatorward winds and upper poleward winds, with air ascending at the equator and descending in the subtropics (Fig. 1.6c). This cell is referred to as the Hadley cell and is thermally direct, as a parcel of air following the cell will transport energy away from the equator. Thus, the Hadley circulation is forced by the meridional gradient of net radiative fluxes and converts the resulting available potential energy into kinetic energy. Another prominent feature is the strong westerly winds in the upper levels at the poleward limit of the Hadley cell, which is referred to as the subtropical jet. These jets are, to a good approximation, in thermal wind balance with the meridional temperature gradient (Fig. 1.6d).

The positions of the Hadley cell and the jet cores and intensity depend strongly on the season. As the insolation maximum moves poleward it is followed by the Hadley cell, while the jet intensity becomes stronger in the winter hemisphere. The extent of the Hadley cell is constrained by conservation of angular momentum and can be deduced by conceptual models (James, 1994). An extension further than the tropics would require very strong westerly winds in the upper levels, which would eventually become unstable, even in the presence of friction. Thus, poleward transport of heat north of the tropics must be taken over by another type of circulation.

A second circulation, the Ferrel cell, is visible poleward (Fig. 1.6c) of the Hadley cell. It is weaker and thermally indirect, implying that it is a sink of kinetic energy and must be forced by mechanical stirring. Indeed, strong eddy activity takes place in this region (Fig. 1.6d), where the eddies are able to transport air parcels poleward in the upper levels. The Ferrel cell is therefore able to transport heat poleward via the transient eddy heat fluxes, reducing the temperature gradient and wind shear generated by the Hadley circulation. The high-frequency transient eddies (periods of less than 10 days) are due to baroclinic instability and convert the available potential energy to eddy kinetic energy.

1.4.2 Ocean

On the large scale, the ocean circulation is driven by momentum fluxes (wind stress) and affected by fluxes of heat and freshwater at the ocean–atmosphere interface. The mean sea surface temperature field (Fig. 1.7a) shows that northern latitudes in the North Atlantic are much warmer than those in the Pacific, which is related to the pattern of the meridional overturning circulation (MOC), as discussed below. The MOC is also the reason that the surface salinity in the North Atlantic is about 2 g kg^{-1} greater than that in the North Pacific (Fig. 1.7b).

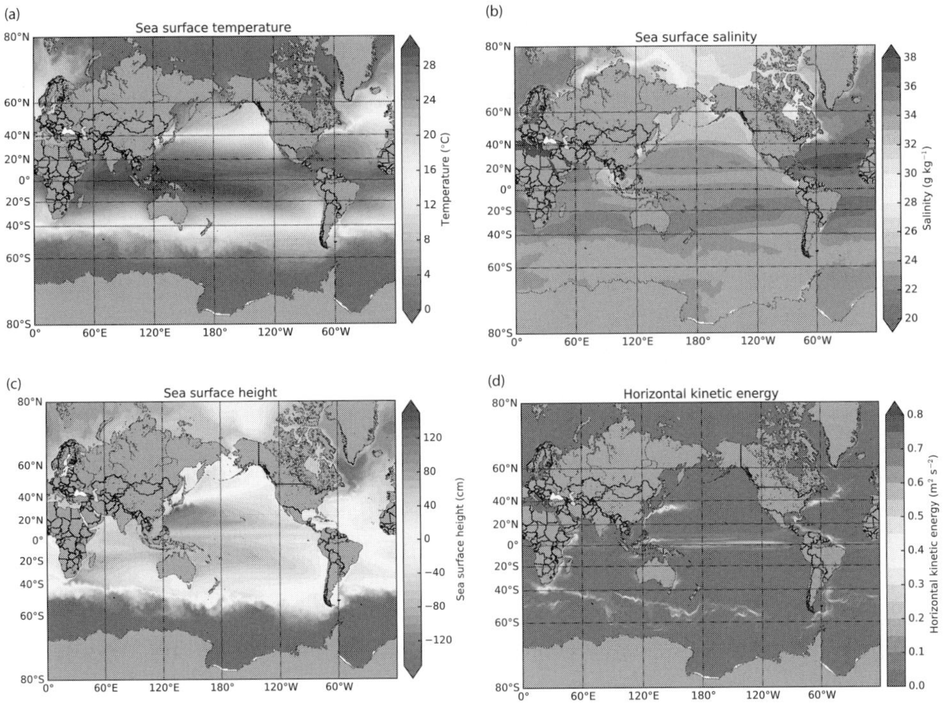

Figure 1.7 (a) Sea surface temperature, (b) sea surface salinity, (c) sea surface height, and (d) surface kinetic energy. All fields are obtained by averaging over the CESM model years 170–199. (A black and white version of this figure appears in some formats. For the color version, please refer to the plate section.)

The dominant ocean surface circulation patterns include midlatitude wind-driven gyres, a complex equatorial circulation system, and the Antarctic Circumpolar Current. The surface currents are approximately tangent to sea surface height curves and hence in Fig. 1.7c one can distinguish the cellular gyre circulation at midlatitudes, with strong currents near the western boundary of each basin (e.g., Gulf Stream, Kuroshio, Agulhas). Much of the kinetic energy is concentrated in these western boundary currents, as well as in the Antarctic Circumpolar Current (Fig. 1.7d).

A textbook sketch of the three-dimensional ocean circulation (Kuhlbrodt et al., 2007) is provided in Fig. 1.8. In the North Atlantic, the Gulf Stream transports relatively warm and saline waters northward, contributing to the relatively mild European climate. The heat is quickly taken up by the atmosphere, making the water denser. When there is strong cooling in winter, the water column becomes unstably stratified, resulting in strong convection in certain areas (e.g., the Greenland Sea and the Labrador Sea). The net result of this process is the formation of a water mass called North Atlantic Deep Water (NADW), which overflows the various ridges that

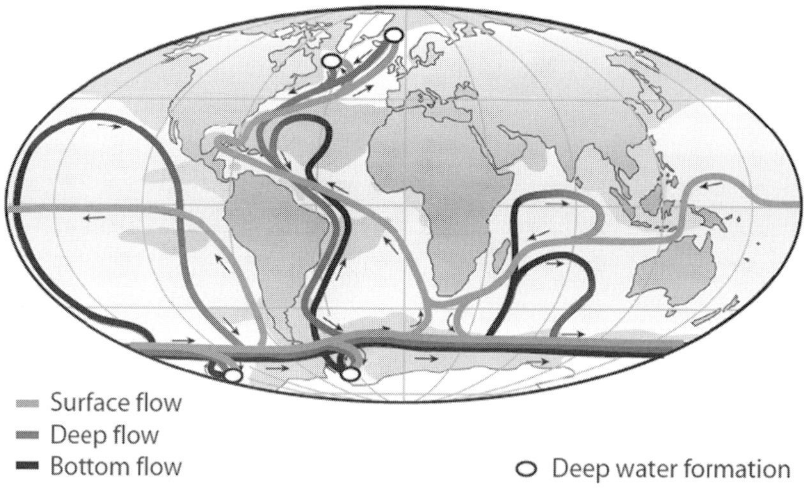

Figure 1.8 Sketch of the three-dimensional ocean general circulation (after Kuhlbrodt et al., 2007).

are present in the topography and enters the Atlantic basin. The NADW flows southward at mid-depth in the Atlantic, enters the Southern Ocean and from there reaches the other ocean basins. Through upwelling in the Atlantic, Southern, Pacific, and Indian Oceans, water is slowly brought back to the surface and the mass balance is closed by transport of water back to the sinking areas in the North Atlantic.

The MOC is a crucial component of the global circulation as it is strongly coupled to meridional heat transport. The MOC is the zonally averaged volume transport (at each latitude-depth) and its strength at 26°N in the Atlantic is now routinely monitored by the RAPID-MOCHA array (Cunningham et al. , 2007). Mean patterns of the MOC in the Atlantic and Indo-Pacific are shown in Fig. 1.9 as determined from the CESM simulation. The Atlantic MOC has its maximum at about 1000 m depth around the separation latitude of the Gulf Stream (40°N) with a strength of about 20 Sv (Fig. 1.9a). Its value at 26°N is in good agreement with that observed from the RAPID-MOCHA array (about 19 Sv). In the Pacific, there is no northern overturning and most of the deep water is formed in the Southern Ocean (Fig. 1.9b).

To summarize, in this chapter we have provided an overview of the mean climate state of the Earth by considering (dynamically consistent) output of a CESM simulation. The dominant balance of the climate system is the radiation balance, which determines to a large extent the global mean surface temperature. Large-scale inhomogeneities in this balance drive the large-scale motions in the ocean

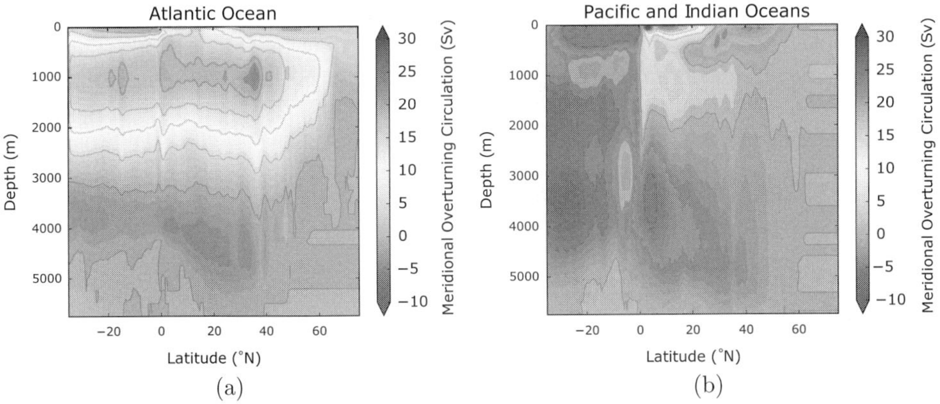

Figure 1.9 (a) Atlantic MOC and (b) Indo-Pacific MOC. All fields are obtained by averaging over the CESM model years 170–199. (A black and white version of this figure appears in some formats. For the color version, please refer to the plate section.)

and atmosphere. These motions, including the large-scale winds and ocean currents, induce equator-to-pole heat transport and affect the surface temperature distribution of the Earth.

2

Climate Variability

Climate processes take place over a broad set of timescales that can range from seconds (the dissipation scale of the three-dimensional atmospheric turbulence) up to tens of thousands of years (the scale of the changes in Earth's orbital parameters affecting the insolation). In this chapter, a short overview is given of climate variability phenomena and the processes causing this variability. In Section 2.1, temporal characteristics of the phenomena are described and a null-hypothesis for sea surface temperature (SST) variations is given. Large-scale patterns of variability, such as atmospheric Rossby waves the North Atlantic Oscillation (NAO), the monsoon variability, the El Niño–Southern Oscillation (ENSO) and the Atlantic Multidecadal Oscillation (AMO) are discussed in Sections 2.2–2.6.

2.1 Phenomena and Null-Hypothesis

An artist's view of the temporal characterization of climate variability is provided in Fig. 2.1. Such a "spectrum" can be generated (Mitchell, 1976) from many different time series of observations of the instrumental record and proxy data, including ice cores and ocean sediments, each with their own length and temporal resolution. The most notable peaks in Fig. 2.1 are the two peaks at one year and one day that are induced by insolation variability. There are also several broad-band signatures, mainly due to large-scale patterns (periods longer than one year) such as ENSO (Wang and Picaut, 2004), Pacific Decadal Oscillation (PDO) (Mantua and Hare, 2002), and the AMO (Schlesinger and Ramankutty, 1994). Another broad-band peak is placed at the monthly scale, and represents the so-called intraseasonal variability, with the Madden–Julian Oscillation (MJO) as a major phenomenon (Zhang, 2005). Finally, the broad range of periods centered around 6–10 days represents the midlatitude synoptic weather variability. Note that Fig. 2.1 is quite a simplification compared to the picture arising from the analysis of composite spectra (Lovejoy, 2014).

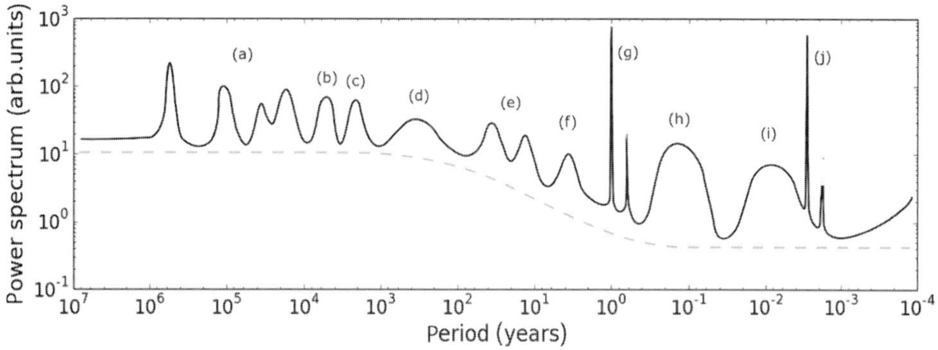

Figure 2.1 Artistic rendering of the energy density spectrum of an idealized climatological time series. This idealization is obtained taking into account the various timescales typically displayed by different climatological data sets, with a broad range of sampling frequencies and time coverage. We can identify various peaks associated with different known processes: (a) Milancović cycles, (b) Heinrich events, (c) Dansgaard–Oeschger cycles, (d) centennial variability, (e) interdecadal variability, (f) ENSO, (g) seasonal cycle and its harmonics, (h) intraseasonal variability, (i) weather, (j) diurnal cycle and its harmonics. Adapted from Dijkstra and Ghil (2005).

Over the complete timescale range, the spectrum shows a continuous background which is usually referred to as a red-noise background (Imkeller and Von Storch, 2001; Marshall and Plumb, 2008). An explanation of such behavior in SST spectra was missing until 1976, when Hasselmann described it in a very simple and straightforward way (Hasselmann, 1976). He proposed to divide the processes into fast atmospheric ones, acting as a forcing over the slower upper ocean mixed-layer dynamics. When the SST is indicated by T, the upper ocean heat balance can be written as

$$\frac{dT}{dt} = -\gamma\,(T - \bar{T}_a) + \eta, \tag{2.1}$$

where γ models the inverse thermal damping timescale of the ocean, $\bar{T}_a$ is the average atmospheric temperature, and η represents the short-term fluctuations in the ocean–atmosphere heat flux.

Hasselmann argued that the atmospheric forcing can be represented as white noise. In this way, (2.1) can be transformed into the stochastic differential equation:

$$d\Theta_t = -\gamma\,\Theta_t\,dt + \sigma\,dW_t, \tag{2.2}$$

where $\Theta_t = T - \bar{T}_a$ is the stochastic temperature anomaly, σ is the white noise intensity, and W_t is a Wiener process. From (2.2) we can see that the variable Θ, modeling the ocean temperature anomaly, is integrating the white noise provided by the atmosphere. The solution to (2.2) is called an Ornstein–Uhlenbeck process, and displays a red spectrum (Imkeller and Von Storch, 2001) of the form

$$P(\omega) = \frac{\sigma^2}{2\pi\gamma} \frac{1}{\omega^2 + \gamma^2}, \tag{2.3}$$

where ω is the frequency and $P(\omega)$ the spectral power density. For large ω, the spectrum falls off with ω^{-2} and for small ω (compared to γ) the spectral power is near constant. In this way, we are able to understand the red noise behavior of SST variability at many locations in the upper ocean.

In the remaining sections of this chapter, we will discuss climate variability phenomena that either have a well-defined pattern (but not a dominant timescale), or both a well defined pattern and timescale.

2.2 Atmospheric Waves and Teleconnections

Beyond the weather timescales there are two types of waves that are particularly important to explain atmospheric (and oceanic) phenomena: Rossby and Kelvin waves. Rossby waves play a fundamental role in atmospheric dynamics. They are responsible for the meanders of the jet streams that govern day-to-day weather and are the cornerstone elements of the theories that explain the influence of the tropics on the extratropical regions. The main restorative force of these waves is the meridional gradient of planetary vorticity, which depends on the latitudinal variation of the Coriolis parameter $f = 2\Omega \sin\theta$. Here, Ω is the Earth's angular velocity and θ indicates the latitude. The displacement of the fluid particles is in the horizontal plane, and not in the vertical one as in normal gravity waves.

For a one-dimensional plane Rossby wave propagating in the zonal direction in a constant-density shallow-water layer with a flat bottom and with uniform zonal wind U, the dispersion relationship is

$$\omega = Uk - \frac{\beta}{k} \tag{2.4}$$

where ω is the frequency of the waves, $\beta = r_0^{-1} df/d\theta$ is the gradient of planetary vorticity, r_0 the radius of the Earth, and k is the zonal wave number. Thus, the phase speed ω/k is always to the west with respect to the mean flow U. In midlatitudes U is large and Rossby waves tend to propagate eastward with respect to the Earth's surface. Rossby waves are dispersive and are such that long waves (small k) tend to propagate faster to the west. Stationary Rossby waves satisfy $k^2 = \beta/U$, and typical values of $U = 10$ ms^{-1} and $\beta \simeq 10^{-11}$ (ms)$^{-1}$ give wavelengths of $\lambda = 2\pi/k \simeq 6000$ km.

The impact of the tropics on midlatitudes is based to first order on the dispersion of Rossby waves. Changes in latent heat associated to displaced rainfall patterns in the tropics induce divergent circulation anomalies that act as sources of Rossby waves that then propagate in arch-like trajectories toward the extratropics (Trenberth et al., 1998). In the case of ENSO (see Section 2.4), the anomalous heat source persists for several months, thus leading to a stationary wave pattern in the extratropics, called a teleconnection.

Atmospheric teleconnections are an important component of climate variability (Glantz et al., 1991). Teleconnections fluctuate over a broad range of scales and form patterns that are extended over large portions of the Earth's surface. They are usually detected through analysis of the covariance matrix of a climatic variable over a certain area using empirical orthogonal functions (EOF) analysis (see Section 3.2.4). The teleconnections associated with NAO (see Section 2.3) and ENSO (see Section 2.4) are main examples of the long-range correlations that can be found in the atmosphere. Many of these patterns can, to first approximation, be explained by means of linear theory. However, especially in the extratropics, where the atmosphere is more active and the eddy forcing is more intense, nonlinearities play an important role. Held et al. (2002) show that, to first order, the atmosphere is a linear system beyond the weather timescale, with spatially stationary patterns given by the resonance of the circulation with the topography and heat sources. In this framework the teleconnections are fluctuations over various timescales of the amplitude of these stationary waves, induced by changes in the forcing intensities.

Kelvin waves are solutions of the shallow water equations for a constant-density fluid with Coriolis parameter f and depth H. Kelvin waves are particularly important in the ocean, where continental boundaries allow the propagation of these waves. Kelvin waves propagate along the boundary with speed $c = \sqrt{gH}$ (that is, as a non-rotating gravity wave), where g is the gravitational acceleration. Nevertheless, Kelvin waves have one characteristic that makes them unique: They can propagate along the coast in one direction only, i.e., in the Southern Hemisphere they have the coast to their left, and in the Northern Hemisphere to their right. This is because the alongshore velocity component is in geostrophic balance (balance between Coriolis acceleration and the horizontal pressure gradient). Moreover, the amplitude of the wave is maximum next to the coast and decreases exponentially in the across-shore direction with an e-folding scale given by the Rossby radius of deformation $R_D = \sqrt{gH}/f$. In the atmosphere and ocean, Kelvin waves can also occur along the equator (where $f = 0$) because it serves as a boundary. Thus, equatorial Kelvin waves in the atmosphere and ocean always propagate from west to east.

2.3 The North Atlantic Oscillation

The NAO is a pattern of atmospheric variability that impacts greatly the weather conditions over Europe, Northern Asia, and North America (Lifland, 2003). When

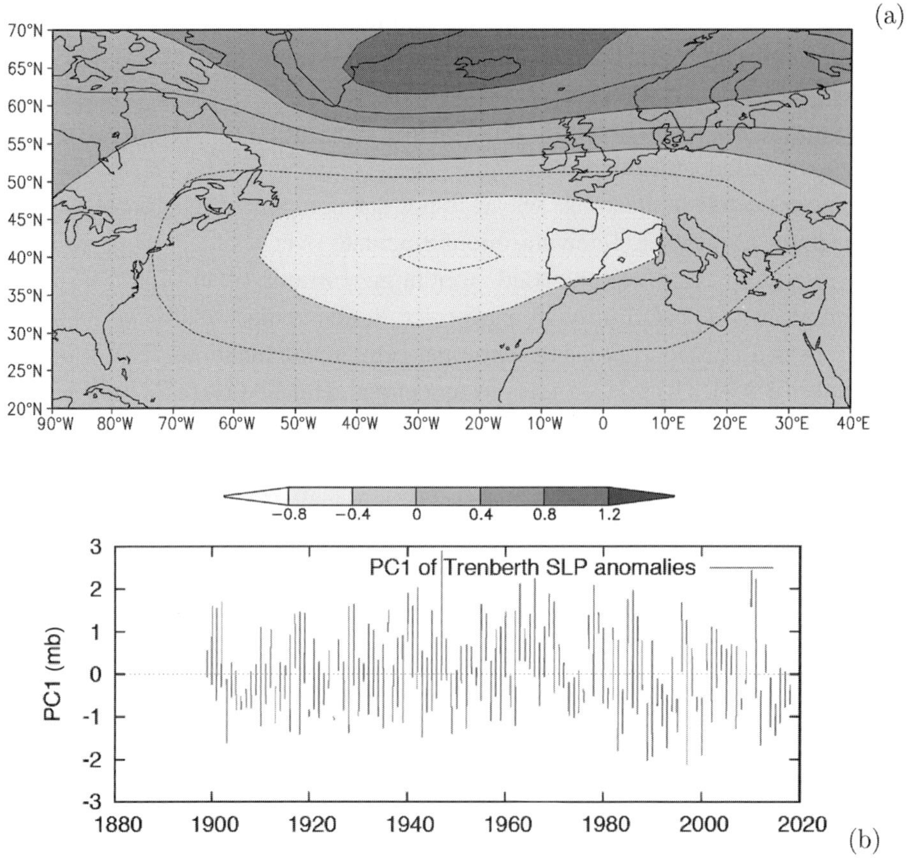

Figure 2.2 (a) Leading EOF of the DJF mean sea level pressure anomalies in the North Atlantic sector (20°N–70°N; 90°W–40°E) over the years 1899–2018. (b) First principle component corresponding to the pattern in (a). Figure created with the Climate Explorer (http://climexp.knmi.nl) using the Trenberth Sea Level Pressure (SLP) data.

monthly mean anomalies of sea level pressure data over the North Atlantic region (20°N–70°N; 90°W–40°E) are analyzed for dominant statistical patterns of variability, the first EOF (see Section 3.2.4) for the winter season (December to February, labeled DJF) shows a pattern as in Fig. 2.2a. This is the pattern of the NAO, with the centers with highest amplitude located at the (land) stations Stykkisholmur (Iceland) and Lisbon (Portugal).

The corresponding principle component (PC, see Section 3.2.4) is shown in Fig. 2.2b. The spectrum of the winter-mean NAO index is slightly red, and the major conclusion is that there is no preferred timescale of NAO variability. However, there is a weak spectral peak between 6–9 years that may be important, as time series of climate variables related to NAO do show peaks in the decadal range of the spectrum (Da Costa and Colin de Verdiere, 2004). The main reason for such high variability is that NAO is mainly an internal atmospheric mode resulting from a collective interaction of different instability patterns. This means that the high level of turbulence present in the mid-latitude atmosphere affects its evolution, resulting in a highly irregular behavior.

During a positive NAO phase, the pressure difference over the dipole is higher than normal, and in the negative phase the opposite occurs. When we deal with a positive NAO phase the higher-than-normal meridional pressure gradient moves the polar jet northwards. Since along the jets we have the passage of the storm systems (storm-track), it means that the weather perturbations will move eastward at latitudes higher than normal, at least in Europe. Thus, Northern Europe will be wetter than normal, and Southern Europe dryer. The opposite holds for the negative NAO phase, in which Southern Europe tends to be wetter than normal. Hence, the socioeconomic effects of NAO are huge and a lot of effort has been devoted to understanding its dynamics and predictability.

2.4 The El Niño–Southern Oscillation

The mean state over the eastern Pacific Ocean SST is the result of the continuous action of easterly trade winds that push equatorial waters from the Peruvian coasts toward the western limits of the basin. This causes both an increase of sea level in the western Pacific boundary and cold water upwelling in the eastern basin (Philander, 1990). The net result is the formation of two distinct areas (Fig. 2.3), one of relatively cold waters in the east (called cold tongue) and one of relatively hot waters in the west (called warm pool).

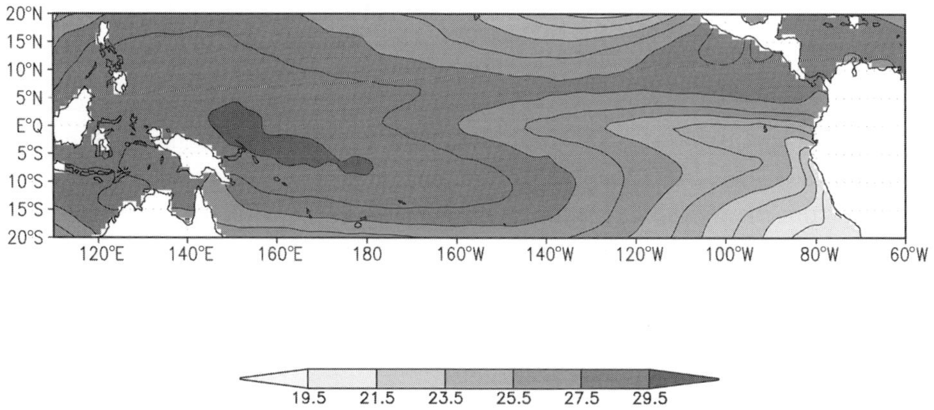

Figure 2.3 SST (° C) of the mean state in the tropical Pacific Ocean. The cold tongue on the right and the warm pool on the left are clearly distinguishable. Data from ERA Interim Reanalysis.

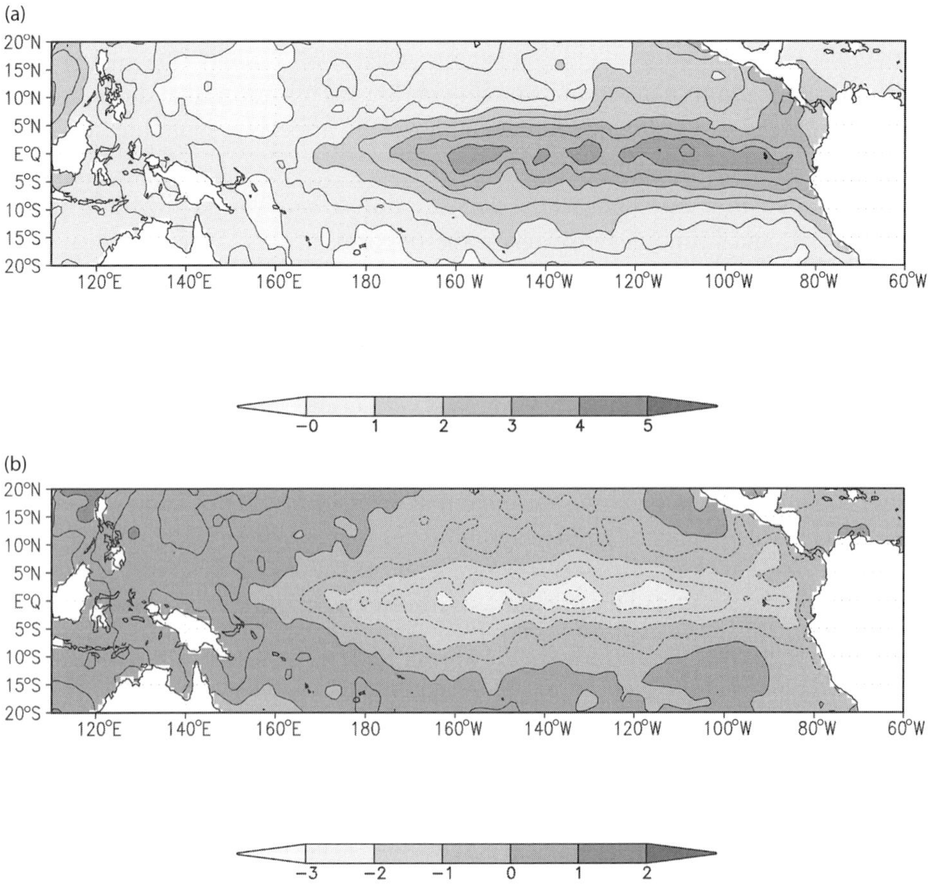

Figure 2.4 Sea surface temperature anomalies (in $^\circ$ C) during a strong El Niño event (a) and during a strong La Niña event (b). Data from ERA Interim Ranalysis.

During an El Niño event, the SST anomaly is typically +5°C in the eastern equatorial Pacific. These SST changes lead to a weakening of the trade winds resulting in a warmer-than-normal cold tongue area since there is less cold water coming from the upwelling and the cold water is further from the surface. If the warming of the cold tongue is above a certain threshold, the resulting configuration is called El Niño (Fig. 2.4a). Conversely, when the trade winds are stronger than normal, the cold tongue will be colder than normal. If the cooling of the cold tongue is below a certain value, the configuration is called La Niña (Fig. 2.4b). The US National Oceanic and Atmospheric Administration (NOAA) definition of El Niño is a three-month average warming of at least 0.5°C, while other organizations prefer different definitions. To stress the fact that it is a coupled ocean–atmosphere phenomenon, one uses the ENSO acronym (El Niño Southern Oscillation).

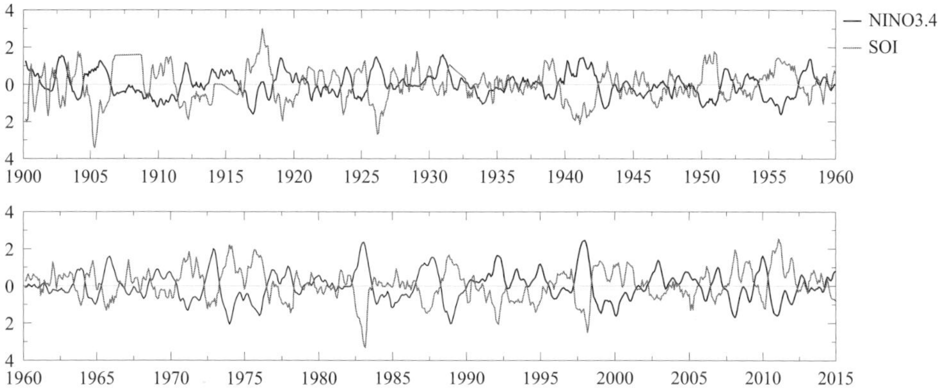

Figure 2.5 Time series of three-month running mean NINO3.4 index (dark line) and SOI (light line) over the years 1900–2014. NINO3.4 index is from HadISST dataset (Rayner et al., 2003) and SOI is provided by NCEP from www.cpc.ncep.noaa.gov.

Unlike NAO, El Niño displays a preferred timescale of about 3–4 years. One measure of the temperature variations of the eastern equatorial Pacific is the NINO3.4 index, defined as the SST anomaly averaged over the region (120°W–170°W)×(5°S–5°N). From the time series (dark curve in Fig. 2.5) for the period 1900–2015, it can be observed that strongest El Niños were those of 1982/1983 and 1997/1998. Changes in the tropical atmospheric circulation are strongly connected to changes in SST, which can be captured by the Southern Oscillation Index (SOI) – the normalized difference of the pressure anomalies between Tahiti (18°S, 150°W) and Darwin (12°S, 131°E). The SOI is plotted as the light curve in Fig. 2.5; when the SOI is negative, the westward trade winds are weak, and vice versa. The strong anti-correlation of NINO3.4 and the SOI is striking from Fig. 2.5.

The canonical or conventional type of ENSO is also called Eastern Pacific (EP) ENSO (Kao and Yu, 2009). However, in the last two decades another type of ENSO has been observed, in which the temperature anomaly near the South American coast is not affected, but an anomaly arises in the central Pacific (Larkin and Harrison, 2005). This variant is called Central Pacific (CP) ENSO (Kao and Yu, 2009), "dateline" ENSO (because the anomaly arises near the dateline) (Yuan and Yan, 2013), or ENSO "Modoki" (Modoki is Japanese for "similar but different")(Cai and Cowan, 2009).

The influence of ENSO on the atmospheric surface temperature is regionally strong. In particular, strong fluctuations of the SST along the equator, a zone in which the atmosphere is very sensitive to the ocean's temperature, greatly affect the atmospheric circulation patterns. This in turn modify the average distribution and intensity of rains and droughts not only over the tropical and subtropical Pacific, but also on remote regions through atmospheric teleconnections. These changes can be so intense as to significantly affect human activities, such as agriculture, that strongly depend on rainfall.

The effects of the CP ENSO are different from those of the conventional EP ENSO. The El Niño Modoki leads to more hurricanes which more frequently make landfall in the Atlantic (Kim et al., 2009). The La Niña Modoki leads to a rainfall increase over northwestern Australia rather than over the eastern part as in a conventional La Niña (Cai and Cowan, 2009).

2.5 Tropical Circulation and Monsoons

In the tropics there is a zone of deep convection, the Intertropical Convergence Zone (ITCZ) (Fig. 2.6), which encircles the Earth and varies seasonally northward and southward following the interface between the winter and summer Hadley cells. These swings are more intense over land, which responds to insolation changes more rapidly than the ocean. Along the main path of the ITCZ numerous convective subsystems are located. The monsoon systems (that is, systems in which the convection is triggered by a strong air–sea contrast) in India and northern South

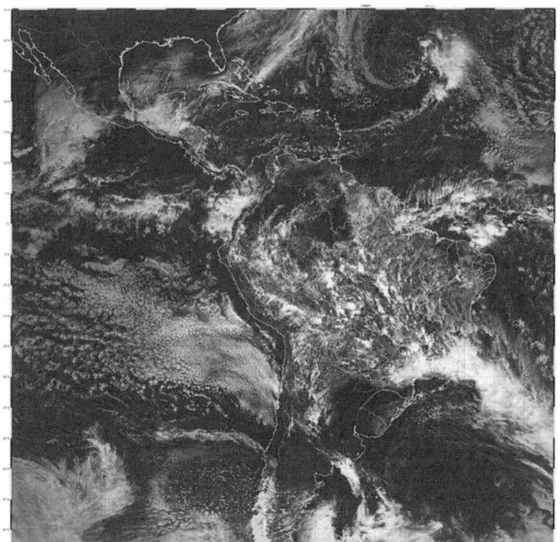

Figure 2.6 Image taken by the GOES 16 satellite of South America on January 30, 2018. The ITCZ is visible as white bands just to the north of the equator in the Pacific and Atlantic Oceans. The SACZ is seen as a diagonal band of cloudiness that extends from the Amazon rainforest to the subtropical southern Atlantic Ocean.

America are examples, as well as the South Pacific Convergence Zone in the Pacific Ocean.

A particularly relevant tropical convective system is the so-called South Atlantic Convergence Zone (SACZ) (Barros et al., 2002), which extends between southern-central Brazil and the subtropical Atlantic Ocean (Fig. 2.6). The SACZ is one of the main components of the South American Monsoon System: it is a convective pattern that extends from the Amazon rainforest to the subtropical South Atlantic ocean, oriented in a northwest–southeast direction (Bombardi et al., 2013; Carvalho et al., 2002, 2004). When active, the SACZ is associated with heavy precipitation over the Amazon rainforest and southeastern Brazil, causing floods and landslides over the densely populated areas. The main pattern of SACZ variability consists of a precipitation dipole with center over the SACZ and over Uruguay, such that when the SACZ is active there is decreased rainfall over Uruguay and vice versa. This pattern varies on several timescales, ranging from day-to-day due to the passage of fronts, to intraseasonal, interannual, and even longer timescales. It has also been associated with the observed summertime rainfall trend in eastern South America during the twentieth century (Barreiro et al., 2002; Junquas et al., 2012; Kodama, 1992; Nogués-Paegle and Mo, 1997).

Due to its socioeconomic importance, there have been several attempts to improve our understanding and predictability of rainfall over the SACZ on seasonal to interannual timescales, particularly focusing on the modulating role of the upper ocean. Studies have shown that the subtropical South Atlantic Ocean may influence the evolution of the SACZ (Barreiro et al., 2002, 2005; Robertson and Mechoso, 2000). For example, Barreiro et al. (2002, 2005) show that even though the region is dominated by internal atmospheric variability, SST anomalies can force a dipole of precipitation anomalies located mainly over the oceanic portion of the SACZ. Subsequent studies have suggested that the air–sea interaction is such that an initially stronger SACZ – due to internal atmospheric variability – induces an oceanic cooling that in turn negatively affects the convective precipitation, resulting in a negative feedback loop (Almeida et al., 2007; Chaves and Nobre, 2004).

Early studies defined monsoon as a regional phenomenon similar to the sea breeze, caused by the seasonal reversal of the wind direction due to the differential heating of the land and the surroundeding ocean. This word came from the Arabic word *mausim*, meaning season, which was used by sailors in the sixteenth century to describe the reversal of winds in the Arabian sea. More modern studies, starting from Saha and Saha (1980), consider the monsoon as a part of the global atmospheric circulation system, and therefore as a global phenomenon. The definition of the monsoon has therefore broadened from the rainy season in the Indian subcontinent, to the so-called "Monsoon Zone," which consists of several "classical monsoons": South Asian (or Indian) Monsoon (SAM), East Asian Monsoon (EAM),

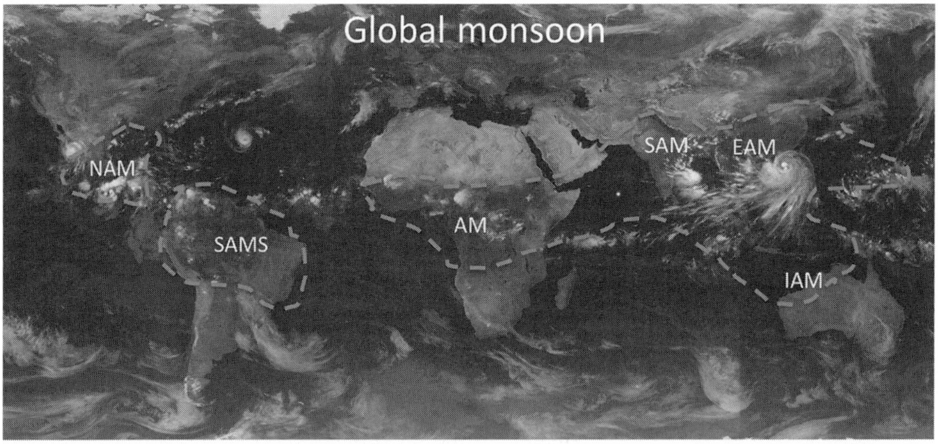

Figure 2.7 Global monsoon: SAM – South Asian Monsoon (includes Indian Summer Monsoon (ISM) during June to September and Winter Monsoon: October to December); EAM – East Asian Monsoon; IAM – Indo-Australian Monsoon; AM – African Monsoon; NAM – North American Monsoon; and SAMS – South American Monsoon System. The ITCZ is one of the main drivers of monsoon, and is shown as a belt of clouds circling the globe and passing through the monsoon regions.

Indo-Australian Monsoon (IAM), and African Monsoon (AM). More recently, the South American Monsoon System (SAMS) and North American Monsoon (NAM) have been defined (Fig. 2.7).

Currently, the Indian Monsoon directly affects the lives and prosperity of more than 1.7 billion people – nearly one-quarter of the world's population. The effects of the Indian Monsoon are enormous on the regional scale, through its effects on the agriculture and economy of the Indian subcontinent. The effects are also important on a global scale, through the monsoon's coupling with other large-scale climatic phenomena, including ENSO and the Indian Ocean Dipole (Gadgil, 2004; Sabeerali et al., 2011; Sankar et al., 2011; Wu et al., 2012, 2003). Numerous studies of the Indian Monsoon during the last century significantly improved the understanding of the monsoon regularities and its variability from millennial to interannual and intraseasonal time scales.

Nowadays, the Indian Monsoon is thought to be driven by several factors, including the land–ocean surface temperature gradient between the Indian subcontinent and surrounding bodies of water (Arabian Sea and Bay of Bengal), the migration of the ITCZ northward during the Northern Hemisphere summer, northward shifting of the westerly subtropical jet stream, the establishment of the tropical easterly jet, and the Tibetan upper troposphere high and low level jet stream, which splits into two monsoon branches. These factors, combined with the topography of the

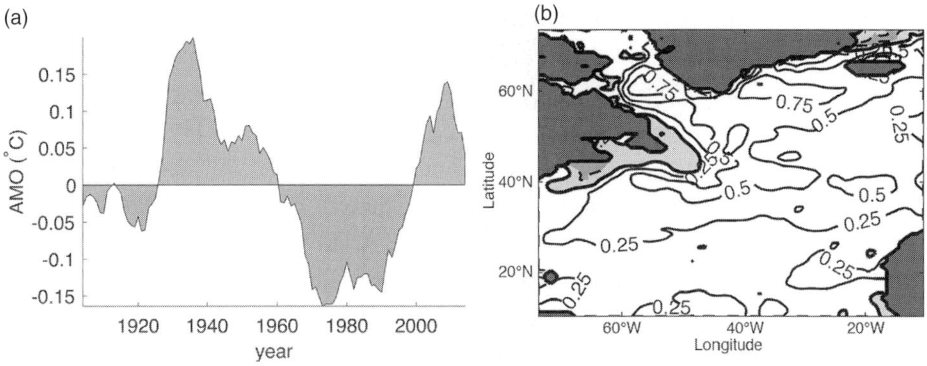

Figure 2.8 (a) The AMO index (the ten-year running mean of detrended Atlantic SST anomalies north of the equator) of the instrumental record (using data from the HadSST2 dataset). (b) An impression of the AMO pattern as shown by the difference in observed average North Atlantic SST between the periods 1950–1964 (warm period) and 1970–1984 (cold period). Units are in degrees Celsius and negative values are shaded. Figure from Dijkstra (2013).

Indian subcontinent, dominated by the Himalayas and Tibetan Plateau in the north, the Western Ghats from the Arabian Sea and Eastern Ghats from the Bay of Bengal, create a low-pressure monsoon trough, and an area of deep convection establishes over the Indian subcontinent (Ananthakrishnan and Soman, 1990; Joseph et al., 1994). However, there are still open crucial questions in the understanding of the present-day monsoon variability, which limit its skillful forecasting (Webster, 1998).

2.6 The Atlantic Multidecadal Oscillation

The North Atlantic SST appears to have a distinct signal of multidecadal variability. The first analyses of this variability were done by Schlesinger and Ramankutty (1994) from spectral analysis of four global-mean temperature records. Delworth and Mann (2000) extended the instrumental record with proxy data stretching back at least 300 years and demonstrated that there is a significant spectral peak in the 50–70-year frequency band. This variability was named the Atlantic Multidecadal Oscillation by Kerr (2000), and an AMO index was defined by Enfield et al. (2001) as the ten-year running mean of detrended Atlantic SST anomalies north of the equator. An AMO index, defined by Enfield et al. (2001) as the ten-year running mean of detrended Atlantic SST anomalies north of the equator, is plotted in Fig. 2.8a. Warm periods existed in the 1940s and from 1995 onward, whereas during the 1910s and 1970s the North Atlantic was relatively cold.

A first impression of the pattern of the AMO was obtained from an analysis of the instrumental SST record in the North Atlantic over the years 1950–1990 (Kushnir, 1994). Fig. 2.8b shows the difference between the average SST during the relatively warm years 1950–1964 and the relatively cool years 1970–1984. There is a negative SST anomaly near the coast of Newfoundland and a positive SST anomaly over the rest of the North Atlantic basin. Although the pattern of multidecadal variability has been characterized more accurately (Deser and Blackmon, 1993; Latif, 1998; Moron et al., 1998) by subsequent analysis of longer and better-quality SST and sea level pressure (SLP), it is still difficult to extract a dominant pattern of multidecadal variability with much confidence, because the observational time series is relatively short (about 150 years of SST and SLP observations).

The effect of AMO on the land-surface temperature is broadly distributed, and not just in the Atlantic region, but also in Africa, southern Asia, and Canada (Muller et al., 2013). Enfield et al. (2001) showed that the AMO has a strong negative correlation with US continental rainfall, with less (more) rain over most of the central USA during a high (low) AMO index period. Sutton and Hodson (2005) found that sea level pressure, precipitation, and temperature over Europe and North America, particularly during June to August, are influenced by the AMO. Positive correlations have also been found between the AMO index and rainfall in the Sahel, the hurricane intensity in the Atlantic, and the strength of the Indian summer monsoon (Feng and Hu, 2008; Gray et al., 1997; Knight et al., 2006; Zhang and Delworth, 2006).

To summarize, in this chapter we have provided examples of patterns of variability in the climate system. The null-hypothesis of SST variability being a red-noise process (with an ω^{-2} spectral decay for large frequencies) caused by the integration of atmospheric (white) noise by the upper ocean gives a rough description of the observed background spectrum. On top of this, it is remarkable that, on the large scale, climate variability appears to be organized in patterns (such as ENSO, NAO, and AMO), which can be determined by elementary statistical analysis. The patterns play an important role in climate variability on the continental scale.

3

Climate Data Analysis

In this chapter we will review the main sources of climate data, which are going to be used in the following chapters, as well as give an overview of linear and nonlinear techniques used to analyze these data.

3.1 Climate Data

Climate system data can be divided in three types: *direct empirical data*, obtained from instrumental measurements, *proxies*, that is indirect empirical data obtained by a number of different sources called archives, and *synthetic data*, obtained from model simulations in which empirical data are incorporated through, e.g., data assimilation.

An example of instrumental measurements is the *Hadley Centre Sea Ice and Sea Surface Temperature* (HadISST) dataset (Rayner et al., 2003) for sea surface temperature (SST). These SST data are provided by the *Met Office Marine Data Bank*, which is a combination of monthly globally complete fields of SST and sea ice concentration from 1871 onwards. In later chapters we also use the *Extended Reconstructed Sea Surface Temperature* (ERSST) (Smith and Reynolds, 2003). This is a global monthly SST analysis derived from the *International Comprehensive Ocean–Atmosphere Dataset* where missing data are filled in by statistical methods. This data set begins in January 1854, continues to the present, and includes anomalies computed with respect to the 1971–2000 monthly climatology. Most of the instrumental data can be found on the Climate Explorer (see https://climexp.knmi.nl).

The long-term climate evolution leaves clear traces on Earth that can be quantified via *proxies or archives*. As an example, every year in the poles, layers of snow are deposited, which compact and form ice. The thickness of these ice layers is an indirect measure of the intensity of the storms in those regions. In addition, the air trapped within the snow in every layer can give information on the chemical composition of the low atmosphere at the time the layer was formed. In a similar way, other proxies provide information about air temperature, volcanic activity, CO_2 concentration, etc.

Depending on the archive we consider, we are able to investigate different timescales at different resolutions. For example, tree ring width is used to estimate surface solar irradiance in a period ranging up to 10^4 years in the past. Using other databases such as ice cores or lake sediments we can reach 10^5 years and through marine sediments we can investigate up to 10^7 years in the past, in the late Cenozoic, a period that corresponds to the presence of the great mammals on the Earth (Mudelsee, 2014).

While proxies are an extremely useful tool in paleoclimatological studies, their interpretation is not without problems. In particular, the relation between the proxy and the observable that it represents can have a large uncertainty. For this reason, the information inferred by a proxy should be, whenever possible, cross-checked with models and/or a proxy of another kind. Most of the proxy data can be found on NOAA's Paleoclimatology site at www.ncdc.noaa.gov/data-access/paleoclimatology-data/datasets.

As we saw seen in Chapter 1, models are very useful for integrating the information from directly measured data. For example, a satellite observing the temperature of the Earth's surface cannot measure, at the same time, the temperature of the whole surface. Because it can measure, at a certain time, an area of a certain width, the temperature in the other regions has to be inferred in some way. This is done by incorporating the measured data in a model using a procedure called *data assimilation*, such as Bayesian data assimilation (Blayo et al., 2014). The data produced in this way is a mixture of observation and synthetic data. In particular, the synthetic part is chosen in a way that displays the best possible accordance with the measured one. These data sets are called *reanalysis data*, and they will be used in several chapters in this book. Because models differ in the ways they assimilate the observed data (Trenberth et al., 2005), a careful model inter-comparison is needed to confirm the significance of the results of the data analysis (trends, variations, etc.) (Simmons et al., 2005, 2010; Wang et al., 2016b).

Two popular reanalysis data sets are those produced by the NCEP/NCAR reanalysis project (Kistler et al., 2001), which is at a resolution of 2.5° in space and daily in time, and ERA Interim reanalysis (Dee et al., 2011), which is available with a resolution down to 0.25° in space and daily in time. The NCEP/NCAR reanalysis starts in 1948, while the ERA Interim starts in 1979. However the latter is derived from a more modern and complete model, and using only recent data (the database starts at the beginning of the satellite era), and thus it can be considered more reliable.

3.2 Linear Analysis Tools

In this section we present several tools commonly used to analyze climatological data. We begin with a brief review of linear techniques, i.e., methods that quantify linear properties of time series. The analysis of the climate data is typically performed by first determining the (total) mean and the seasonal mean, defining the (seasonal) climatology. In most cases, the anomalies with respect to the climatology are analyzed.

Assume that we have a monthly sampled database, with time series of an observable $y_i(t)$ recorded at the geographical location i. The seasonal climatology $c_i(t)$ is computed from $y_i(t)$ by averaging the values of each month over all years, i.e.,

$$c_i(t) = \frac{1}{Y-1} \sum_{n=0}^{Y} y_i(t + 12n), \tag{3.1}$$

where Y is the number of years. This gives a time series of length 12 months. The anomaly time series $z_i(t)$ is then the difference

$$z_i(t) = y_i(t) - c_i(t). \tag{3.2}$$

More advanced ways to compute the climatology use an empirical orthogonal function (EOF) decomposition (Section 3.2.4) or Fourier spectral analysis (Section 3.2.2).

3.2.1 Correlation Analysis

In time series analysis, the *cross-correlation*, $\rho_{X,Y}(\tau)$, correlation coefficient, or Pearson correlation coefficient is a measure of similarity of two time series, X and Y, synchronously sampled, as a function of a time-shift (or *lag*), τ, of one relative to the other. It is defined as

$$\rho_{X,Y}(\tau) = \frac{\mathrm{cov}(X(t), Y(t+\tau))}{\sqrt{\sigma_X^2 \sigma_Y^2}} = \frac{E[(X(t) - \mu_X)(Y(t+\tau) - \mu_Y)]}{\sigma_X \sigma_Y}, \tag{3.3}$$

where $\mathrm{cov}(X, Y)$ is the covariance of the two series, μ_X and μ_Y are their mean values, and σ_X^2 and σ_Y^2 are their variances. The correlation coefficient is symmetric, $\rho_{X,Y}(\tau) = \rho_{Y,X}(-\tau)$. An alternative measure of correlation between two variables is the *Spearman* correlation, which is equal to the Pearson correlation between the rank values of those two variables (i.e., the relative position labels: first, second, third etc.). While Pearson's correlation detects linear relationships, Spearman's correlation detects monotonic (linear or not linear) relationships.

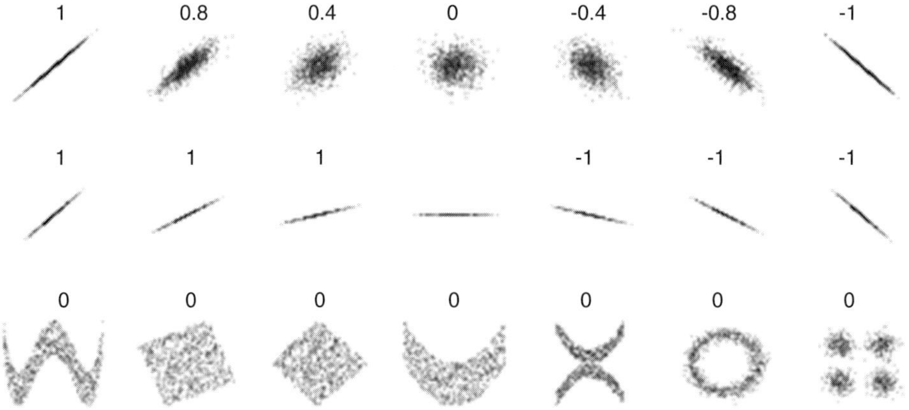

Figure 3.1 Examples of Y vs X plots revealing different types of relationships, which are not necessary captured by the linear cross-correlation coefficient (indicated over the plots). Image credit: DenisBoigelot/Wikimedia Commons

From the definition of the covariance, ρ is bounded between 1 and -1. In particular 1 (-1) means perfect linear correlation (anticorrelation), which implies $X(t) \propto Y(t+\tau)$ ($X(t) \propto -Y(t+\tau)$). Instead, if the two series are independent, then

$$E[(X(t)-\mu_X)(Y(t+\tau)-\mu_Y)] = E[(X(t)-\mu_X)]E[(Y(t+\tau)-\mu_Y)] = 0, \quad (3.4)$$

and $\rho_{X,Y}(\tau) = 0$. However, the converse is not necessarily true because, as shown in Fig. 3.1 which presents several examples, the correlation coefficient detects only linear dependencies. In the case that X and Y both are normally distributed (top row), zero correlation indeed detects mutual independence. Notice in the middle row of Fig. 3.1 that, when the variables are linearly related with minimal noise, the cross-correlations reflect the sign of the slope only, and for a completely horizontal line the correlation is not defined as the variance of Y is zero. Many shapes which possess an obvious (but nonlinear) relationship between the variables yield to a zero correlation providing the slope of a linear regression between them is zero, as shown in the bottom row.

The cross-correlation is widely used for determining the time delay between two signals, e.g., the maximum (or minimum) of $\rho_{X,Y}(\tau)$ indicates the optimal temporal lag τ that renders the time series X and Y best aligned. As will be discussed in Chapter 4, this property has been extensively used to define *climate networks.* An example of the maps of optimal lag times computed from surface air temperature (monthly NCEP/NCAR CDAS1 reanalysis, see Section 3.1) is presented in Fig. 3.2. The lag time is defined in the interval $[-6,6]$ months, but similar maps can be obtained when the lag is defined in the interval [0–12] months (Tirabassi and Masoller, 2013). Panel (a) displays the lag times between the time series at a

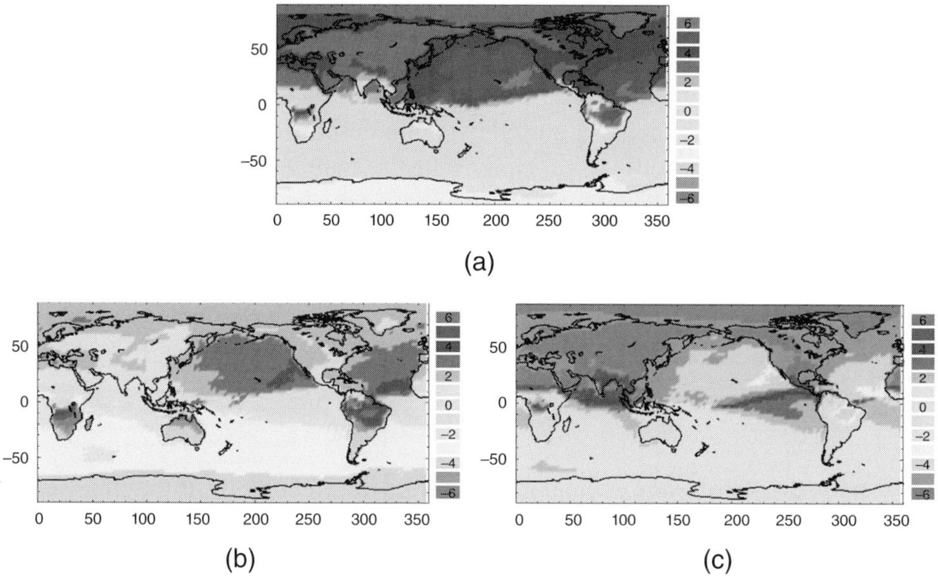

Figure 3.2 Lag times computed from cross-correlation analysis of monthly surface air temperature reanalysis. The lag is defined in the interval [−6,6] months. Panel (a) displays the lag times between the time series at a reference point located in El Niño basin and all the other time series; in panel (b) the reference point is located in the Indian Monsoon area; and in panel (c) it is in central Australia. (A black and white version of this figure appears in some formats. For the color version, please refer to the plate section.)

reference point located in the El Niño basin and all the other time series; in panel (b) the reference point is located in the Indian Monsoon area; and in panel (c) it is in central Australia. The maps reveal clear characteristic patterns, signatures of climatic regions. We can also note the memory effect induced by the oceans, and the almost perfect six-month symmetry between the Northern and Southern Hemispheres.

While the lag analysis can be extremely useful, care must be taken when choosing the interval of lag times within which the maximum or minimum of the cross-correlation is searched for (Martin et al., 2013). Another issue that must be carefully considered is the *statistical significance* of cross-correlation values (Section 3.4). As discussed by Palus (2007) with the illustrative example of the cross-correlation between the number of Republicans in the US Senate in the years 1960–2006 and the number of sunspots in the same period: with appropriated surrogate data the observed “high” correlation (0.52) is not significant.

The *auto-correlation* is the cross-correlation of a time series with itself, i.e.,

$$c(\tau) = \frac{\text{cov}(X(t), X(t+\tau))}{\sigma_X^2}. \tag{3.5}$$

The auto-correlation of a time series provides information about the "memory" of the processes that generated it. While a time series that is generated by a white-noise process is uncorrelated ($x(t)$ and $x(t + \tau)$ are independent), many natural phenomena, and in particular in our climate, are described by processes in which the state at time t depends on previous states. Examples include the Ornstein–Uhlenbeck process (as discussed in Chapter 2) and *autoregressive processes*. In these cases, the presence of auto-correlations needs to be taken into account when analyzing the significance of cross-correlation values (see Section 3.4).

For climate data, if the series does not have a specific periodicity, the auto-correlation usually decays exponentially with increasing τ as

$$c(\tau) \sim \exp\{-\tau/\tau_d\} \tag{3.6}$$

where τ_d is the *decorrelation time* of the time series. In contrast, if the time series under examination displays an oscillatory behavior with a well-defined frequency, then the auto-correlation will present an oscillating behavior enveloped in an exponential decay.

Figure 3.3 displays two examples of auto-correlation functions for two time series of zonal wind anomalies at 850 hPa and two latitudes: the solid line represents the auto-correlation of an equatorial time series, while the dashed one is of an midlatitude one. In the inset the two time series are shown. In the extratropics the

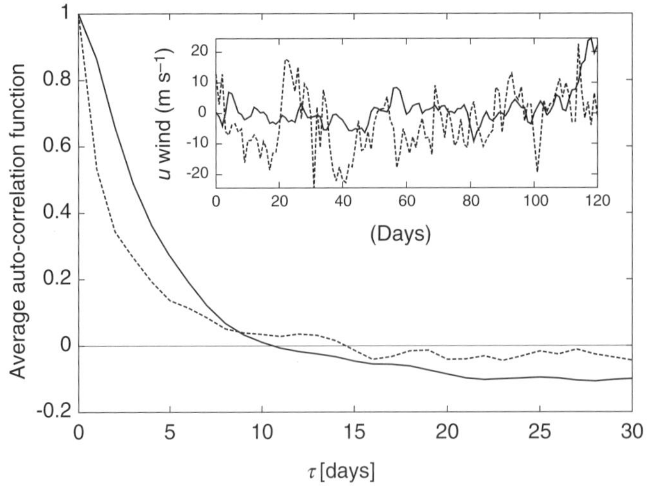

Figure 3.3 Auto-correlation function of two time series (in the inset) of zonal wind at 850 hPa anomalies at the equator (solid) and at midlatitude (dashed). In the midlatitude, where the atmosphere is more turbulent, the fluctuations of the time series are faster and more intense. This is reflected by the auto-correlation, that drops to zero quicker than the equatorial one. Source: Tirabassi (2015).

atmosphere is more energetic and turbulent with respect to the tropics, which are dominated by very regular trade winds. These features are reflected by the aspect of the two series, one being much more noisy than the other. This in turn reflects into the auto-correlation functions, with the tropical one having a slower decay (longer decorrelation time) than the extratropical one.

3.2.2 Spectral Analysis

Assuming a wide-sense-stationary random process, a well-known relationship links the auto-correlation of a time series with its power spectrum: the auto-correlation function has a spectral decomposition given by the power spectrum of that process, a result that is known as the Wiener–Khinchin theorem. To investigate in detail the spectral properties of a time series, the Fourier decomposition is a natural choice. In the case of a discrete signal the Fourier power spectral density is usually called a *periodogram* and is computed via a fast Fourier transform (FFT) algorithm. The periodogram gives information about the timescales involved in the processes under study. In fact, given a spectrum $P(\omega)$, the energy carried by the time series at the ω frequency is $E(\omega) = |P(\omega)|^2$.

Under the null-hypothesis (Chapter 2), the time series displays a *red spectrum* (Imkeller and Von Storch, 2001) and the largest fraction of the energy is contained in the slow timescales (low frequencies). For this reason, the spectrum of an observable needs to be tested against the red spectrum given by an autoregressive process of order 1, often indicated by AR(1), with same mean, variance, and auto-correlation at lag 1 as the original time series. In this way, eventual peaks at low frequencies can be explained as just a product of integrated noise, not representing real cycles. An example of such a situation is presented in Fig. 3.4, which that displays the spectrum of the time series of NINO3.4 index (solid line) together with the 5% highest and lowest percentiles (dashed lines) of the power spectrum distribution of 100 red-noise processes with the same auto-correlation. As we can see, only in an interval of frequencies is the spectrum larger than the 5% largest values of the red spectrum, ranging roughly from one to ten years with significant peaks around 3–4 years.

3.2.3 Wavelets

Let us imagine that a time series represents a harmonic oscillation with a frequency that varies with time, $X(t) = \sin(\omega(t)\ t)$. A Fourier spectral analysis would not be able to capture this feature of the time series, but an alternative methodology, known as *wavelet analysis*, can provide this information.

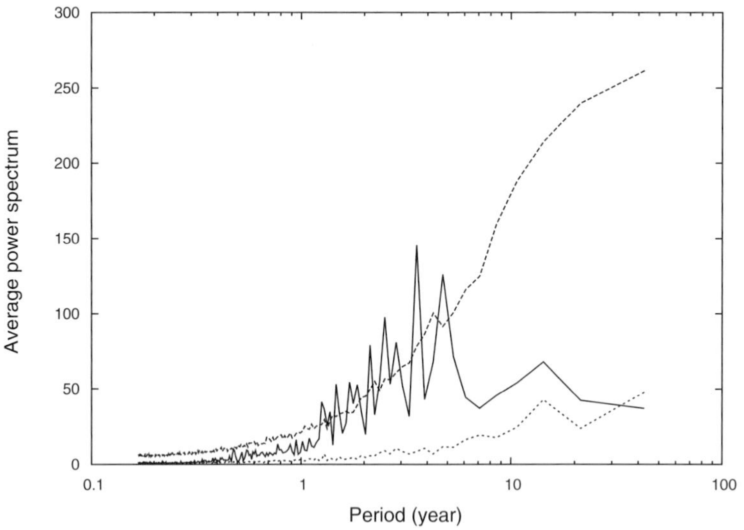

Figure 3.4 Power spectrum of NINO3.4 monthly time series (solid), expressed as a function of the period instead of the frequency, with (dashed) the 5% highest and lowest percentiles of the power spectrum distribution of 100 red-noise processes with same lag-1 auto-correlation and variance of the NINO3.4 time-series. Source: Tirabassi (2015).

The wavelet method uses an expansion of the time series in a series of *localized waves*, both in time and in frequency, which are called *wavelets* (Daubechies, 1992). Wavelets have a well-defined frequency and an amplitude that is significantly different from 0 only during a short amount of time. An example of a basic wavelet is a monochromatic plane wave enveloped in a Gaussian profile; other wavelets are obtained by stretching and translations of the basic wavelet. Figure 3.5 displays the wavelet spectrum for the time series of NINO3.4. Along the time axis several intervals of periodicity between three and six years can be seen, which characterize the typical range of variability of El Niño (see also Fig. 3.4).

3.2.4 Empirical Orthogonal Functions

If, instead of a single time series, the data to analyze are composed of observables at different locations, one might be interested in the identification of the spatial patterns that the data display, and among them, to select those that carry the greater amount of variance of the field. To do so we can use the EOF technique, also called *principal component analysis* (PCA).

Let us suppose we have a database of synchronous time series $X_{t,i}, i = 1, \ldots, N$ recorded at N different geographical locations (as we will see, the spatial sampling

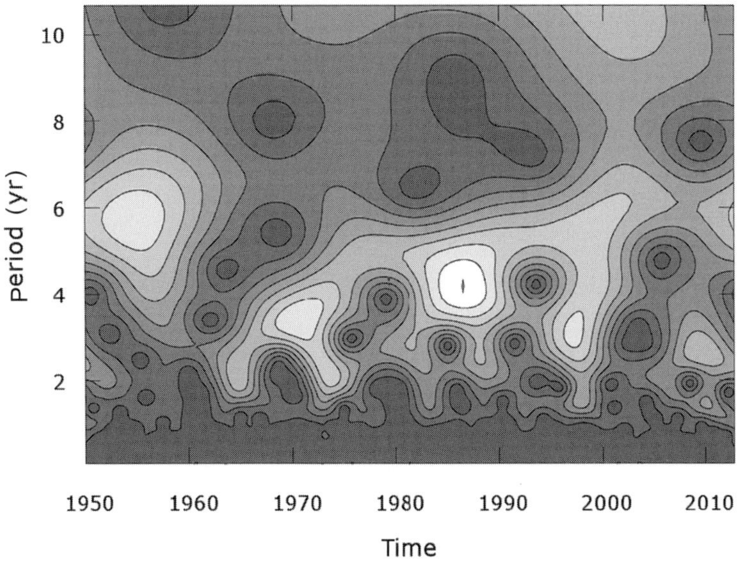

Figure 3.5 Wavelet spectrum of NINO3.4 time plane wave series (lighter colours indicate higher values).

does not need to be regular). If each time series $\vec{X}_i$ is normalized to zero mean and unit variance we can write their covariance matrix as

$$\mathbf{C} = \mathbf{X}^T\mathbf{X}. \tag{3.7}$$

Since $\mathbf{C}$ is symmetric, the eigenvectors form an orthogonal basis on which we can project the data matrix $\mathbf{X}$:

$$X_{t,i} = \sum_{j=1}^{N} P_{t,j}A_{ji}, \tag{3.8}$$

where each time series $\vec{P}_j$ is called a principal component (PC) and each $\vec{A}_j$ is an EOF of the field $\mathbf{X}$. The EOFs have the same dimension as the spatial dimension of the database, and they thus represent a spatial pattern. Looking at (3.8), we can then see in the PCs the time evolution of the maps represented by the EOFs (Fig. 3.6). The sum of the oscillations of all the EOFs give the original field $\mathbf{X}$. Since the trace of $\mathbf{C}$ is equal to the variance of X, the coefficient

$$\lambda_j = \frac{C_{jj}}{\mathrm{Tr}(\mathbf{C})} \tag{3.9}$$

represents the variance fraction explained by each statistical mode of variability. Usually it is common to consider the first modes only, because they can explain alone a great part of the field variance. However, among all the main modes,

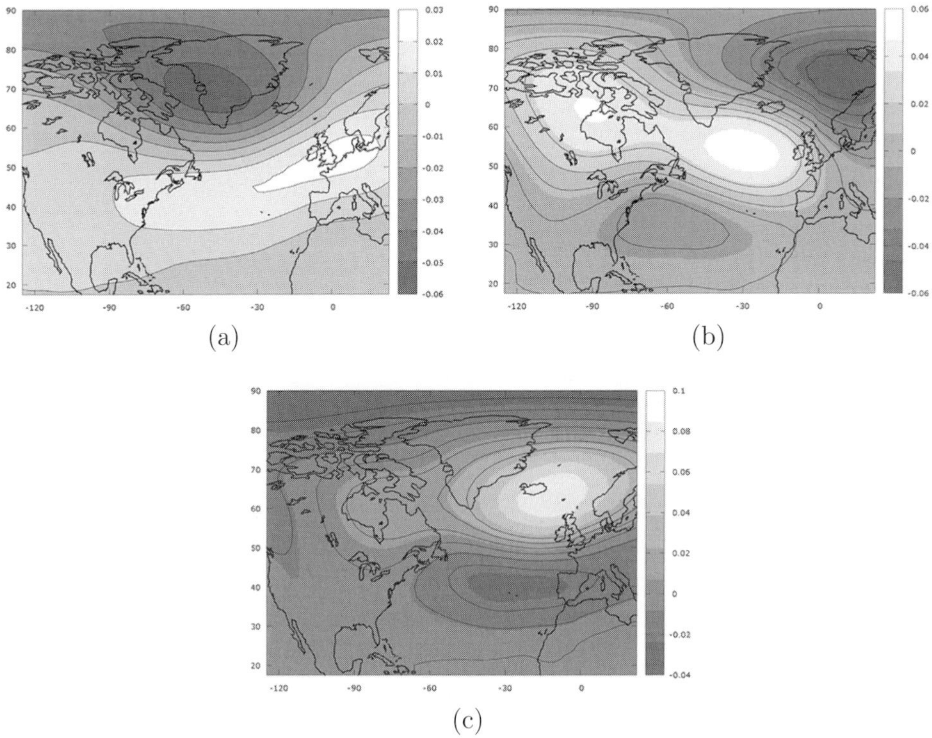

Figure 3.6 (a)–(c) first three EOFs for geopotential height at 500 mbar. Data from NCEP/NCAR monthly reanalysis. Source: Tirabassi (2015).

the first one is the most physically informative, since it is not constrained by the orthogonality requirement imposed by the symmetry of **C**. To overcome this restriction, advanced approaches can be used to derive alternative EOFs-like principal modes (Navarra and Simoncini, 2010; Runge et al., 2015).

As an example, Fig. 3.6 displays the first three EOFs of the North Atlantic geopotential height field at 500 mbar, as derived from NCEP/NCAR monthly reanalysis. As we can see, the first EOF displays a dipole centered over high and midlatitudes. This pattern is closely related to the North Atlantic Oscillation (NAO).

3.3 Nonlinear Analysis Tools

A wide range of nonlinear tools for the analysis of complex signals is available in the literature. For example, a nonlinear extension of traditional EOF decomposition has been proposed and applied to climate data (Mukhin et al., 2015). Here we focus on several tools that are used in various chapters of this book; general and comprehensive reviews can be found elsewhere (Bradley and Kantz, 2015; Kantz and Schreiber, 2003).

3.3.1 Entropy

In the framework of *information theory* (Cover and Thomas, 2006), the standard way to quantify the degree of *unpredictability* or *disorder* of a time series is by means of *Shannon entropy* (Shannon, 1948), defined as:

$$H = -\frac{1}{\log M} \sum_{n=1}^{M} p_n \log p_n. \tag{3.10}$$

The $p_n, n = 1, \ldots, M$ form the probability distribution of the values in the time series, M is the number of bins, and the entropy is normalized to the maximum entropy, $\log M$, of the uniform distribution. If log base 2 is used, the units of entropy are bits. The entropy can be understood as the "quantity of surprise one should feel upon reading the result of a measurement" (Hlavackovaschindler et al., 2007).

An important drawback of this measure is that, when the probability distribution is estimated from the histogram of data values, the measure depends on how the phase space is partitioned, i.e., the bins used to compute the histogram. In Section 3.3.5 a symbolic method of analysis will be presented that allows computing a probability distribution without the need to define a partition.

An entropy analysis of surface air temperature (SAT) time series is presented in Fig. 3.7, where two monthly reanalysis, NCEP/NCAR CDAS1 and ERA Interim (see Section 3.1), are compared (Arizmendi et al., 2017). Before computing the entropy, each time series is normalized to zero mean and unit variance. Well-defined spatial structures are uncovered, with the extratropics having higher entropy values with respect to the tropics, which can be expected due to stronger SAT variability. In the western Pacific there is a difference between the two reanalyses, which is due to the presence of extreme values (outliers).

3.3.2 Mutual Information

The mutual information (M_I) of two stochastic variables measures the mutual dependence between the two variables. More specifically, it quantifies the amount of information obtained about one variable, through the knowledge of the other variable. Considering two time series, X and Y, which have associated marginal probability distribution functions $p(x)$ and $p(y)$, and a joint one, $p(x,y)$, M_I is defined as:

$$M_I = \sum_x \sum_y p(x,y) \, \log\left(\frac{p(x,y)}{p(x)p(y)}\right). \tag{3.11}$$

In this sum, states with zero probability of occurrence are ignored. If a log base 2 is used, the units of entropy are in bits.

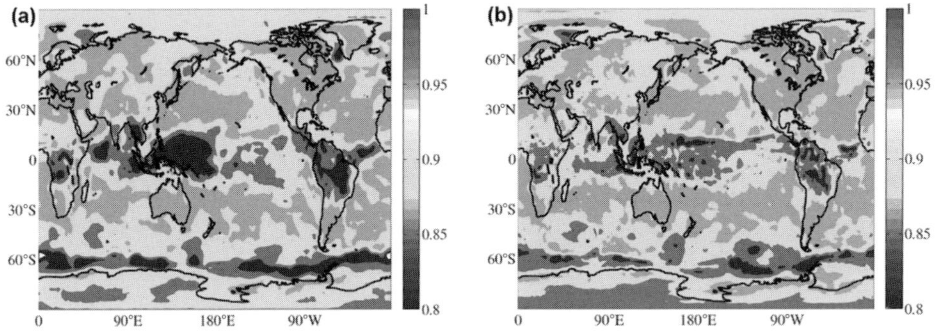

Figure 3.7 Shannon entropy of monthly surface air temperature time series. (a) NCEP CDAS1 reanalysis, (b) ERA Interim reanalysis. The number of bins M is the same for all time series within a reanalysis database, but is adjusted in each database to take into account the different length of the time series: in panel (a) $M = 40$; in panel (b) $M = 20$. Source: Arizmendi et al. (2017). (A black and white version of this figure appears in some formats. For the color version, please refer to the plate section.)

We can think of M_I as the nonlinear partner of the cross-correlation. In fact, two variables are independent if and only if $p(x, y) = p(x)p(y)$, so that and $M_I = 0$. Nevertheless, when estimating the probabilities from finite time series one can obtain a small positive value of M_I even if two series are independent(Cellucci et al., 2005; Steuer et al., 2002). Therefore, the statistical significance of M_I values needs to be carefully analyzed (Section 3.4). The M_I is symmetric and can also be computed with a time lag. However, unlike the cross-correlation, the M_I is always positively defined.

3.3.3 Event Synchronization

An alternative approach to measure synchronization and to infer time delays between signals, originally proposed by Quian Quiroga et al. (2002), is known as event synchronization (ES). It is based on the relative timings of events in a time series (defined, e.g., as threshold crossings, local maxima or minima, etc.). The degree of synchronization is obtained from the number of quasi-simultaneous events.

The method starts by extracting two event series from two time series, X and Y. An event l that occurs at X at time t_l^x is considered to be synchronized with an event m that occurs at Y at time t_m^Y within a time lag $\pm\tau_{lm}^{xy}$ if $0 < |t_l^x - t_m^y| < \tau_{lm}^{xy}$, where $\tau_{lm}^{xy} = \min\{t_{l+1}^x - t_l^x, t_l^x - t_{l-1}^x, t_{m+1}^y - t_m^y, t_m^y - t_{m-1}^y\}/2$. Here, $l = 1, 2, \ldots, s_x$, $m = 1, 2, \ldots, s_y$, where s_x and s_y are the number of events in the X and Y time series respectively. Then, the number of times that an event occurs in X shortly after Y is counted: $c(x|y) = \sum_{l=1}^{s_x} \sum_{m=1}^{s_y} J_{xy}^{lm}$, and vice versa: $c(y|x)$. Here, J_{xy}^{lm} is given by

$$J_{xy}^{lm} = \begin{cases} 1, & \text{if } 0 < t_l^x - t_m^y \le \tau_{lm}^{xy}, \\ 1/2, & \text{if } t_l^x = t_m^y, \\ 0, & \text{else.} \end{cases} \tag{3.12}$$

Then, the *strength of synchronization* is defined as

$$Q_{xy} = \frac{c(x|y) + c(y|x)}{\sqrt{(s_x - 2)\,(s_y - 2)}}, \tag{3.13}$$

and is a nonlinear measure of correlation between two time series.

3.3.4 Directionality and Causality Measures

The cross-correlation and the mutual information are both symmetric measures and thus provide no information about the *direction* of interrelationships. In the literature several techniques for investigating information transfer and causality have been proposed (Baccala and Sameshima, 2001; Balasis et al., 2013; Eichler et al., 2003; Feldhoff et al., 2012; Granger, 1969; Rosenblum and Pikovsky, 2001; Runge et al., 2015; Sameshima and Baccala, 1999; Sugihara et al., 2012). Here we focus on an information-theoretic measure based on the transfer entropy, and on Granger causality.

We consider two time series, x_i and y_j, which are abbreviated as i and j. A *directionality index* (DI) that can yield insight into the direction of information transfer is (Hlavackovaschindler et al., 2007; Palus and Stefanovska, 2003):

$$\mathrm{DI}_{ij}(\tau) = \frac{\mathrm{TE}_{ij}(\tau) - \mathrm{TE}_{ji}(\tau)}{\mathrm{TE}_{ij}(\tau) + \mathrm{TE}_{ji}(\tau)}, \tag{3.14}$$

where $\tau > 0$ is the *characteristic timescale of information transfer* and $\mathrm{TE}_{ij}(\tau)$, $\mathrm{TE}_{ji}(\tau)$ are the *transfer entropies*, defined as:

$$\mathrm{TE}_{ij}(\tau) \equiv M_I(i; j|i_\tau),$$
$$\mathrm{TE}_{ji}(\tau) \equiv M_I(j; i|j_\tau).$$

Here, $i_\tau = i(t-\tau), j_\tau = j(t-\tau)$, and $M_I(i;j|k)$ is the *conditional mutual information* (CMI) that measures the amount of information shared between two time series $i(t)$ and $j(t)$, given the effect of a third time series, $k(t)$, over $j(t)$. It can be rewritten as:

$$M_I(i;j|k) = \sum_{m,n,l} p_{ijk}(m,n,l) \log \frac{p_k(l)p_{ijk}(m,n,l)}{p_{ik}(m,l)p_{jk}(n,l)}. \tag{3.15}$$

The transfer entropy $\mathrm{TE}_{ij}(\tau)$ is obtained when $k(t)$ is replaced by the past of the time series $i(t)$ to account for the information transfer time τ, and quantifies the amount of information shared between $i(t)$ and $j(t)$, given the influence of $i(t-\tau)$

over $j(t)$. In the same way, $\mathrm{TE}_{ji}(\tau)$ is obtained when $k(t)$ is replaced by the past of the time series $j(t)$ to account for the information transfer time τ, and quantifies the amount of information shared between $i(t)$ and $j(t)$, given the influence of $j(t-\tau)$ over $i(t)$.

By comparing $\mathrm{TE}_{ij}(\tau)$ and $\mathrm{TE}_{ji}(\tau)$, the DI quantifies the *net* direction of information flow: if $\mathrm{DI}_{ij}(\tau)$ is positive, we can infer *a net directionality from i to j* as $\mathrm{TE}_{ij}(\tau) > \mathrm{TE}_{ji}(\tau)$. Conversely, if $\mathrm{TE}_{ij}(\tau) < \mathrm{TE}_{ji}(\tau)$, then $\mathrm{DI}_{ij}(\tau) < 0$ which gives *a net direction of information flux from j to i*.

The directionality index is antisymmetric (by definition, $\mathrm{DI}_{ij} = -\mathrm{DI}_{ji}$) and is normalized such that $-1 \leq \mathrm{DI}_{ij} \leq 1$ with limiting values

$$\mathrm{DI}_{ij} = 1 \text{ if and only if } \mathrm{TE}_{ij} \neq 0,\ \mathrm{TE}_{ji} = 0 \tag{3.16}$$

(i.e., the information flow is $i \to j$ and there is no back coupling $j \to i$), and

$$\mathrm{DI}_{ij} = -1 \text{ if and only if } \mathrm{TE}_{ji} \neq 0,\ \mathrm{TE}_{ij} = 0 \tag{3.17}$$

(i.e., the information flow is $j \to i$ and there is no back coupling $i \to j$).

Naturally, $\tau > 0$ is a parameter that has to be tuned appropriately to the timescales involved in the series. If τ is too small, $\mathrm{DI}_{ij}(\tau)$ will capture short timescale directionality, and may fail if the time series behave too similarly on those timescales as they do if they are subjected to the same external forcing. On the other hand, if τ is too large, larger than the decorrelation time of the time series, the effect of the past i over j (and of j over i) will be negligible and $\mathrm{DI}_{ij}(\tau)$ will be a small and in principle random value. As will be discussed in Section 3.4, statistical testing allows to assess the significance of $\mathrm{DI}_{ij}(\tau)$.

Another way to infer the direction of interaction between two processes X and Y is by using *Granger causality* (Granger, 1969), which relies in the possibility of modeling the two processes as autoregressive processes. Let us suppose we study the influence of the time series X over the time series Y. We model Y as an autoregressive processes of order d forced by the elements of X:

$$Y_t = \sum_{i=1}^{d} a_i Y_{t-i} + \sum_{i=1}^{d} b_i X_{t-i} + \epsilon_t, \tag{3.18}$$

where $\vec{a}$ and $\vec{b}$ are vectors of constant coefficients and ϵ_t are the noise residuals.

The idea is to test the hypothesis $\vec{b} \neq 0$ against the null hypothesis $\vec{b} = 0$. To do so we proceed in three steps. In the first one we fit with a linear regression $\vec{a}$ and $\vec{b}$, and compute the associated variance of the residuals, $\sigma^2_{\text{coupled}}$. In the second step the fit is repeated, setting $b_i = 0$, and again the variance of the residuals is computed, namely $\sigma^2_{\text{uncoupled}}$. The last step involves comparing the two residual variances: If the variance in the coupled case is smaller than the one in the uncoupled case, it means that the predictive power of the coupled model is higher. Then, the Granger causality estimator (GCE) is calculated as (Mokhov et al., 2011):

$$GCE = \frac{\sigma^2_{\text{uncoupled}} - \sigma^2_{\text{coupled}}}{\sigma^2_{\text{uncoupled}}}. \tag{3.19}$$

From this equation it is clear that if $GCE > 0$ the information given by X allowed for a more precise prediction of Y. Thus, X is said to be Granger causal of Y.

To test the statistical significance of this result, it is possible to use a variable F computed from GCE having a known distribution. In particular we can write the quantity F as

$$F = \left(\frac{N - 2d - 1}{d}\right) \frac{\sigma^2_{\text{uncoupled}} - \sigma^2_{\text{coupled}}}{\sigma^2_{\text{coupled}}}, \tag{3.20}$$

where N is the length of the time series. The quantity F is distributed according to the Fisher–Snedecor distribution with d and $N - 2d - 1$ degrees of freedom (Seber and Lee, 2012).

Let us now suppose that we have not just a couple of time–series X and Y but a whole ensemble of couples of series X_i and Y_i, each pair being an independent realization of the same process described by (3.18). In this case we face a problem of multiple-trial Granger causality. The procedure to compute the GCE is as above, the only difference being in the fit procedure. Writing (3.18) as

$$\vec{Z} = \mathbf{W}\vec{\gamma} + \vec{\epsilon}, \tag{3.21}$$

where $\vec{Z}$ is the time series Y^T_{t+1}, $\vec{\gamma} = \{a_1, \ldots, a_D, b_1, \ldots, b_D\}^T$, and $\mathbf{W}$ is a $2D \times T$ matrix containing the elements of X and Y, we can first obtain

$$\mathbf{W}^T\vec{Z} = \mathbf{W}^T\mathbf{W}\vec{\gamma} + \mathbf{W}^T\vec{\epsilon}. \tag{3.22}$$

Then, by averaging over all the trials (indicated by brackets), we obtain

$$\vec{\gamma} = \left(\langle \mathbf{W}^T\mathbf{W}\rangle\right)^{-1} \langle \mathbf{W}^T\vec{Z}\rangle, \tag{3.23}$$

where we have supposed $\vec{\gamma}$ to be the same in all the trials.

As a final note, the order d of the autoregressive model in (3.18) is chosen in order to minimize the function,

$$S_i(d) = \frac{N}{2} \ln\left(\frac{\sigma^2_{\text{uncoupled}}}{\sigma^2_{\vec{Y}}}\right) + \frac{d}{2} \ln(N), \tag{3.24}$$

which is a good compromise between obtaining a good fit and avoiding over-fitting (Schwarz, 1978).

The Granger methodology has been successfully applied to "disentangle" interactions in many complex systems from observed time series, such as finance, neuroscience, and climate. Granger causality was employed by Mokhov et al. (2011) to understand the interplay between ENSO and the Indian Monsoon. A mutual influence was found, although the interaction in one direction is different from the interaction in the opposite direction. In particular, using nonlinear regressions as a generalization of (3.18), it was shown that the influence of ENSO over the Indian Monsoon is nonlinear, while the influence of the Indian Monsoon on ENSO is mainly linear.

In Tirabassi et al. (2017), two time series that serve as measures of ENSO and the Indian Monsoon (the NINO3.4 index and the All India Rainfall index) were analyzed. Apart from the Granger causality estimator, also the *partial directed coherence* (PDC) (Baccala and Sameshima, 2001; Sameshima and Baccala, 1999) and *directed partial correlation* (DPC) (Eichler et al., 2003) techniques were used. While PDC is a frequency-domain representation of the concept of Granger causality, DPC is an alternative approach for quantifying Granger causality in the time domain. The three measures were found to give consistent results, but the DPC allowed for a better discrimination of the strength of the weaker Indian Monsoon–ENSO interaction.

3.3.5 Ordinal Analysis

A tool of time series analysis able to identify nonlinear correlations and recurrent oscillatory patterns in complex data sets is known as *ordinal analysis* (Bandt and Pompe, 2002). The method takes into account the relative ordering of the data values in a time series and provides an alternative way to compute the probability distribution p_i associated to the time series.

For example, let us imagine we have a time series of anomaly surface air temperature (SATA) records T_i, daily sampled. If the value of a data point at a certain day is larger than the one of the day before, we assign to that day the symbol "1," otherwise we assign the symbol "0." In this way we have reduced the time series to a sequence of two symbols. If we compare the relative ordering of 3 (4) consecutive values, then 6 (24) different symbols (known as *ordinal patterns*) can be defined. The frequency of occurrence of the different symbols gives the ordinal probability distribution p_i that can be used for computing the information-theory measures that were presented in the previous sections. For example, the entropy computed with (3.10) using ordinal probabilities is known as *permutation entropy*.

Using combinatory calculus it is easy to see that the number N of patterns with length K grows as $N = K!$. Thus N quickly diverges with the number of data points

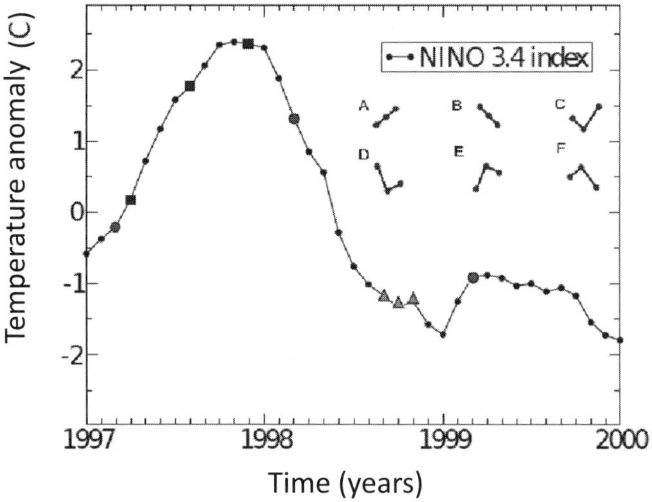

Figure 3.8 Example of how ordinal patterns allow us to select the timescale of the analysis. Three ordinal patterns of length three are indicated in the time series of the NINO3.4 index (monthly averaged). Triangles: intraseasonal pattern, squares: intraannual pattern, and circles: interannual pattern. Source: Deza (2015).

used to define the patterns. This is a severe limitation if we want to study finite length time series. For a standard monthly reanalysis time series, the typical length of the data is $L \sim 10^3$. To estimate correctly the joint probability distribution for the mutual information, one should not go below the heuristic value of ten data points per possible symbol. The number of possible different symbols in a joint probability calculation is N^2. Thus we have a rule of thumb for N of the kind $N \sim \sqrt{L/10}$. For a monthly time series this limits the analysis to $K = 3$ or $K = 4$.

The drawback of using this symbolic method to analyze a time series is that we lose the information about the real magnitude of the data values. However, this ordinal methodology has an important advantage for the analysis of climate data: It allows us to select a specific timescale for the analysis because we can compare data values using a lag, i.e., we can compare the value at a certain day/month with the value at a certain time in the past (for example, the same day of the previous month, or the same day of the previous year).

To fix the ideas, consider monthly sampled climatic data such as the NINO3.4 index, as shown in Fig. 3.8. The time interval of the ordinal patterns can be varied by considering not only three consecutive months (which we will refer to in the next chapter as *intraseason* timescale; e.g., January, February, March; February, March, April; etc.) shown as triangles – but also in three consecutive years (which we will refer to as *interannual* timescale; e.g., January 2010, January 2011, January 2012; February 2010, February 2011, February 2012; etc.), shown as circles, or any other timescale, such as the one shown as squares, where the ordinal pattern is formed with three equally spaced data points covering a one-year period. In this way, ordinal analysis allows us to study timescale-dependent phenomena, which are not captured with probabilities computed from single-time histograms of data values.

A comparison of mutual information values computed by considering the actual values of the data points, and their ordinal representation, is presented in Fig. 3.9. We analyze monthly SATA from the NCEP/NCAR reanalysis and display in color code the M_I between a time series recorded in a reference point (indicated with a cross and located in the eastern Pacific region) and all the time series in the database. The probabilities are computed with six bins, in panel (a) using histograms of values, while in panels (b)–(d) using ordinal patterns (OPs) of length three. In panel (b) the OPs are defined by three consecutive months, in panel (c) by three equally spaced months covering a period of one year, and in panel (d) by three months in consecutive years. In this figure only M_I values that are considered significant (according to a simple threshold criterion that will be discussed in the next section) are shown. We note that an increase in the OP spacing (therefore an increase in the timescale of the analysis) results in an increase in the number of regions having significant M_I values. In particular, we note that in the intraseason timescale (panel b), the only significant M_I values are with neighboring regions, but the geographical area having significant M_I values grows with the timescale of the analysis. We also note that the map obtained by using probability distributions that are computed from the histograms of anomaly values (panel a) and from the probabilities of interannual OPs (panel d) appear quite similar, which suggests that the M_I values in panel a capture climatic phenomena that have a characteristic timescale of three or more years.

3.4 Statistical Testing

A crucial part of statistical data analysis is to test a *null hypothesis* that there is no significant relationship between two observables. Typically it is assumed that observations are sampled from a distribution H with mean μ (null hypothesis). Then, the *p-value* is computed, which is the probability of obtaining a result equal to, or more extreme than, the observed result, assuming that the null hypothesis is true. A p-value is a number between 0 and 1 and the smaller the p-value, the stronger is the evidence against the null hypothesis. Although p-values are easy to use and thus extremely popular, care should be taken when interpreting p-values because a p-value only measures the compatibility of an observation with a hypothesis, not the truth of the hypothesis (Altman and Krzywinski, 2017; Wasserstein and Lazar, 2016).

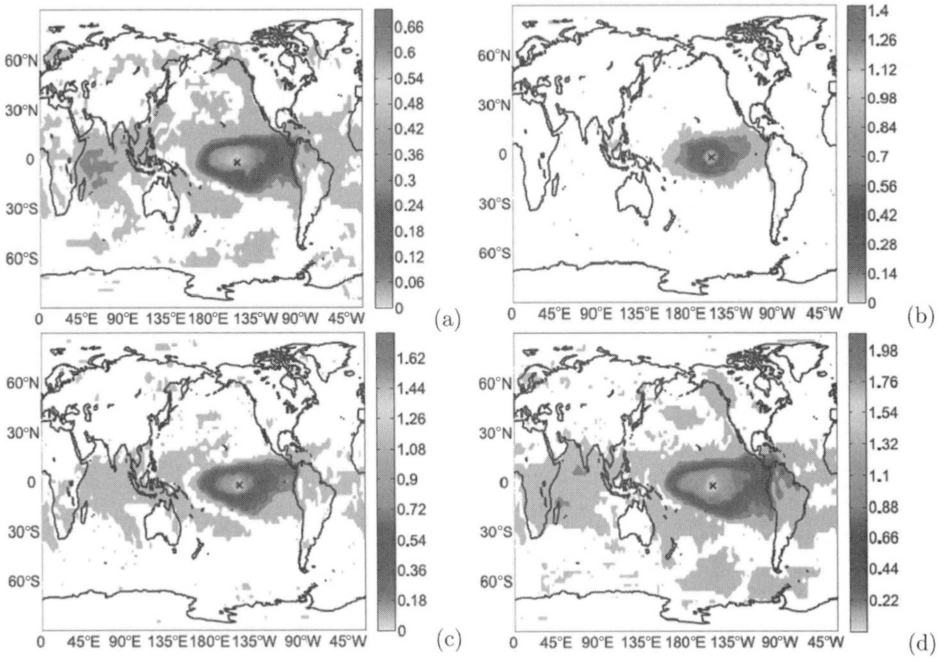

Figure 3.9 Mutual information (M_I) values of monthly SATA from NCEP/NCAR reanalysis. The value of M_I between the time series recorded at a given reference point (indicated with a cross) and all the time series recorded at different points is shown in color code. The probability distributions used in panel (a) are computed from the histograms of anomaly values (using six bins to compute the histogram), while in (b)–(d) OPs are used: in (b), the patterns are formed by three consecutive months, in (c) by three equally spaced months covering a one-year period, and in (d) by three months in consecutive years. Only significant M_I values are shown: a M_I value is considered significant if it is larger than $\mu+3\sigma$, with μ and σ being the mean value and the standard deviation of the distribution of M_I values computed from surrogates. Source: Deza (2015). (A black and white version of this figure appears in some formats. For the color version, please refer to the plate section.)

Another way to estimate the likelihood of an observation, considering the null hypothesis, is by using a *significance threshold*, defined in terms of the reference value, μ, and the standard deviation, σ, of the distribution H: An observation x is considered not consistent with the null hypothesis if it differs from the expected mean μ more than a threshold $n\sigma$, i.e., $|x - \mu| > n\sigma$. The larger the number n, the more strict the significance criterion is. This criterion was used in Fig. 3.9 to infer significant values of mutual information with the distribution H computed from *surrogates*.

The simplest null hypothesis is that the distribution H is a Gaussian of μ and σ adjusted to the observed data. However, complex phenomena are usually not described by Gaussian distributions. A more appropriate approach is to generate surrogate data, i.e., synthetic data aimed at modeling the observed data. Algorithms to generate surrogates can be classified into two groups (Mudelsee, 2014): (1) model-based surrogates: models that reproduce main statistical properties (e.g., histogram, auto-correlation) of the observed data are used to generate surrogates; and (2) model-free surrogates: generated from the observed data by a suitable transformation that removes the property whose significance is being tested. Some of the methods used are (Lancaster, 2018) as follows:

- *Random shuffle*: The simplest approach, surrogates are generated simply by random permutations of the original time series. The permutations maintain the probability distribution of the data values, but destroy correlations among them.
- *Fourier* surrogate (also known as *random phases* surrogate): In order to preserve the auto-correlation function (and thus the power spectra), surrogate data are created by the inverse Fourier transform of the modules of Fourier transform of the original data with new (uniformly random) phases. If the surrogates must be real time series, the Fourier phases must be antisymmetric with respect to the central value of data. This method, however, does not conserve the probability distribution of the data values.
- *Amplitude adjusted Fourier transform* (AAFT) surrogate: This method attempts to preserve both the auto-correlation structure and the probability distribution of data values. This is done by transforming the PDF in a Gaussian before applying the Fourier surrogate, to take advantage of the property of Gaussian distributions under Fourier transforms, and then applying the inverse transformation. The drawbacks of this method are: it is time consuming; the nonlinear transformations might modify the auto-correlations; and a long time series is in general needed, which is often not available in climatology.
- *Bootstrap surrogate*: This is generated by randomly resampling with replacement from the original set of data, using blocks of data of length given by decorrelation time (3.6) of the original time series.

Fig. 3.10 presents an example of statistical testing of mutual information M_I (3.11) and DI (3.14) values of two time series of SATA, recorded in two regions in the Pacific Ocean (referred as i) and in the Indian Ocean (referred as j) (Deza et al., 2015). Histograms of the M_I and DI values ($\tau = 1$ month) computed from 100 bootstrap surrogates are shown; the thick vertical lines indicate the significance thresholds, and the thin lines indicate the M_I and DI values computed from the original time series. The M_I threshold is defined as $\mu + 3\sigma$ with μ and σ being the mean and standard deviation of the M_I distribution computed from bootstrap surrogates; the DI thresholds are defined as $\mu \pm 3\sigma$ with positive (negative) value corresponding to the $i \to j$ ($j \to i$) direction; here, μ and σ are the mean and the

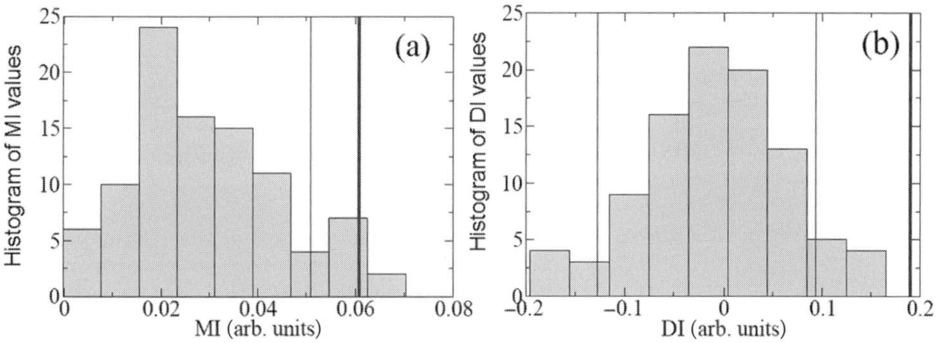

Figure 3.10 Histograms of the (a) M_I and (b) DI values computed from 100 bootstrap surrogates: the thin lines indicate the significance thresholds ($\mu \pm 3\sigma$) and the thick lines indicate the M_I and DI values computed from the original data. In this example both values are significant. Source: Deza (2015).

standard deviation of the DI distribution computed from bootstrap surrogates. In this example both values are significant; as expected, there is significant evidence of interdependency between the time series in the two regions, and significant evidence of a net direction of information transfer from the ENSO region to the Indian Ocean region.

To summarize, in this chapter we have described different climatological data sets and presented various linear and nonlinear tools of time series analysis, which will be used in the following chapters to define and to characterize climate networks. Linear methods include correlation analysis, spectral analysis, and EOFs; nonlinear methods include symbolic ordinal analysis, event synchronization, and the use of information-theory measures such as entropy and mutual information.

Many other analysis tools have been used to investigate climate data. See Donner et al. (2011) for an overview of methods based on recurrence plots; Zappala et al. (2016, 2018) for methods based on the Hilbert transform; Mudelsee and Bermejo (2017) for the analysis of extreme distributions, and Balasis et al. (2013) for information transfer and causality.

While our focus is the analysis of climatological data, most of the tools presented here have been applied to a broad range of signals across interdisciplinary fields.

4

Climate Networks
Construction Methods and Analysis

Systems composed by a large number of interconnected units can be modeled as complex networks, where the nodes represent the individual units and the links among the nodes represent the interactions among them. Many real-world systems such as the internet, ecological, social, and biological systems, and even the human brain can be described by complex networks. Early work in the 1990s discovered several important network properties, such as nontrivial topologies (Barabási and Albert, 1999; Watts and Strogatz, 1998). Over the years, complex networks have attracted a huge amount of research, and new concepts and measures have yielded new insight into hidden features and structures (Albert and Barabási, 2002; Barabási, 2009; Caldarelli, 2007; Newman, 2010; Strogatz, 2001).

As discussed in previous chapters, our climate is made up of a huge number of nonlinear subsystems with mutual nonlinear interactions and feedback loops active on a wide-range of spatial scales – from several meters to thousands of kilometers – and timescales – from several hours to many years. Therefore, modeling the Earth climate system by using complex networks can clearly advance our understanding of climate phenomena (Donges et al., 2009a,b; Tsonis and Roebber, 2004; Tsonis and Swanson, 2006, 2008; Yamasaki et al., 2008). In this chapter we review basic concepts of network theory and methods for constructing climate networks, and in the following chapters we will present several advances gained by using the network approach.

4.1 Complex Networks

A *network* or *graph* G is composed of two sets: the set of *nodes* or *vertices*, V, and the set of *edges* or *links*, E. Each edge connects two elements of V, i.e., two nodes, and the set E defines the *network topology*, which can be regular or non-regular. Large non-regular networks are usually referred to as complex networks. Another classical example is the *random network*, or *Erdös–Rényi network*, constructed by drawing links between each pair of nodes with a fixed probability p. The network definition is very general, which makes them applicable to the description of numerous systems. Edges are assigned to represent some relationship between the nodes they connect, as for example some physical interaction, or some statistical dependency.

4.1.1 General Definitions and Properties

The set of links that represent relationships between nodes is described by the so-called *adjacency matrix*, $\mathcal{A}$, defined as

$$\mathcal{A}_{ij} = 1, \quad \text{if } i \text{ and } j \text{ are linked nodes; } \mathcal{A}_{ij} = 0, \quad \text{otherwise.} \tag{4.1}$$

If the network contains N nodes the adjacency matrix will be a $N \times N$ matrix. The above definition is appropriate if the relations among the nodes are non-directional (that is, having i connected to j implies that also j is connected to i). In this case the adjacency matrix is symmetric. If the relationships are directional, then we have $\mathcal{A}_{ij} = 1$ only if there is a link from i to j, giving a nonsymmetric adjacency matrix.

Additionally, the connections between nodes can have associated "weights" which quantify the strengths of the relationships. In this case the network is described by the weighted adjacency matrix, $\mathcal{W} = \{w_{ij}\}$, with $w_{ij} \neq 0$ if i and j are linked nodes and $w_{ij} = 0$ otherwise. $\mathcal{W}$ can also be symmetric or asymmetric.

One of the main features of a complex network topology is that the number of links varies from node to node. The *degree* of a node (also called *degree centrality*) is the number of links the node has (i.e., the number of *neighbors*):

$$k_i = \sum_j \mathcal{A}_{ij}. \tag{4.2}$$

For directed networks, incoming and outgoing degrees can be defined ($K_I(i) = \sum_j \mathcal{A}_{ji}$, $K_O(i) = \sum_j \mathcal{A}_{ij}$). Similarly, for weighted networks, *weighted degrees*, also called *strength* or *strength centrality*, can be defined: $s_i = \sum_j w_{ij}$. Analogous incoming and outgoing strengths can be defined for directed networks.

The *degree distribution*, $P(k)$, is the probability that a randomly selected node has k links. It is an important characteristic of a network, and in many networks $P(k)$ follows a power law. These networks are called *scale-free* (Barabási and Albert, 1999), and are characterized by a large inhomogeneity in the degree values: Many nodes have a low degree, whereas a few of them, the *hubs*, are connected to a very large number of other nodes.

The degree is just one of the many quantities that can be computed to analyze the properties of a network (Caldarelli, 2007). Another popular measure is the *clustering coefficient*, C_i, which measures the fraction of a node's neighbors that are neighbors also among themselves. More specifically (for undirected networks):

$$C_i = \frac{2R_i}{k_i(k_i-1)} = \frac{1}{k_i(k_i-1)} \sum_{j=1}^{N} \sum_{l=1}^{N} A_{ij} A_{jl} A_{li}, \tag{4.3}$$

where k_i is the degree of node i and R_i is the number of connected pairs in the set of neighbors of node i, which was expressed in the previous formula in terms of the adjacency matrix. C_i takes values between 0 and 1.

The assortativity a_i of a node i is the average degree of its neighbors, that is, of all the nodes to which node i is linked. It can be computed from the adjacency matrix as

$$a_i \equiv \frac{1}{k_i} \sum_{j=1}^{N} A_{ij} k_j. \tag{4.4}$$

The assortativity characterizes the tendency of a node to be connected to nodes with high degree. Note that a node can have low degree but at the same time high assortativity. As with the degree values, the assortativity coefficient can range from 0 to $N-1$.

In weighted networks one can normalize the weights as $p_{ij} = w_{ij} / \sum_k w_{ik}$, where the sum is over all nodes k connected to node i. Then a *network entropy* at node i can be defined:

$$H_i = -\sum_j p_{ij} \log p_{ij}, \tag{4.5}$$

which is a measure of the diversity of values of the link weights attached to node i.

In connected networks one can define a *network distance* between two nodes as the minimum number of links one has to cross to go from one to the other. This also identifies the *shortest paths* among these two nodes (there may be more than one). The *betweenness coefficient*, B_i, of node i is the fraction of shortest paths crossing a node over the number (N_S) of all the possible shortest paths in the network (Newman, 2010):

$$B_i = \frac{\sum_{jk} g_{jk;i}}{N_S}, \tag{4.6}$$

where $g_{ij;k}$ is the number of shortest paths between nodes j and k passing through i, and the sum is over all node pairs (j,k). Related to the features of the shortest paths is the *small-world* property (Watts and Strogatz, 1998) that in simple terms describes the fact that despite their often large size, in most networks there is a rather short path between any two nodes. The most popular manifestation of the small world concept is the "six degrees of separation" concept, uncovered by the social psychologist Milgram (1967), who found that there was a path of acquaintances with a typical length of about six between most pairs of people in the United States (Albert and Barabási, 2002; Kochen, 1989). Random Erdös–Rényi networks have short distances between their nodes. In the original work by Watts and Strogatz (1998), the name *small-world network* was given to structures in which paths are short but simultaneously clustering is high, which excludes random networks.

4.1.2 Functional and Structural Networks

In many situations the interactions among the elements of a complex systems are unknown, or only partially understood, and thus a complete knowledge of the network topology is lacking. Technical limitations, or the nature of the network itself, sometimes make it impossible to know how the nodes are actually interconnected between themselves. A fundamental and challenging problem is then how to optimally infer, from the observed data, the network that more closely resembles (or mimics) the actual connectivity of the system. The inferred network is usually referred to as the *functional* network, while the real network of physical interconnections is referred to as the *structural* network. Two paradigmatic examples are brain networks and climate networks.

The procedure to infer a functional brain network is through the statistical study of the correlations between the time series recorded from different brain regions, which, in turn, are the nodes of the network. First, a covariance matrix representing the coordinated activity between all pairs of nodes is calculated, and then a functional network is obtained by thresholding the covariance matrix: if two nodes display a high correlation they are considered linked, otherwise not. In this way, from the filtered covariance matrix an adjacency matrix is obtained.

A functional network does not necessarily represent real physical connectivity: When two or more nodes are influenced by the same source (for example, they are under the influence of the same external forcing), their dynamics might be highly correlated even if they are not interacting directly. As one aims to find a functional network that resembles as close as possible the genuine physical connectivity of the system, it is crucial to develop reliable techniques capable of detecting real interdependencies (Van Wijk et al., 2010). In recent years numerous approaches for optimal inference of functional networks have been proposed (see, e.g., Tirabassi et al. (2015b) and references therein).

In Section 4.2 we explain how the concept of functional networks has been applied to the climate system.

4.2 Construction of Climate Networks

Climate networks are built through bivariate statistical analysis of interdependencies between pairs of time series that describe climatological variables in different geographical regions of Earth, which, in turn, are the nodes of the network. The procedure to construct a climate network is displayed in Fig. 4.1, adapted from Donges et al. (2015).

In **step 1**, a spatial grid is chosen to define the nodes of the network. **Step 2** includes the selection of the climatological variable to be analyzed (surface air temperature [SAT], sea surface temperature [SST], geopotential height, precipitation, etc.), pre-processing of time series (for example, anomaly time series are typically normalized to unit variance), and the computation of a statistical similarity measure (SSM), $\mathcal{S}_{ij}$, to quantifying interdependencies between pairs of time series (see the following and Sections 3.2 and 3.3). In **step 3** the construction of the climate network typically involves some thresholding criterion to select the statistically significant SSM values (see Section 3.4). Considering, for example, that the significant links have $\mathcal{S}_{ij}$ values above a given threshold, W, then the adjacency matrix is determined as

$$\mathcal{A}_{ij} = \mathcal{H}(\mathcal{S}_{ij} - W), \tag{4.7}$$

where $\mathcal{H}$ is the Heaviside function. In **step 4**, the obtained network is investigated by using various analysis tools. We note that the interpretation of the node's strength will be affected by the positions of the grid points chosen in step 1. For example, if a grid uniform in latitude and longitude coordinates is used, nodes closer to the Earth's poles have less space between them than those in the equator. A convenient way to take into account this effect is to replace the *degree* by the *area-weighted degree*, also known as the *area-weighted connectivity* (AWC) (Tsonis and Swanson, 2006):

$$AWC_i = \frac{\sum_j w_j \mathcal{A}_{ij}}{\sum_j w_j}, \tag{4.8}$$

where $w_j = \cos(\lambda_j)$, with λ_j being node j latitude, is the factor taking into account the decreasing area associated with grid cells of higher latitude. Thus AWC_i represents the fraction of the total area that node i is connected to. An alternative to the use of area-weighted connectivity is to consider an equal-area map projection (as in Sections 4.4, 6.5, and 7.3), so that each grid point already represents the same area of Earth. Finally, in **step 5**, the results of this analysis are interpreted in terms of dynamical processes of the Earth system (e.g., winds, ocean currents, etc.).

As will be discussed in the following chapters, a wide range of climate network studies have demonstrated that the correlation structures of a global or regional

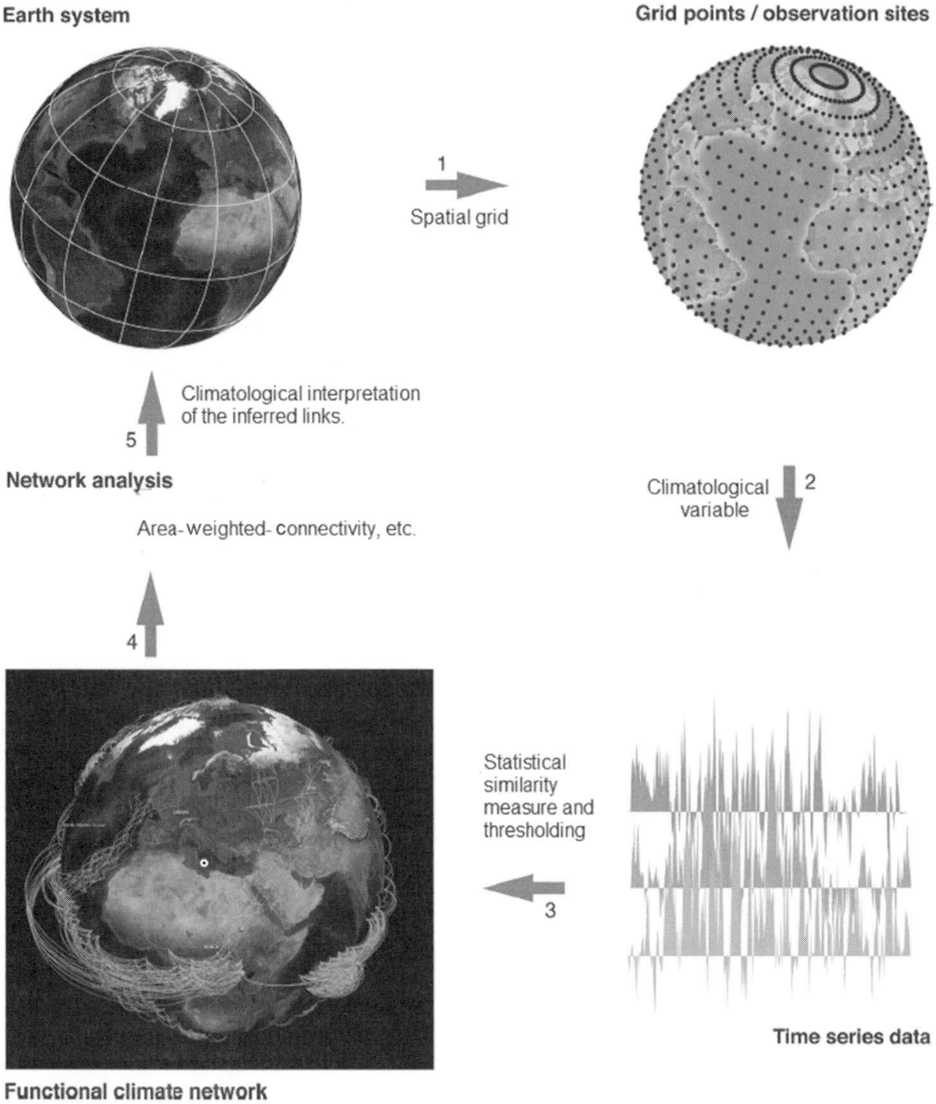

Figure 4.1 Methodology used to extract a climate network from climatic signals. Source: Donges et al. (2015). (A black and white version of this figure appears in some formats. For the color version, please refer to the plate section.)

climatological field (such as SAT, pressure, etc.) uncovered by using the network approach yield complementary and novel insights on spatial as well as temporal patterns, which account for a large fraction of the fields' variance (Donges et al., 2015).

Regarding similarity measures, used in step 2 of network building, many quantities have been used. The first and one of the most used (Tsonis and Roebber, 2004) is the Pearson correlation (see Section 3.2.1) at zero lag ρ_{ij} between two time series of the same climatic variable X but at two locations i and j. From (3.3):

$$\rho_{ij} = \rho_{X_i,X_j}(\tau = 0) = \frac{\mathrm{cov}(X_i(t), X_j(t))}{\sqrt{\sigma_{X_i}^2 \sigma_{X_j}^2}} . \tag{4.9}$$

The Pearson correlation at a specific lag time τ can also be used, if one has information on a relevant timescale providing the value of the delay τ. A procedure to select a meaningful value of τ, adapted to each pair of locations, and select suitable weights for them was introduced by Yamasaki et al. (2008) and further extended by Wang et al. (2013, 2016a) to distinguish between positive and negative correlations. In the method, one considers the normalized anomalies of some climatic variable at locations s, say $\Theta_s(t)$, and computes a lagged correlation function: $X_{s_1,s_2}(\tau \geq 0) = \langle \Theta_{s_1}(t)\Theta_{s_2}(t+\tau) \rangle$, and $X_{s_1,s_2}(\tau \leq 0) = X_{s_2,s_1}(-\tau)$, with averages over time. Two optimal time lags, $\tau^{\pm}$, are defined for each pair (s_1, s_2) as the lags at which $X_{s_1,s_2}(\tau)$ are maximal and minimal. Then, positive and negative link weights are assigned to the pair of nodes (s_1, s_2) as

$$W^{+}_{s_1,s_2} = \frac{\max(X_{s_1,s_2}) - \mathrm{mean}(X_{s_1,s_2})}{\mathrm{std}(X_{s_1,s_2})} , \tag{4.10}$$

$$W^{-}_{s_1,s_2} = \frac{\min(X_{s_1,s_2}) - \mathrm{mean}(X_{s_1,s_2})}{\mathrm{std}(X_{s_1,s_2})} , \tag{4.11}$$

where max and min indicate the maximal and minimal cross-correlation, $X_{s_1,s_2}(\tau^{\pm})$, and mean and std represent the mean and the standard deviation over values of τ. When s_1 is to the west of s_2 and the time lag is positive, the link direction is to the east. Finally, positive and negative networks can be constructed by establishing links between pairs of nodes with weights larger than some significant threshold. Sections 6.4 and 7.1 contain examples of the analysis of negative and positive links, respectively. Note that the calculations in (4.10)–(4.11) involve a rather nontrivial modification of the original correlations, so that it may be difficult to interpret the weights $W^{\pm}$. Martin et al. (2013) note that this may lead to incorrect interpretation of network calculations, especially if in the search for the optimal τ several physically different timescales are scanned.

Nonlinear measures of similarity between time series at different locations have also been used, constructed both from histograms of the climatic variable (Section 3.3.2) and from ordinal patterns (Section 3.3.5). Later in this chapter and in the rest of the book we present several examples of networks constructed in this way. Finally, we mention that synchronization measures such as event synchronization (Section 3.3.3) have also been employed (see examples in Section 6.6).

As in other types of networks, in climate networks the links can be unweighted or weighted (by the corresponding SSM value). When considering the weights of the links, the presence of strong auto-correlations can hinder the interpretation of the results, and a way to overcome the problem is to normalize the SSM matrix (Guez et al., 2014). Links can also be undirected or directed (if an asymmetric SSM measure is used). Different techniques (lag correlation and others) can allow inferring directed links (e.g., Gozolchiani et al., 2011). The next section discusses, as examples, climate networks inferred from SAT anomalies (SATA), by using different statistical similarity measures.

4.2.1 Undirected Network Inferred from SATA

As an example of the general procedure discussed in the previous section, here we analyze the climate network inferred from (NCEP/NCAR monthly reanalysis covering the period January 1949 to December 2006, see Section 3.1) by using, as a similarity measure, the symmetric mutual information (M_I, see Section 3.3.2) computed from the probabilities of the ordinal patterns (see Section 3.3.5). The resulting network is analyzed in terms of the map of AWC (4.8), and the maps that display the significant MI values.

Figure 4.2 displays the AWC when the ordinal patterns are formed by comparing SATA in the same month during four consecutive years (Barreiro et al., 2011). In this figure the networks obtained with three significance thresholds are shown. It can be seen that in this "interannual" timescale (see Fig. 3.8) the dominant atmospheric connections are located in the tropical Pacific and Indian Ocean

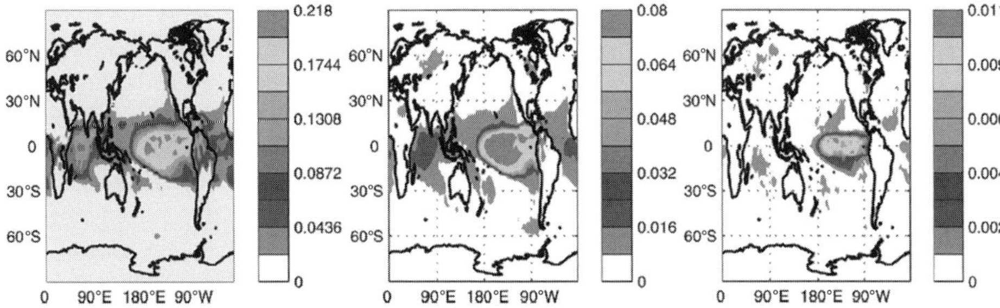

Figure 4.2 Area-weighted connectivity in color code, when the climate network is inferred by using the mutual information computed from the probabilities of ordinal patterns defined by comparing SATA (NCEP/NCAR monthly reanalysis) in four consecutive years. The color code is such that the white (red) regions indicate the geographical areas with zero (greatest) connectivity. The significance threshold increases from left to right. Source: Barreiro et al. (2011). (A black and white version of this figure appears in some formats. For the color version, please refer to the plate section.)

areas, mainly associated to the El Niño phenomenon. One can also see that, as expected, the connectivity of the network decreases as the M_I significance threshold is increased. For the highest threshold considered (shown in c, here the threshold is selected such that the density of the network is 0.1% of the total possible links) the El Niño–Indian Ocean tele-connection is significantly weakened with respect to the lower threshold network, shown in panel a. Here the threshold is chosen equal to the maximum M_I value obtained from surrogate (shuffled) data, which gives a network with 2.7% of the total links.

Figure 4.3 displays the effect of the time-scale used to define the ordinal patterns in the AWC map (Deza et al., 2013); the maps of significant M_I values in the ENSO region were shown in Fig. 3.9.

In Fig. 4.3 one can see that the equatorial Pacific has large connectivity only when considering interannual timescales. In addition, one can see that the AWC when the M_I is computed from the probability distribution function of SATA values, panel (a), (i.e., without taking into account the temporal ordering of the data points) looks as a "superposition" of spatial structures which are present only in some of the maps shown in panels (b)–(d). See, for example, the highly connected green spot in the Labrador Sea, which is also seen in Fig. 4.3b and to a lesser extent in Fig. 4.3c, but is not present in Fig. 4.3d. The Labrador Sea is one of the most important regions of deep-water formation in the north Atlantic. The formation of this water occurs in wintertime and depends on the passage of extratropical storms that cool the surface. The passage of storms is in turn related to the state of the North Atlantic Oscillation (NAO). As a result, there is a clear connection of the Labrador Sea with the rest of the north Atlantic mainly on seasonal timescales, and it is mostly independent of ENSO activity.

Thus, when transforming the anomalies time series into series of symbols (ordinal patterns), varying the spacing between data values allows us to identify and disentangle the timescales of different processes and interactions.

4.2.2 Directed Network Inferred from SATA

Here we use the directionality index (DI) presented in Section 3.3.4 to infer the directionality of the links of the climate network constructed from the analysis of SATA (NCEP/NCAR daily reanalysis covering the period from January 1949 to December 2013). The results presented here were reported by Deza et al. (2015). A critical comparison of linear and nonlinear methods for inferring directed climate networks was performed by Hlinka et al. (2013).

Figure 4.4 displays the directionality of the links of two nodes in the tropics (indicated with triangles) when the parameter τ that represents the characteristic

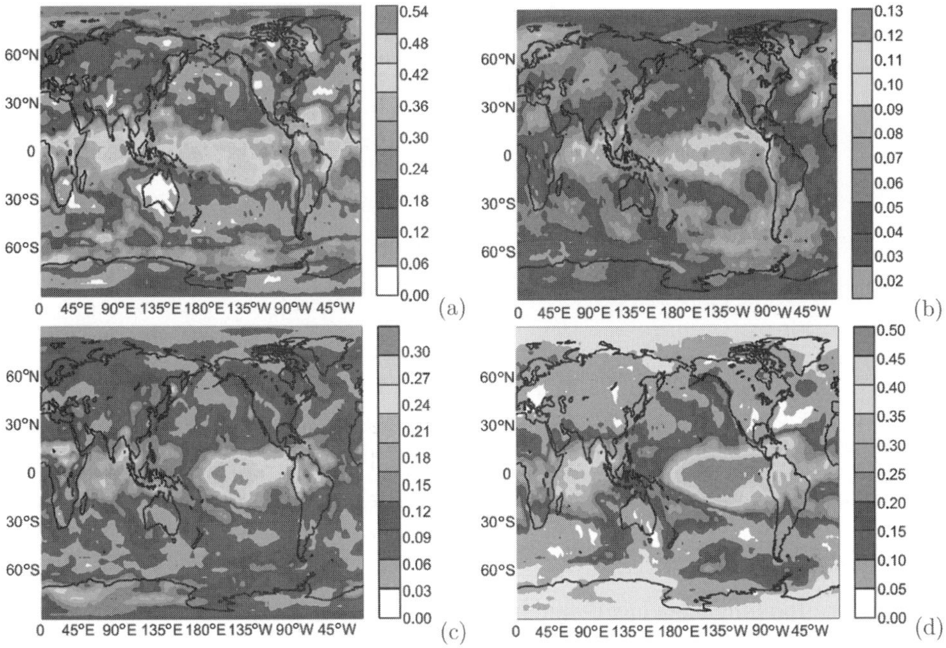

Figure 4.3 Area-weighted connectivity in color code, when the mutual information M_I is computed from (a) the histograms of anomaly values and (b)–(d) from ordinal probabilities: in (b), the are patterns formed with three consecutive months, in (c) by three equally spaced months covering a one-year period, and in (d) by three months in consecutive years. Only significant M_I values are shown: an M_I value is considered significant if it is larger than $\mu + 3\sigma$, with μ and σ being the mean value and the standard deviation of the distribution of M_I values computed from surrogates. The maps of significant M_I values in the ENSO region were shown in Fig. 3.9. Source: Deza (2015). (A black and white version of this figure appears in some formats. For the color version, please refer to the plate section.)

timescale of information transfer is equal to 30 days. Here the color code indicates the DI value: outgoing links are shown in red, while the incoming links are shown in blue. A double significance analysis using bootstrap surrogates (see Section 3.4) is performed: for a D_I value to be considered significant, the corresponding MI values also have to be significant.

Figure 4.4a shows, as expected, the central Pacific influenced by the eastern Pacific (in blue) and influencing the global network, with many regions in the tropics and in the extratropics in red. Reciprocally, Fig. 4.4b shows that the blue links come to the node in the Indian Ocean from a well-defined region in the central Pacific Ocean. In addition, few red outgoing links connect the node in the Indian Ocean to other regions. A main drawback of the DI is that it does not distinguish indirect from direct information transfer. Therefore, the red areas influenced by the

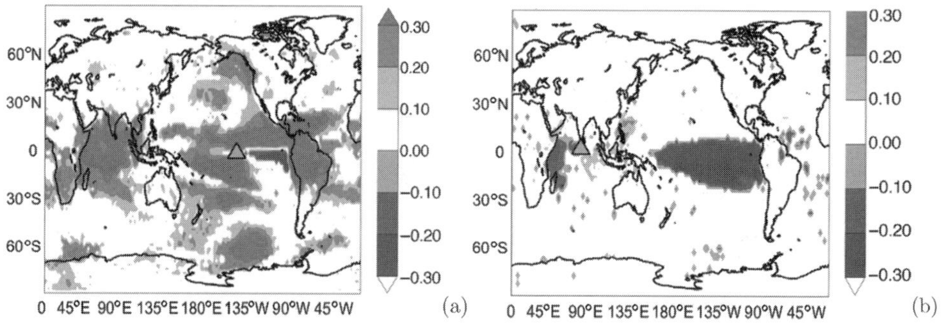

Figure 4.4 Directionality of the links in a node in the central Pacific (a) and in a node in the Indian Ocean (b), indicated with a triangle. The color code indicates the directionality index: outgoing links are shown in red while incoming links are shown in blue. The time scale of information transfer is $\tau = 30$ days. Source: Deza et al. (2015). (A black and white version of this figure appears in some formats. For the color version, please refer to the plate section.)

node in the Indian Ocean can be an artifact in the sense that these regions might be indirectly influenced by the ENSO region. A possible way to overcome this problem could be separating correlations into direct and indirect components, as proposed by Zhou et al. (2015).

Next, we keep $\tau = 30$ days and analyze the directionality of the links in various regions on the equatorial Pacific. As Fig. 4.5 shows, the influence of the Pacific Ocean is almost global. The DI captures the fact that – even if there are feedbacks and the Pacific is affected by extratropical perturbations and other ocean basins (e.g., the tropical Atlantic) – the influence is *effectively* from the Pacific to the rest of the world.

Moreover, the maps show that the largest influence is exerted by the equatorial Pacific, close to the dateline. This is clear in the extratropical atmosphere, as well as in the tropical north Atlantic. On the other hand, the connection to the Indian Ocean and south Atlantic is not so sensitive to the point considered over the equatorial Pacific. As the reference point moves further west from the dateline the influence decreases substantially, only maintaining weak connections to the tropical north Atlantic and Indian Oceans. The methodology can thus be applied to find the best region to construct an index that describes the Pacific influence over the area where climate anomalies are studied.

Notice that all maps show a blue tongue of incoming links to the east of the point considered; it is seen first in Fig. 4.5a and extends westward until covering the whole Pacific Ocean in Fig. 4.5d. This feature is interpreted as being due to the equatorial cold tongue and the fact that easterly trade winds blow over the equator, thus advecting air from the east to the west of the point.

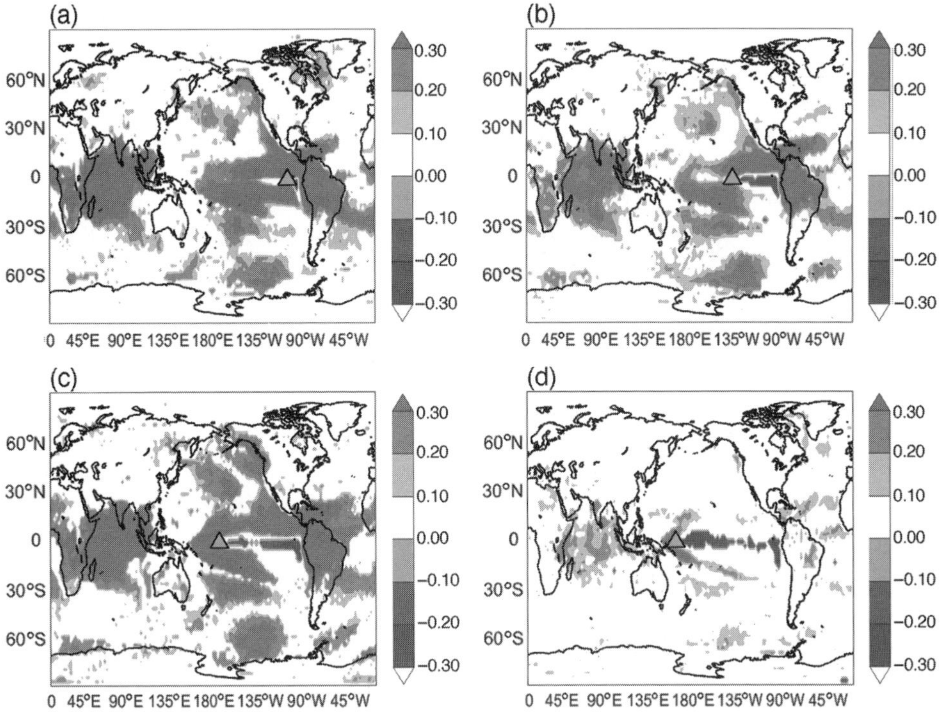

Figure 4.5 Directionality over the equatorial pacific with $\tau = 30$ days. It is observed that most of the regions over the central and eastern Pacific Ocean have strong effects over a large part of the world; especially over the tropical areas and the rest of the Pacific Ocean. Source: Deza et al. (2015). (A black and white version of this figure appears in some formats. For the color version, please refer to the plate section.)

The influence of the time of information transfer τ on these results is described by Deza et al. (2015).

4.3 Climate Communities

An important characteristic of complex networks in general, and of climate networks in particular, is the presence of subsets of highly interconnected nodes, which are nevertheless only weakly linked to the rest of the network. These sets of nodes the referred to as *communities* (Fortunato, 2010; Newman, 2010). Communities are very interesting in the framework of complexity reduction or system coarse-graining because nodes that belong to the same community can be thought of as a coherent part of a large macro-system, while connections between nodes in different communities can be coarse-grained into links among communities. In this way one can build a network of communities that, with reduced complexity, contains relevant "macroscopic" information, extracted from of the original network.

Detecting the underlying community structure of a complex network is a challenging task. Many methods have been developed for it (Fortunato, 2010; Fortunato and Hric, 2016; Newman, 2010), usually involving the maximization of some quantity (as for example the modularity parameter (Newman, 2010; Newman and Girvan, 2004)), indicative of the quality of a partition of the network into communities, or of its significance with respect to some null model. The situation is complicated in networks with nodes representing spatially located regions (Barthélemy, 2011), as in the case of brain functional networks and climate networks. Here, the effects of space in the topology of the network are particularly important, and can dominate methods for community detection, thus obfuscating more subtle but less trivial community structure.

Spectral methods for community detection (Fortunato, 2010; Newman, 2010) use the eigenvalues and eigenvectors of the Laplacian or related matrices. The Laplacian matrix, $\mathcal{L}$, of an undirected network with adjacency matrix $\mathcal{A}$ is

$$\mathcal{L}_{ij} = k_i \delta_{ij} - \mathcal{A}_{ij}, \tag{4.12}$$

where k_i is the degree of node i. For example, to partition a network into two significant subgraphs, one uses the sign of the components of the eigenvector associated to the second smallest eigenvalue (the Fiedler vector (Fortunato, 2010; Newman, 2010)). To obtain a partition in a larger number of communities, several eigenvectors are grouped with algorithms such as k-means clustering (Fortunato, 2010).

Among other limitations (Fortunato and Hric, 2016), spectral methods usually require specifying *a priori* the number of communities to be found. *Infomap* (Rosvall and Bergstrom, 2008) is a popular community-detection algorithm that does not require specifying the number of communities to be detected. The main idea is to consider a random walk in the network and to find the optimal partition such that the random walk remains most of the time inside individual communities, with only a few jumps between them, as indicated by the information-theoretical content of the optimal code used to encode the random walker's trajectories. The optimization algorithm is efficiently implemented and freely available: details of the method and code implementations can be found at www.mapequation.org.

As an example, Fig. 4.6, taken from Tirabassi and Masoller (2016), displays a set of climate communities inferred by using the *Infomap* algorithm. The analysis was performed over SATA (NCEP/NCAR monthly reanalysis covering the period from January 1948 to May 2012). In this case the network was defined by using symbolic analysis: ordinal patterns with seasonal timescale (i.e., comparing SAT anomalies in three consecutive months, see Section 3.3.5). The transition probabilities (TPs) that describe the statistics of the symbolic sequence in each geographic region were

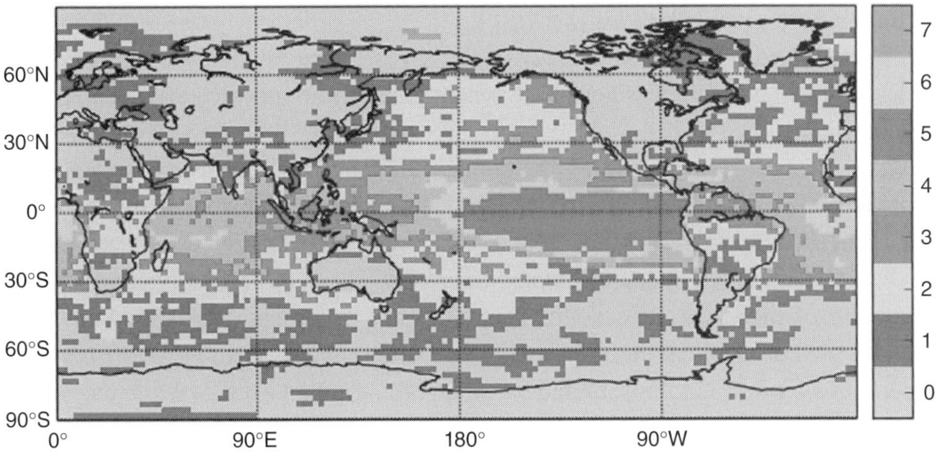

Figure 4.6 Climate communities inferred with the *Infomap* algorithm. The adjacency matrix is obtained from symbolic statistical analysis of SAT anomalies. Regions depicted with the same color belong to the same community. Four macro-communities are identified: extratropical continents and oceans, tropical oceans, and El Niño basin. Source: Tirabassi and Masoller (2016). (A black and white version of this figure appears in some formats. For the color version, please refer to the plate section.)

computed, and regions with similar (dissimilar) TPs were strongly (weakly) linked. In this way the information of the evolution of the SAT anomalies at the seasonal timescale was encoded into transition probability matrices.

As can be seen in Fig. 4.6, the *Infomap* algorithm divides the world into eight areas, labeled with different colors. These areas share similar symbolic transition probabilities. The continents in the two hemispheres are in the same community and a large coherent area is detected in the ENSO basin, while the oceans are divided into tropical and extratropical. We note that the communities group regions where similar processes dominate the SAT variability. For example, the community analysis (1) identifies the central equatorial Atlantic as having a similar behavior to El Niño; (2) separates the behavior of SAT over the maritime continent from that of the Indian and tropical Atlantic Oceans, consistent with a different rainfall regimes; and (3) considers the tropical north and south Atlantic as belonging to the same community, which is consistent because SAT is strongly controlled by air–sea heat fluxes.

It is important to remark that such community structure cannot be inferred from networks that are constructed from correlation analysis (by using Pearson coefficient or mutual information). The classic tools are not useful to identify communities formed by regions with similar climate, because they do not provide direct connections among extratropical regions (see, for example, the maps of mutual information values shown in Fig. 3.9). In order to belong to the same community, two nodes must be part of the same group of highly interconnected nodes, and in the correlation approach, where the links are prominently local, direct teleconnections across hemispheres are scarce and weak.

We confirm this by presenting the communities obtained by using the standard approach, in particular, measuring dynamical similarities with the Pearson correlation coefficient and using the threshold $W = 0.5$ to extract significant cross-correlation values. In this case *Infomap* returns 8604 communities, but only 20 are composed of more than two nodes (Tirabassi and Masoller, 2016). As shown in Fig. 4.7, which displays the largest 16, only communities 0 and 1 correspond to coherent structures, namely El Niño basin, and the tropical oceans, while the others appear to be just noise.

The structure of the network and the number of communities depend of the threshold used to select the links that represent significant similarities. Decreasing the threshold leads to a more connected network, while increasing the threshold results in a sparser network. The number of communities depends on the number of links, which in turn depends on the threshold. In order to uncover a coherent, well-defined community structure, the threshold has to be carefully chosen.

Figure 4.8 displays the effect of the threshold in the average degree and in the number of communities, and we can see that there is a negative correlation between the number of communities and the average degree. As the threshold

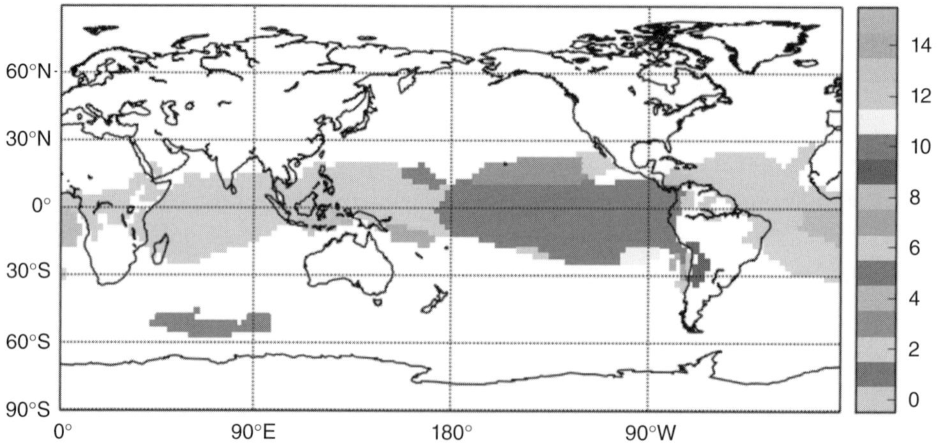

Figure 4.7 Community structure uncovered by *Infomap* algorithm when the network is constructed using the Pearson cross-correlation coefficient as a measure of dynamical similarity. Source: Tirabassi and Masoller (2016). (A black and white version of this figure appears in some formats. For the color version, please refer to the plate section.)

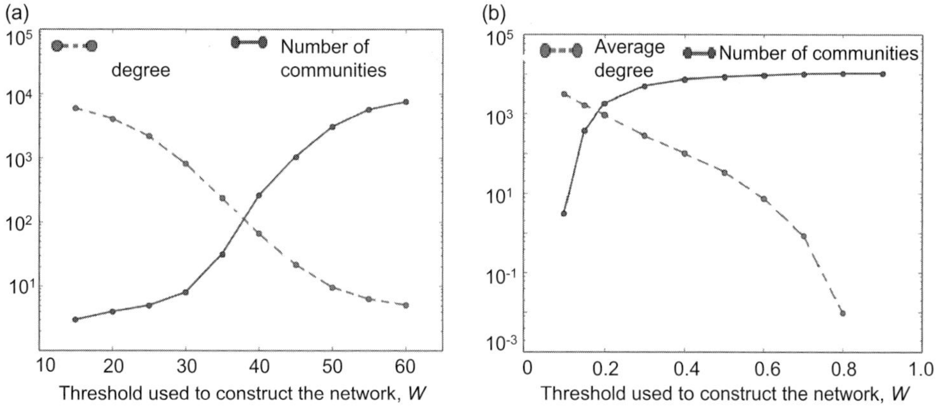

Figure 4.8 Dependence of the network average degree and of the number of *Infomap* communities with the threshold value when the network is obtained from (a) symbolic statistical analysis of SAT anomalies and (b) from cross-correlation analysis of SAT anomalies. Source: Tirabassi and Masoller (2016).

increases, relevant links are removed and the fragmentation of the network leads to the increase of the number of communities. Therefore, one needs to select a threshold that provides a good compromise between the need to limit the small-communities-proliferation and the need to include in the network only the relevant, significant links.

As can be seen in Fig. 4.8, with both approaches (symbolic ordinal analysis and cross-correlation) the number of communities increases as the threshold is increased, but the change is less abrupt when comparing the transition probabilities computed with ordinal analysis (note the logarithmic vertical scale). With cross-correlation, an abrupt increase of the number of communities occurs at low values of the threshold, which means that a low threshold needs to be used in order to limit the number of communities; however, the obtained set of communities will not be informative because when using a low threshold, the network includes too many links that do not represent significant similarities.

4.4 Flow Networks

Previous sections in this chapter have used climate networks constructed from statistical correlations or dependencies between the different geographic nodes (*functional networks*). An alternative approach is to construct the climate networks by taking into consideration the actual physical transport of air, water, or other substances between different geographical locations. In this section we discuss the main tools for the construction and characterization of these networks inferred from transport processes, which are known as *flow networks*. Whereas for the functional climate networks the input variables are climatic variables measured or computed at each spatial point, to evaluate transport we need as input the velocity field of the region of interest in the atmosphere or the ocean during a given time interval. Alternatively, fluxes of different substances (water vapor, ozone, salt, etc.) can be used.

The connection between these two network approaches (functional correlation climate networks and flow networks) was investigated by Tupikina et al. (2016), who proposed a theoretical approach to verify relations between the correlation matrix of flowing quantities and climate network measures (see also Molkenthin et al., 2014). The analysis, developed for correlations of a scalar field that satisfies an advection–diffusion dynamic in the presence of forcing and dissipation (temperature, or geophysical flows, for example) revealed that correlation networks are not sensitive to steady sources and sinks and also uncovered a profound impact of the signal decay rate on the network topology. The theory developed provided an elegant framework for the analytical calculation of degree and clustering for a meandering flow resembling a geophysical ocean jet.

4.4.1 Network Description of Lagrangian Transport in Fluid Flows

Among the many tools recently developed to quantify and understand transport processes in fluid flows (Haller, 2015; Mancho et al., 2006, 2013; Peacock and Dabiri, 2010; Samelson and Wiggins, 2006; Wiggins, 2005), there is one family in which portions of moving fluid are explicitly tracked, the so-called *set-oriented* methods (Dellnitz et al., 2009; Froyland and Dellnitz, 2003; Froyland, 2005; Froyland et al., 2007, 2010, 2012; Levnajić and Mezić, 2010; Santitissadeekorn et al., 2010; Tallapragada and Ross, 2013).

One of the basic tools in this type of method is the *Perron–Frobenius or transfer operator*, which quantifies the amount of fluid transported from some initial region to other regions under time evolution. A discretized version of that operator is the *transport matrix* (also known as *transfer matrix*) indicating which part of the fluid domain is connected with which, and by what amount of flow (Fig. 4.9). In this matrix form, the transfer operator can be interpreted as an adjacency matrix that defines a transportation or *flow network*, an analogy that has been recognized as useful (Dellnitz et al., 2006; Nilsson-Jacobi et al., 2012; Preis et al., 2004; Rossi et al., 2014; Santitissadeekorn and Bollt, 2007; Ser-Giacomi et al., 2015b; Speetjens et al., 2013; Thomas et al., 2014). Within this network interpretation, the network analysis tools discussed in the previous section become available to extract information about the transport processes. Recent work has related the connectivity of the functional climate network inferred by using correlation analysis to the underlying fluid flow network (Molkenthin et al., 2014; Tupikina et al., 2016).

Flow networks describing the material fluid flow among different locations are directed, weighted (Newman, 2004), spatially embedded (Barthélemy, 2011), and time-dependent (Holme and Saramäki, 2012). Since fluid flow is a process occurring in continuous space, a discretization procedure involving a coarse-graining of space is needed to have access to the techniques of network theory. In the following we enumerate the steps needed to construct the discrete transport network starting from the continuous flow.

First, we have to define the network *nodes*. As in the functional networks defined in the previous sections, network nodes are associated to geographical locations: We subdivide the fluid domain of interest in a large number N of boxes, $\{B_i, i = 1, \ldots, N\}$, so that node j represents the fluid box B_j. Although it is not strictly necessary, it is convenient to consider boxes that have the same area (for two-dimensional flows) or volume (for three dimensions). This is conveniently done by using grid points regularly placed in an area-preserving or sinusoidal projection (see examples in Sections 6.5 and 7.3). Each box will contain exactly the same amount of fluid.

Second, we need to establish the connections between nodes (i.e., boxes in the fluid domain). We establish a directional *link* between two nodes when an exchange of fluid occurred from one to the another during a given time interval. The *weight* of this link will be proportional to the amount of fluid transported. When the input data is the spatio-temporally dependent velocity field in a region, this amount of fluid can be obtained from a Lagrangian point of view by following trajectories of ideal fluid particles and keeping a record of their initial and final positions (i.e., starting and ending nodes) during the time interval considered.

More formally we initialize at time t_0 a large number of particles, say N_i, each at an arbitrary position $\mathbf{x}_0$ inside each node B_i of the network. Then we compute each trajectory by integrating the given velocity field $\mathbf{v}(\mathbf{x}, t)$ for a fixed time τ until the final position $\mathbf{x}$ at $t_0 + \tau$. This defines the *flow map* $\Phi_{t_0}^{\tau}$:

$$\mathbf{x}(t_0 + \tau) = \Phi_{t_0}^{\tau}(\mathbf{x}_0), \tag{4.13}$$

which moves single fluid particles or, by considering the action of the flow map on all the points contained in a fluid region A, it also gives the action of the flow on whole sets: $A(t_0 + \tau) = \Phi_{t_0}^{\tau}(A(t_0))$.

Applying the flow map to the discrete boxes B_i, we have an estimation of the flow among each pair of nodes. More explicitly, given the collection of boxes $\{B_i, i = 1, \ldots, N\}$, we represent the transport between them by the discrete version of the Perron–Frobenius operator $\mathbf{P}(t_0, \tau)$ whose matrix elements are given by (Dellnitz et al., 2009; Froyland, 2005; Froyland and Dellnitz, 2003; Froyland et al., 2007, 2010; Santitissadeekorn et al., 2010):

$$\mathbf{P}(t_0,\tau)_{ij} = \frac{m\left(B_i \cap \Phi_{t_0+\tau}^{-\tau}(B_j)\right)}{m(B_i)}, \tag{4.14}$$

where $m(A)$ is a measure assigned to the set A. In our case it is the amount of fluid it contains, i.e., simply its area or volume. Other measures referring for example to heat or salt content could be implemented in specific applications.

The number of particles transported from box B_i to box B_j gives an estimation of the flow among these boxes, and a numerical approximation to (4.14) is then:

$$\mathbf{P}(t_0,\tau)_{ij} \approx \frac{\text{number of particles from box } i \text{ to box } j}{N_i}. \tag{4.15}$$

Equation (4.14) states that the flow from box B_i to box B_j is the fraction of the contents of B_i which is mapped into B_j. If a non-uniform distribution of some conserved tracer is initially released in the system such that $\{p_i(t_0), i = 1, \ldots, N\}$ is the amount of such tracer in each box $\{B_i\}$ at the initial instant t_0, the matrix $\mathbf{P}(t_0,\tau)$ gives the evolution of this distribution after a time τ as $p_j(t_0+\tau) = \sum_{i=1}^{N} p_i(t_0)\mathbf{P}(t_0,\tau)_{ij}$. Writing $\{p_i\}$ as row vectors: $p(t_0+\tau) = p(t_0)\mathbf{P}(t_0,\tau)$. A probabilistic interpretation of (4.14) is that $\mathbf{P}(t_0,\tau)_{ij}$ is the probability for a particle reaching box B_j, under the condition that it started from a uniformly random position within box B_i. The matrix $\mathbf{P}(t_0,\tau)$ is row-stochastic, i.e., it has non-negative elements and $\sum_{j=1}^{N} \mathbf{P}(t_0,\tau)_{ij} = 1$, but not exactly column stochastic. The quantity $\sum_{i=1}^{N} \mathbf{P}(t_0,\tau)_{ij}$ measures the ratio of fluid present in box B_j after time τ with respect to its initial content at time t_0. This ratio will be unity, and the matrix doubly stochastic, if the flow $\mathbf{v}(\mathbf{x},t)$ is incompressible.

Because of the time dependence of the velocity field, the results of the Lagrangian simulations will depend on both the initial time t_0 and the duration of the simulation τ. Once these parameters are fixed, we can build a network described by a transport matrix $\mathbf{P}(t_0,\tau)$ that characterizes the connections among each pair of nodes from initial time t_0 to final time $t_0+\tau$. We interpret $\mathbf{P}(t_0,\tau)$ as the adjacency matrix of a weighted and directed network, so that $\mathbf{P}(t_0,\tau)_{ij}$ is the weight of the link from node i to node j. The replacement of the continuous flow system by a discrete network introduces discretization errors. Even if the integration is done accurately, the initial and final locations of the transported particles are only specified up to a precision Δ, given by the linear side of the boxes. This implies that our network approach does not display explicitly fluid structures smaller than the box length-scale Δ.

The network constructed in this way characterizes the final locations of all fluid elements a time τ after their release at time t_0, but gives no information

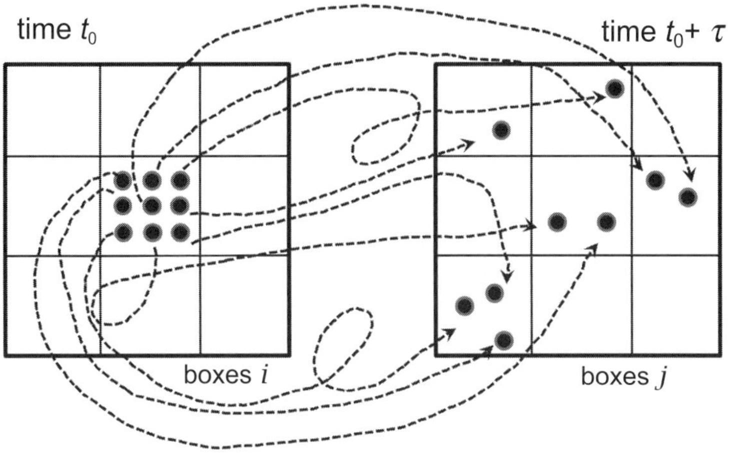

Figure 4.9 Construction of the transport matrix $\mathbf{P}(t_0, \tau)_{ij}$ from tracer's advection, following (4.15).

on particle locations at intermediate times. Also, since each of the matrices $\mathbf{P}(t_0 + k\tau, \tau)$, for $k = 0, 1, \ldots, n-1$, is a stochastic matrix, one can consider the discrete-time Markov chain in which an initial vector giving occupation probabilities $p(t_0) = (p_1(t_0), \ldots, p_N(t_0))$ for the different boxes is evolved in time as $p(t_n) = p(t_0)\mathbf{P}(t_0, \tau)\mathbf{P}(t_1, \tau)\ldots\mathbf{P}(t_{n-1}, \tau)$, where $t_k = t_0 + k\tau$. This time evolution will not be exactly equal to the true evolution $p(t_n) = p(t_0)\mathbf{P}(t_0, n\tau)$, but a Markovian approximation to it in which the memory of the precise particle positions inside each cell is lost after a time τ. The Markovian approximation may be reasonable in some circumstances and in fact it has been successfully used in geophysical flow problems (Dellnitz et al., 2009; Froyland et al., 2012, 2014; Ser-Giacomi et al., 2015a,c). As far as only the initial and final states of the transport processes are considered, the matrix $\mathbf{P}(t_0, \tau)$ contains the full transport information, independently of any Markovian hypothesis. The Markov approximation, however, will be used later, when analyzing discretized approximations to long-time fluid pathways.

4.4.2 Dispersion, Mixing, and Network Entropies

Important properties of geophysical flows depend on their dispersion characteristics, i.e., how far away the fluid can be transported during some time, and how diverse the target regions are. Mixing of fluid with different characteristics, another process of great geophysical relevance, will occur at a particular place if fluid from different origins arrives there at a particular time. In this section we show that information on these processes, of great impact on climate, can be obtained from the transport matrix defining the flow network (Ser-Giacomi et al., 2015b).

In dynamical systems approaches to flow processes, a standard way to quantify dispersion is by means of the finite-time Lyapunov exponent (FTLE). It is defined as (Shadden et al., 2005)

$$\lambda(\mathbf{x}_0, t_0, \tau) = \frac{1}{2|\tau|} \log \Lambda_{max} \tag{4.16}$$

where Λ_{max} is the maximum eigenvalue of the strain tensor:

$$C(\mathbf{x}_0, t_0, \tau) = \left(\nabla\Phi_{t_0}^{\tau}(\mathbf{x}_0)\right)^T \nabla\Phi_{t_0}^{\tau}(\mathbf{x}_0), \tag{4.17}$$

constructed from the Jacobian matrix $\nabla\Phi_{t_0}^{\tau}(\mathbf{x}_0)$ of the flow map. M^T means the transpose of the matrix M. For $\tau > 0$ this is the forward FTLE. By running the flow map backwards in time ($\tau < 0$), we get the backwards FTLE field, which quantifies the strength of mixing into a particular location. The interpretation of (4.16) in a two-dimensional flow is that an initial circle of infinitesimal diameter δ located at $\mathbf{x}_0$ at t_0 will become an ellipse of major axis $e^{\tau\lambda(\mathbf{x}_0,t_0,\tau)}\delta$ after being advected by the flow during a time τ. The minor axis will be a decreasing function of τ, contracting at an exponential rate related to a negative Lyapunov exponent that can be computed from the second eigenvalue of $C(\mathbf{x}_0, t_0, \tau)$.

An obvious quantity in the network interpretation suitable to be related to dispersion and mixing, and thus to the forward and backwards Lyapunov values, is the *degree* of a node. Since the flow network is directed, we should distinguish between the in-degree $K_I(i)$, i.e., the number of links pointing to a particular node i, and the out-degree $K_O(i)$, the number of links pointing out of it.

A first problem in relating these network properties to the actual physics of dispersion and mixing is that their values are dependent on the spatial scales chosen for discretization (there is also a dependency on the numbers N_i of particles used to compute the transport matrices, but it disappears for large N_i). This problem is easy to solve by recalling that every box has an associated area. Dealing first with the out-degree case for definiteness, $K_O(i)$ is proportional to the total area of all nodes that received some contents from the initial node i. This quantity has a well-defined meaning that can be related to the continuous flow dynamics with only a minor dependence on the discretization procedure. By using a map projection and grid in which all boxes have the same area Δ^2, the area corresponding to the out-degree of node i is $K_O(i)\Delta^2$. If a non-area preserving projection is used, then one has to replace degrees by an area-weighted version of them (Section 4.1.1). We can use generic ideas of chaotic dynamics to obtain heuristically a more precise relationship between two quantifiers of dispersion: the degree and the Lyapunov exponent. In regions dominated by hyperbolic structures, each of the fluid boxes will be stretched into a long, thin filament after a sufficiently long time τ. If we want to compute the number of boxes reached by it, it is enough to consider its length, since the width quickly becomes smaller than the box size Δ. Let us consider an initial line of length $L(t_0) \approx \Delta$ inside the initial box B_i. A small segment of it, of length $dl(t_0)$ at position $\mathbf{x}_0 \in B_i$ will become elongated by a factor given by the local FTLE: $dl(t_0+\tau) = dl(t_0)e^{\tau\lambda(\mathbf{x}_0,t_0,\tau)}$. Integrating over the initial positions along the line we get an estimation of the final length $L(t_0+\tau)$ of the filament. A better estimation $\bar{L}(t_0+\tau)$ of this length can be done by averaging over positions transverse to the line, to take into account different locations of the initial line in the box:

$$\bar{L}(t_0+\tau) \approx \frac{1}{\Delta}\int_{B_i} d\mathbf{x}_0 e^{\tau\lambda(\mathbf{x}_0,t_0,\tau)}, \tag{4.18}$$

where the longitudinal and transverse integrations have been combined into the integration of $\mathbf{x}_0$ over area B_i. The area of the boxes covered by the filament is $\mathcal{A}(t_0+\tau) \approx \bar{L}(t_0+\tau)\Delta$ so that the out-degree of the initial box will be

$$K_O(i) = \frac{\mathcal{A}(t_0+\tau)}{\Delta^2} \approx \frac{1}{\Delta^2}\int_{B_i} d\mathbf{x}_0 e^{\tau\lambda(\mathbf{x}_0,t_0,\tau)} \equiv \left\langle e^{\tau\lambda(\mathbf{x}_0,t_0,\tau)}\right\rangle_{B_i}. \tag{4.19}$$

Thus, we have a useful relationship between a natural quantity in the network description of fluid flows and a standard characterization of dispersion in the dynamical systems approach to such flows: The degree of a node associated to a box is the average or coarse-graining of the stretching factor $e^{\tau\lambda}$ in that box.

Expression (4.19) suggests defining

$$H_i^0(t_0,\tau) \equiv \frac{1}{\tau}\log K_O(i) \tag{4.20}$$

so that

$$\left\langle e^{\tau\lambda(\mathbf{x}_0,t_0,\tau)}\right\rangle_{B_i} = e^{\tau H_i^0(t_0,\tau)}. \tag{4.21}$$

From the convexity of the exponential function, we have $H_i^0(t_0,\tau) \geq \langle\lambda(\mathbf{x}_0,t_0,\tau)\rangle_{B_i}$. The previous expressions are reminiscent of the properties of the topological entropy of a dynamical system, as giving the exponential growth in time of the length of a material line (Tél and Gruiz, 2006). Pushing forward the analogy, we can define a sequence of Rényi-like entropies (Rényi, 1970) associated to a particular node i:

$$H_i^q(t_0,\tau) \equiv \frac{1}{(1-q)|\tau|}\log\sum_{j=1}^{N}\left(\mathbf{P}(t_0,\tau)_{ij}\right)^q, \tag{4.22}$$

which we call *network entropies*. Due to their dependence on the finite size of the partition, they are related to the ϵ-entropies discussed by Boffetta et al.

(2002). Note, however, that here the transport matrix involves only two states of the trajectories, separated by an interval of time τ which remains finite, and the dependence on the initial location, box B_i, is kept. The entropies H_i^0 and H_i^1 should be understood as defined by the limits $q \to 0$ and $q \to 1$, respectively. Applying l'Hôpital's rule to the definition of the network entropy of order $q = 1$, one gets:

$$H_i^1(t_0, \tau) = -\frac{1}{\tau} \sum_{j=1}^{N} \mathbf{P}(t_0, \tau)_{ij} \log \mathbf{P}(t_0, \tau)_{ij}. \tag{4.23}$$

This gives the amount of information (per unit of time) gained by observing the position of a particle at time $t_0 + \tau$, knowing that it was initially (time t_0) somewhere in box B_i. This quantity is precisely the discrete finite-time entropy studied by Froyland and Padberg-Gehle (2012). The product τH_i^1 is the network entropy defined in (4.5).

All the network entropies measure the *diversity* in the amounts of fluid received by the nodes connected to a given box, but weighting them in different ways: In H_i^0 all nodes are counted equally independently of the amount of fluid they receive, so that it informs only about the degree, as seen in (4.20); for increasing values of q, nodes receiving more fluid are weighted with increasing strength. Although the network entropies have been introduced here in the particular context of flow networks, we note that they can be defined for any weighted network, giving generalizations of the degree to quantify the unevenness of the weight distribution toward the nodes connected to a given one.

One can extend the arguments leading to (4.21) to obtain a relationship between the set of network entropies and moments of stretching factors. The result is (Ser-Giacomi et al., 2015b):

$$e^{(1-q)\tau H_i^q(t_0,\tau)} \approx \left\langle e^{(1-q)\tau\lambda(\mathbf{x}_0,t_0,\tau)} \right\rangle_{B_i} . \tag{4.24}$$

For $q = 0$ we re-obtain (4.21). In the limit $q \to 1$ we get $H_i^1(t_0, \tau) \approx \langle \lambda(\mathbf{x}_0, t_0, \tau) \rangle_{B_i} = \lambda_i(t_0, \tau)$. The arguments above can be repeated to get the same relationship (4.24) between network entropies in the backward time direction and backward Lyapunov exponents.

All these expressions are similar to the ones presented, for example, by Paladin and Vulpiani (1987) relating Rényi entropies and generalized Lyapunov exponents defined from moments of the stretching factor $e^{\lambda\tau}$. But here the moments are not by averaging along a dynamic trajectory but inside a box B_i. In the same way as the value of any of the network entropies at node i characterizes the inhomogeneity in the fluxes sent from i to other nodes, the difference between the different entropies (different q) at a single node i characterizes the inhomogeneity of the FTLE inside box B_i. This is a way in which small-scale features present in the Lagrangian trajectories get statistically represented in the network description. Relationships such as (4.24) are not exact for finite Δ and τ, but are expected to become more accurate for increasing τ and decreasing Δ. Section 7.3 will contain a validation of the formulae above for a flow network describing the surface circulation of the Mediterranean Sea.

It is important to remark that so far we have assumed that the boxes $\{B_i\}$ have equal areas. Expression (4.22) would need corrections in a more general case. See, for example, the case of H_i^1 in Froyland and Padberg-Gehle (2012).

4.4.3 Communities in Flow Networks

We can apply to flow networks the different methods suited to detect community organization in weighted and directed networks. Since here links are assigned according to fluid flow between nodes, the interpretation of the communities revealed by these methods will be that of regions interchanging little fluid with the surroundings, but intense internal mixing inside.

We display here two examples. First, Fig. 4.10 shows a spectral community partition of the global ocean obtained from four left eigenvectors of the transport matrix describing surface ocean transport during 1000 years forward in time (velocity fields obtained from the OFES model). See details in Froyland et al. (2014). Second, at very different spatial and temporal scales, Fig. 4.11 shows

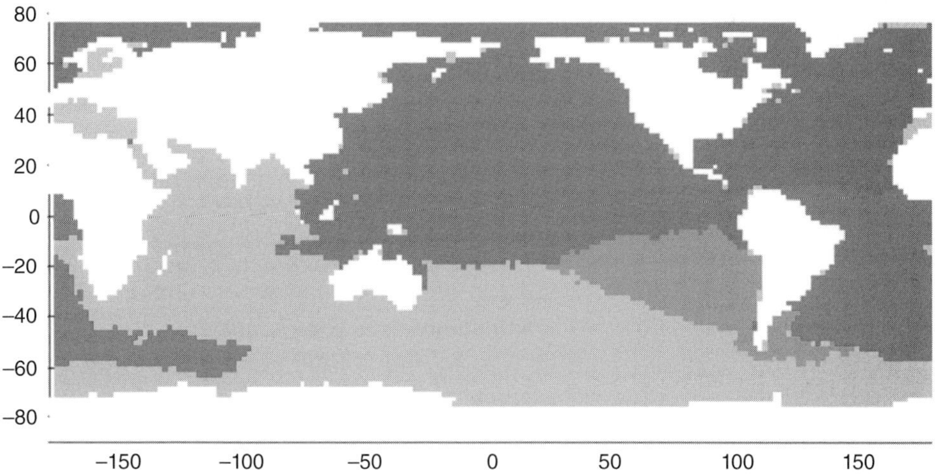

Figure 4.10 A spectral partition of the global ocean into five communities, using four left eigenvectors of the transport matrix describing surface ocean transport during 1000 years. Source: Froyland et al. (2014). (A black and white version of this figure appears in some formats. For the color version, please refer to the plate section.)

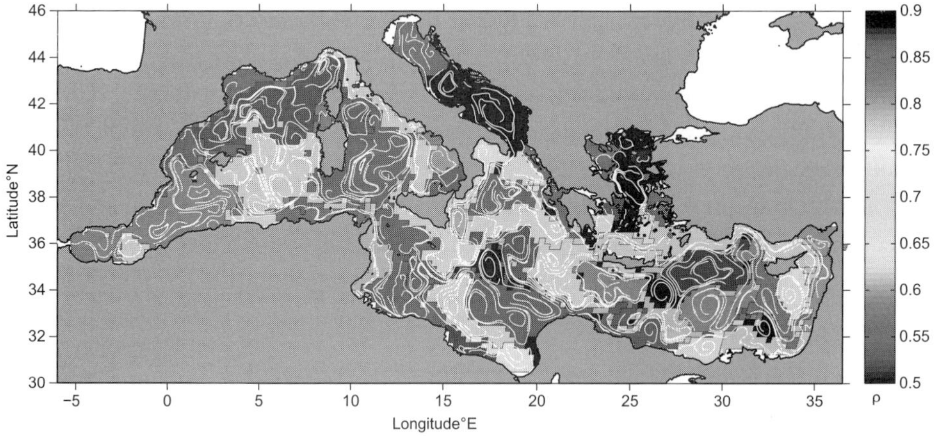

Figure 4.11 A partition of the Mediterranean Sea into 65 communities by Infomap, from the transport matrix describing surface transport for one month starting on January 1 2011. Each community is colored with the value of its coherence ratio ρ (4.27)) White lines are average streamlines during the month considered. Source: Rossi et al. (2014). (A black and white version of this figure appears in some formats. For the color version, please refer to the plate section.)

communities computed with the Infomap algorithm applied to the flow network of surface Mediterranean circulation during one month (January 2011, velocity data from the NEMO model). See additional details in Rossi et al. (2014) and Ser-Giacomi et al. (2015b). Note the detection of communities with rather different sizes, features not easy to extract with spectral methods.

A standard way to assess the quality of a network partition is by computing a modularity parameter (Newman, 2010; Newman and Girvan, 2004). But this involves comparison with a random null model that in the case of flow networks has no obvious meaning. Then, alternative quantifiers with a direct interpretation in terms of fluid connectivity should be developed. Here we define a *coherence ratio* and a *mixing parameter*.

If coherent regions A are understood as almost-invariant areas of fluid, this means they are mapped by the flow nearly into themselves after a time τ:

$$\Phi_{t_0}^{\tau}(A) \approx A \,. \tag{4.25}$$

To measure how well this is achieved, one can introduce the *coherence ratio* (Froyland, 2005; Froyland and Dellnitz, 2003):

$$\rho_{t_0}^{\tau}(A) = \frac{m(A \cap \Phi_{t_0+\tau}^{-\tau}(A))}{m(A)} \tag{4.26}$$

where, as before, $m(C)$ is the area of set C, but it can be generalized to other measures. We have $\rho_{t_0}^{\tau}(A) \leq 1$ and values close to unity indicate that A is a truly almost-invariant set.

In our discrete set-up, we consider sets A made of our boxes $\{B_i, i = 1, \ldots, N\}$: $A = \cup_{i\in\mathcal{I}} B_i$, where $\mathcal{I}$ is the set of indices identifying the boxes B_i making A. The coherence ratio is now (Froyland, 2005; Froyland and Dellnitz, 2003)

$$\rho_{t_0}^{\tau}(A) = \frac{\sum_{i,j\in\mathcal{I}} m(B_i)\mathbf{P}(t_0,\tau)_{ij}}{\sum_{i\in\mathcal{I}} m(B_i)} . \tag{4.27}$$

This quantity is used to color the different communities in Fig. 4.11.

For a partition of the fluid domain into p communities: $\mathcal{P} = \{A_1, \ldots, A_p\}$, a global quality figure of the partition is

$$\rho_{t_0}^{\tau}(\mathcal{P}) \equiv \frac{1}{p}\sum_{k=1}^{p} \rho_{t_0}^{\tau}(A_k), \tag{4.28}$$

where again a good partition would be indicated by a value close to 1. When communities are of very different sizes it may be appropriate to weight the average in (4.28) with these sizes.

Physically we can say that $\rho_{t_0}^{\tau}(\mathcal{P})$ represents the fraction of tracers that at time $t_0 + \tau$ are found in the same community where they were released at time t_0. The definition involves the initial and final positions, but gives no information on the particle trajectories in between. Note that coherence ratios measure fluid exchanges between communities, but do not quantify how strong the internal mixing is.

The second quantifier we use is a *mixing parameter* devised to assess how strongly the flow mixes fluid inside communities. To define the mixing parameter $M_{t_0}^{\tau}(A)$ inside a set A we first define a transport matrix conditioned to represent just the transport occurring inside A (more precisely, transport by trajectories that start and end in A):

$$\mathbf{R}(t_0,\tau|A)_{ij} = \frac{\mathbf{P}(t_0,\tau)_{ij}}{\sum_{k\in\mathcal{I}} \mathbf{P}(t_0,\tau)_{ik}}, \quad i,j\in\mathcal{I}. \tag{4.29}$$

As before, $\mathcal{I}$ is the set of indices identifying the boxes B_i making A. The mixing parameter is a normalized version of the sum inside A of the entropies associated to the transition probabilities in $\mathbf{R}(t_0,\tau|A)$:

$$M_{t_0}^{\tau}(A) = \frac{-\sum_{i,j\in\mathcal{I}} \mathbf{R}(t_0,\tau|A)_{ij}\log \mathbf{R}(t_0,\tau|A)_{ij}}{Q_A \log Q_A} . \tag{4.30}$$

Q_A is the number of boxes in A. The maximum value, $M_{t_0}^{\tau}(A) = 1$, is reached when fluid is dispersed from each box in A to all the others uniformly ($\mathbf{R}_{ij} = 1/Q_A, \forall i,j \in \mathcal{I}$). A global quantification of the internal mixing in a community partition $\mathcal{P} = \{A_1, \ldots, A_p\}$ is given by

$$M_{t_0}^{\tau}(\mathcal{P}) = \frac{\sum_{k=1}^{p} m(A_k) M_{t_0}^{\tau}(A_k)}{\sum_{k=1}^{p} m(A_k)}. \tag{4.31}$$

Here, we have weighted the different communities according to their size.

Examples of community partitions of flow networks and their quality figures will be given in Section 7.3.

4.4.4 Optimal Paths in Flow Networks

Prominent connectivity patterns in networks can be revealed by introducing the concept of paths (Newman, 2010; Newman et al., 2006). While its definition is simple and intuitive for static or aggregated networks, for the temporal case paths between nodes can suddenly appear and disappear in time (Holme and Saramäki, 2012; Kempe et al., 2000; Kim and Anderson, 2012; Kostakos, 2009; Lentz et al., 2013; Masuda et al., 2013; Starnini et al., 2012; Tang et al., 2010a,b). Thus, there has recently been a focus toward the definition and characterization of paths in temporal networks. The concept of shortest path in static network analysis has been generalized to include information on the time necessary to establish a space–time connection between nodes. This was the motivation behind the development of fastest path analysis, specially for unweighted and undirected networks (Holme and Saramäki, 2012; Starnini et al., 2012). Although it is relevant to study the time required to build a path among two nodes, it is equally crucial to understand how to quantify the importance and the distribution of such paths. This issue becomes essential when one tries to exploit this information in order to define and evaluate global network properties, for example betweenness centrality.

The analysis is restricted to a time interval $[t_0, t_M]$ in which $M + 1$ snapshots of the state of the network are taken at times $t_l = t_0 + l\tau$, $l = 0, 1, \ldots, M$, with τ the time between them. We consider a temporal, directed, and weighted network of N nodes. Its time-dependent connectivity is described by a set of weighted adjacency matrices $\mathbf{A}^{(l)}$, $(l = 1, \ldots, M)$, in which the matrix element $\mathbf{A}^{(l)}_{IJ} \geq 0$ specifies the strength of connectivity from I to J during the time interval $[t_{l-1}, t_l]$. A convenient way to analyze the system is using *time-ordered graphs* (TOGs) (Kim and Anderson, 2012). Formally, the TOG can be considered a static network of $N \times (M + 1)$ nodes with directed and causal links. For each snapshot l, a group $V(t_l)$ of N nodes replicating the nodes of the original network can be defined. Links are then established only from nodes at successive times, i.e., from $i_{l-1} \in V(t_{l-1})$ to $j_l \in V(t_l)$ with the weights given by those in the original temporal network: $\mathbf{A}^{(l)}_{i_{l-1}\, j_l}$.

We now consider a flow or transport process by releasing random walkers in each node of the network. Their motion is assumed to be Markovian and is defined by single-step transition probabilities proportional to the entries in the adjacency matrices. Specifically the probability of reaching node k_l at time t_l under the condition of being at k_{l-1} at time t_{l-1} is:

$$\mathbf{T}^{(l)}_{k_{l-1}k_l} \equiv \frac{\mathbf{A}^{(l)}_{k_{l-1},k_l}}{K^{(l)}_O(k_{l-1})}. \tag{4.32}$$

Here $K^{(l)}_O(k) = \sum_j \mathbf{A}^{(l)}_{kj}$ is the *out-strength* of node k during time step l, so that $\sum_j \mathbf{T}^{(l)}_{kj} = 1$. Note that, with this normalization, $\mathbf{T}^{(l)}_{ij}$ is nothing but the transfer matrix $\mathbf{P}(t_{l-1}, \tau)_{ij}$ as defined in Section 4.4. In this section we prefer to follow this different notation to avoid confusion of transfer matrices $\mathbf{T}$ with other probabilistic quantities which will be denoted with the letter P.

A generic M-step path μ between two nodes I and J is defined as a $(M+1)$-uplet $\mu \equiv \{I, k_1, \ldots, k_{M-1}, J\}$ providing a sequence of nodes crossed to reach J at time t_M from I at time t_0. Under the Markovian assumption the probability for a random walker to take the path μ under the condition of starting at I is:

$$(p^M_{IJ})_\mu = \mathbf{T}^{(1)}_{Ik_1}\left[\prod_{l=2}^{M-1} \mathbf{T}^{(l)}_{k_{l-1}k_l}\right]\mathbf{T}^{(M)}_{k_{M-1}J}. \tag{4.33}$$

The most probable path (MPP) η^M_{IJ} is the path that maximizes (4.33) with respect to the intermediate nodes $k_2, \ldots, k_{M-1}$. Its probability is denoted by $P^M_{IJ} = \max_\mu\{(p^M_{IJ})_\mu\}$. The exact maximization of (4.33) can be obtained iteratively by noting that in the first step the maximum probability to reach a given node k_1 is simply $P^1_{Ik_1} = \mathbf{T}^{(1)}_{Ik_1}$ and then using the recurrence

$$P^{l+1}_{Ik_{l+1}} = \max_{k_l}\left(P^l_{Ik_l}\mathbf{T}^{(l+1)}_{k_l k_{l+1}}\right) \tag{4.34}$$

for $l = 1, 2, \ldots, M-1$ until reaching $k_M = J$. This type of iterative optimization is similar (taking logarithms of the probabilities involved) to the one used to find optimal configurations of directed polymers in random media (Halpin-Healy and Zhang, 1995) and can be considered as an adaptation of the classical Dijkstra algorithm (Dijkstra, 1959) to the layered and directed structure of the TOG. The computational cost of the maximization is strongly reduced by calculating first *accessibility matrices* (Lentz et al., 2013) and restricting the maximization search to the set of nodes that are *accessible* from I and for which J results *accessible* as well.

To assess whether the most probable path alone is a good representation of the transport dynamics, we introduce the quantity $\lambda^M_{IJ} \equiv P^M_{IJ}/\sum_\mu (p^M_{IJ})_\mu$. It corresponds to the fraction of probability carried by the MPP between I and J with respect to the sum of the probabilities of all paths connecting these nodes after M steps.

Note that the sum in the denominator can be efficiently computed as the entry (I, J) in the matrix product $\prod_{l=1}^M \mathbf{T}^{(l)}$. Depending on the network under investigation, the MPP can actually carry a significant fraction of the total connection probability. When this is not the case, we can relax the definition of MPP and define a subset of *highly probable paths* (HPPs) that carry most of the probability. In particular, we want to identify paths characterized by individually carrying a probability larger than a fraction ϵ of the MPP probability, i.e., larger than ϵP_{IJ}^M, with $0 \leq \epsilon \leq 1$. Exhaustively searching for all such paths becomes computationally prohibitive except for the smallest N and M values. Here we compute the set $\mathcal{Q}_{IJ}^M$ of all paths of M steps between I and J that are constructed by joining the MPP from I to an intermediate k_l and the MPP from k_l to J. In principle there would be $(M-2) \times N$ such paths, one for every choice of the intermediate k_l, but this number is in fact much smaller when considering that k_l should be accessible from I (Lentz et al., 2013), that J should be accessible from k_l itself, and that many of the resulting paths turn out to be repeated. Out of these we consider the subset $\mathcal{K}_{IJ}^M(\epsilon) = \{\mu \in \mathcal{Q}_{IJ}^M \mid (p_{IJ}^M)_\mu > \epsilon P_{IJ}^M\}$. This set contains the MPP, and although it may miss some of the paths with probability larger than ϵP_{IJ}^M, we expect it to contain a sufficiently representative sample of them. This can be checked by calculating $\lambda_{IJ}^M(\epsilon) \equiv \sum_{\nu \in \mathcal{K}_{IJ}^M(\epsilon)} (p_{IJ}^M)_\nu / \sum_\mu (p_{IJ}^M)_\mu$, the fraction of probability carried by this set of HPPs.

The search for HPPs could be improved by defining new subsets of them $\mathcal{K}_{IJ}^M(r,\epsilon)$. The parameter r is introduced to determine the number of constraints imposed in the search for these relevant paths. Given the initial (I) and final (J) points, we fix r nodes at intermediate times and look for paths between I and J made of segments which are MPPs connecting these intermediate nodes. For $r = 0$ we recover the MPPs and for $r = 1$ we have the paths in the set $\mathcal{K}_{IJ}^M(\epsilon)$ defined above. Evaluation of these sets of HPPs can be computationally costly for high values of r, since the algorithm scales exponentially with r. Nevertheless, interesting results can be obtained considering already low-order HPPs, i.e., $r = 1$ and $r = 2$.

4.4.5 MPP-Betweenness

Equipped with the above definitions we can now characterize network properties that are dependent on optimal paths in different ways. One of these is the concept of *betweenness centrality*, which is generally defined as the proportion of shortest paths passing through a node. We introduce here a definition based on the number of *most probable paths* crossing a node. Specifically, we define the betweenness of node K after M steps as $\mathcal{B}_K^M = \sum_{IJ} g_{IJ;K}^M / N_M$, where the sum is over all pairs of initial nodes I and final accessible nodes J, N_M is the total number of connected pairs of nodes at time step M (computable from accessibility matrices, see Lentz et al. (2013)), and $g_{IJ;K}^M$ is the number of times the node K appears in the most probable path connecting I and J. Fixing the time interval M corresponds to considering paths with the same temporal duration. In this way we ignore connections that are occurring at shorter or longer times (Kim and Anderson, 2012) and that can be significantly more probable. It is possible to overcome this limitation by performing a multistep analysis: we can look at all MPPs with M in a given interval $[M_{min}, M_{max}]$ and choose the MPP, $\eta_{IJ}^{[M_{min},M_{max}]}$, with the highest probability. The multistep analysis leads to an alternative definition of betweenness, i.e., a multistep MPP-betweenness $\mathcal{B}_K^{[M_{min},M_{max}]}$ which is calculated considering the multistep MPPs instead of the fixed-M one.

4.5 Event Synchronization Networks

The event synchronization (ES) method (see Section 3.3.3) provides an alternative way to construct a network from climate observations: event synchronization networks (ESNs).

ES was first used by Malik et al. (2010, 2012) to analyze spatial and temporal patterns of extreme rainfall during the Indian Summer Monsoon (ISM; see Section 6.6.2). Considering rainfall time series, which are not smooth and continuous (as those of temperature or pressure fields) and often contain a high-frequency component, it is important to choose an appropriate method to infer similarity of dynamical behavior in different geographical sites. Malik et al. (2010) used the ES method to study precipitation data, defining, at each grid site, an extreme event series of rainfall and computing the strength of synchronization, Q_{ij}, between each pair of grid sites i and j. By normalizing as $0 \leq Q_{ij} \leq 1$, where $Q_{ij} = 1$ means complete synchronization, and $Q_{ij} = 0$ means absence of synchronization, a correlation-like matrix was obtained, which represented the strength of synchronization of the extreme rainfall events between pairs of grid sites. The ES network is then constructed by selecting only pairs of sites that show strong correlations, and the time lags between the events are used to define the direction of the links.

This approach has also been used to predict extreme rainfall events in the Central Andes of South America (Boers et al., 2014). From real-time satellite-derived rainfall data, Boers et al. (2014) were able to predict more than 60% (90% during El Niño conditions) of rainfall events above the 99th percentile in the Central Andes, and were also able to detect a linkage between polar and tropical regimes, as the underlying responsible mechanism.

To summarize, in this chapter we have presented an overview of various approaches for constructing climate networks that will be used in the following chapters: "functional networks" constructed from correlation analysis (linear or nonlinear, by using cross-correlations or mutual information), which can be undirected or directed (the link directionality can be defined from correlation time lags, measures of information transfer, or Granger causality). We have also introduced "flow networks," constructed from the analysis of transport processes in flows, and "event synchronization networks," constructed from the analysis of correlations and time delays between extreme events. As the construction and the analysis of climate networks involve managing a huge amount of data, in the following chapter we present efficient computational tools, and then, in the following chapters, we discuss new insights of atmospheric and ocean dynamical processes, which are gained by using the network approach.

5

Computational Tools for Network Analysis

In this chapter we provide an overview of the different numerical tools available for the efficient reconstruction and analysis of climate networks. The computational challenge is described in Section 5.1 and is followed by sections in which serial (Section 5.2) and parallel (Section 5.3) tools are presented.

5.1 Computational Problem

In Chapter 4, we covered a number of techniques to reconstruct networks from climate data (from observations or model simulations). Assume that, for a particular problem, climate data are given in the form of a time series, represented by an $N \times M$ matrix, where N is the number of locations and M is the number of temporal measurements (for example, daily or monthly values). Then, for example, to reconstruct a Pearson correlation climate network (PCCN) one needs to calculate and process of the order of $N^2/2$ correlation values. Such computations become challenging for large N. For example, with a network of 10^6 nodes, this would result in 5×10^{11} correlation calculations. A further challenge is the memory needed for such a computation. Only to keep the calculated correlation matrix in memory for further processing, about 4×10^3 GB of memory is required (consider 8 bytes of memory for each of the 5×10^{11} matrix items), which is not available in the vast majority of current computing platforms.

In addition, analyzing the resulting network (graph) is nontrivial and also computationally challenging. Considering a graph G with V nodes and E edges (links between nodes). A typical step in an algorithm to analyze G involves visiting each $v \in V$ and its neighbors $\bar{V} \subset V$ (the set of vertices connected to v by an edge $e \in E$), then their consecutive neighbors, and so on. Processing such steps is normally done within a computational complexity of the order of $n(V)$ and/or $n(E)$ squared or cubed, where $n(V)$ indicates the number of elements in the set V. For example, the computation of the clustering coefficient, which measures the degree to which network vertices tend to cluster together, has a time complexity of order $n(V)^3$.

Regarding the reconstruction and analysis of climate networks, there are hence two types of user profiles: those using relatively small networks (small $n(V)$) and those using large networks (large $n(V)$). For the small networks, there are various available software tools for graph analysis, some of them providing implementations of single-machine algorithms such as BGL (Mehlhorn and Näher, 1995), LEDA (Siek et al., 2002), NetworkX (Hagberg et al., 2008), Stanford Network Analysis Platform (SNAP, see http://snap.stanford.edu), and `igraph` (Csardi and Nepusz, 2006). However, most of the existing libraries do not address the processing and memory challenges involved in the construction of graphs with large $n(V)$ from statistical measures of time series.

Indeed most researchers tend to develop their own tools to build correlation matrices beforehand, and thereafter they transform these matrices into appropriate graph data structures that can be handled by the existing libraries of graph analysis. An exception is the software package `pyunicorn` (see http://tocsy.pik-potsdam.de/pyunicorn.php) (Donges et al., 2015), which couples Python modules for numerical analysis with `igraph`. It can carry out both tasks: the construction of climate networks and the analysis of the resulting graphs. As it has been used frequently by researchers in climate networks, we will describe it in more detail in Section 5.2 and show its performance for a typical (relatively small) climate network problem.

The networks that have mostly been handled in climate research applications have only a limited (10^4) number of nodes. As a consequence, coarse-resolution observational and model data have been used with a focus only on large-scale properties of the climate system. Climate is, however, known for its multi-scale interactions and hence one would like to explore the interaction of processes over the different scales (see Chapter 1). Data are available through high-resolution ocean/atmosphere/climate model simulations, but they lead to networks with at least 10^5 nodes. Single-core software is bounded by memory and speed, making it impossible to construct large-node climate networks, and consequently these methods are inappropriate to analyze them. For example, the computation of a clustering coefficient for a network with $V = 10^6$ nodes would be very challenging, if even possible, with existing single-core software.

The most popular approach to tackling such computational challenges is exploiting parallelism for both the construction and the analysis of those massive graphs through the design of efficient algorithms for parallel computing platforms. In this regard, contributions have been made to the development of algorithms that exploit parallel computing machines, such as in the Parallel BGL (Gregor and Lumsdaine, 2005) and CGMgraph (Chan et al., 2005). However, due to structure irregularity and sparsity of real-world graphs, including those built using climate data, there are few parallel implementations that are efficient, scalable, and can deliver high performance. Other factors that contribute to this inefficiency include a manifested irregularity of data dependencies in those graphs, as well as the poor locality of data, making graph exploration and analysis highly dominated by memory latency rather than processing speed (Lumsdaine et al., 2007). Another attempt with NetworKit (see http://networkit.iti.kit.edu) has been a remarkable step toward providing parallel software tools capable of analyzing large-scale networks. Yet the networks analyzed by this software had at most 4×10^7 edges, which is still lower that what is intended to be studied in climate networks. More recently, a parallel network toolbox, Par@Graph, was developed (Ihshaish et al., 2015) and applied to reconstruct and analyze networks using data with large N. In Section 5.3 we will present details of this software and show its performance for a typical (large-node) climate network problem.

5.2 Serial Tools: `pyunicorn`

The Python software package `pyunicorn` (available at https://github.com/pik-copan/pyunicorn) implements concepts and techniques from complex network theory, as well as nonlinear time series analysis, and integrates these approaches to develop new innovative methodologies, such as functional networks, and network-based time series analysis (Donges et al., 2015).

5.2.1 Description

With a focus on complex network theory, `pyunicorn` is a valuable complement to traditional nonlinear time series analysis tools such as TISEAN (Hegger et al., 1999). Its main mode of operation is to import, generate, and export complex networks from time series or fields thereof, and to compute appropriate measures on these networks in order to derive insights about the causal structure and dynamical regimes of underlying processes. Fig. 5.1 shows two examples for the software architecture, one (Fig. 5.1a) for recurrence networks (Donner et al., 2010; Marwan et al., 2009) and the other (Fig. 5.1b) for a mutual information climate network (MICN).

Although the development of `pyunicorn` has mostly accompanied advances in climate research, the generality of the network approach and its extensions, such as spatiotemporal and time-delayed embedding, node-weighted measures, coupled functional networks, and recurrence networks, render this software package widely applicable in numerous fields, such as medicine, neuroscience, sociology, economics, and finance. The `pyunicorn` library is fully object-oriented and its inheritance and composition hierarchy reflects the relationships between the analysis methods in use (Fig. 5.1). As is usually the case with Python libraries, `pyunicorn`

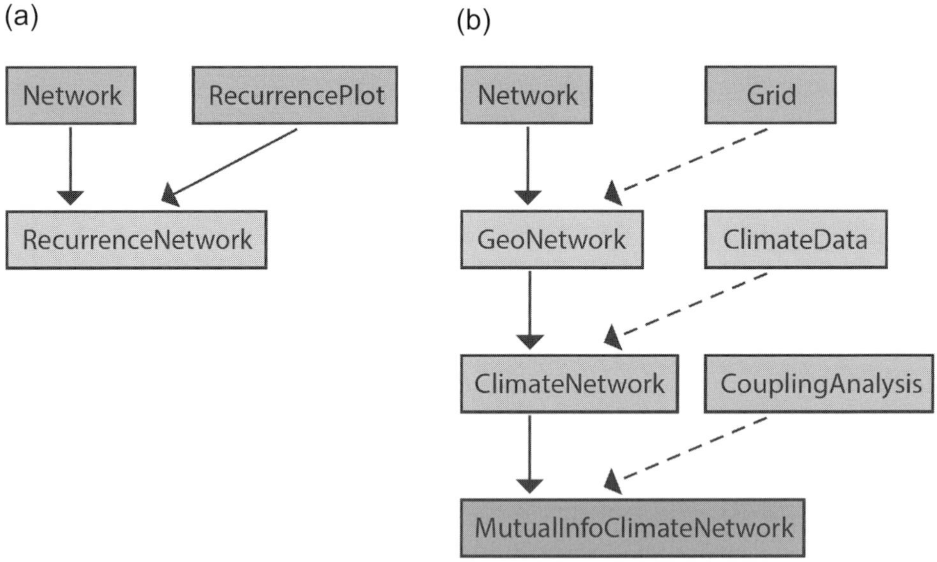

Figure 5.1 Examples for the software architecture of `pyunicorn` displayed as a unified modeling language (UML) diagram of class relationships: ancestry of the `pyunicorn` class (a) `timeseries.RecurrenceNetwork` and (b) `climate.MutualInfoClimateNetwork`. Inheritance (class B inherits from class A, solid arrows) and object composition relationships (class B contains class A, dashed arrows) are indicated. Source: Donges et al. (2015).

tries to provide simple interfaces and clear architecture, while delegating the heavy lifting to specialized tools for performance. The basic network measures and generative models are inherited from `igraph`. Wherever possible, numerically intensive computations are expressed as combinations of highly optimized linear algebra methods from `Numpy` and `Scipy`, and implemented in embedded C otherwise. Thus all costly computations are performed in compiled C, C++, or FORTRAN code.

Since `pyunicorn` internally represents networks as sparse adjacency matrices, it can handle relatively large data sets. However, when the number of nodes is larger than order 10^3, as for example when reconstructing climate networks from high-resolution climate model simulations output, this software may be bounded by the single machine's memory and speed.

5.2.2 Performance

For analyzing the performance of `pyunicorn`, data were used from the Zebiak-Cane (ZC) model (Zebiak and Cane, 1987), which is an intermediate complexity

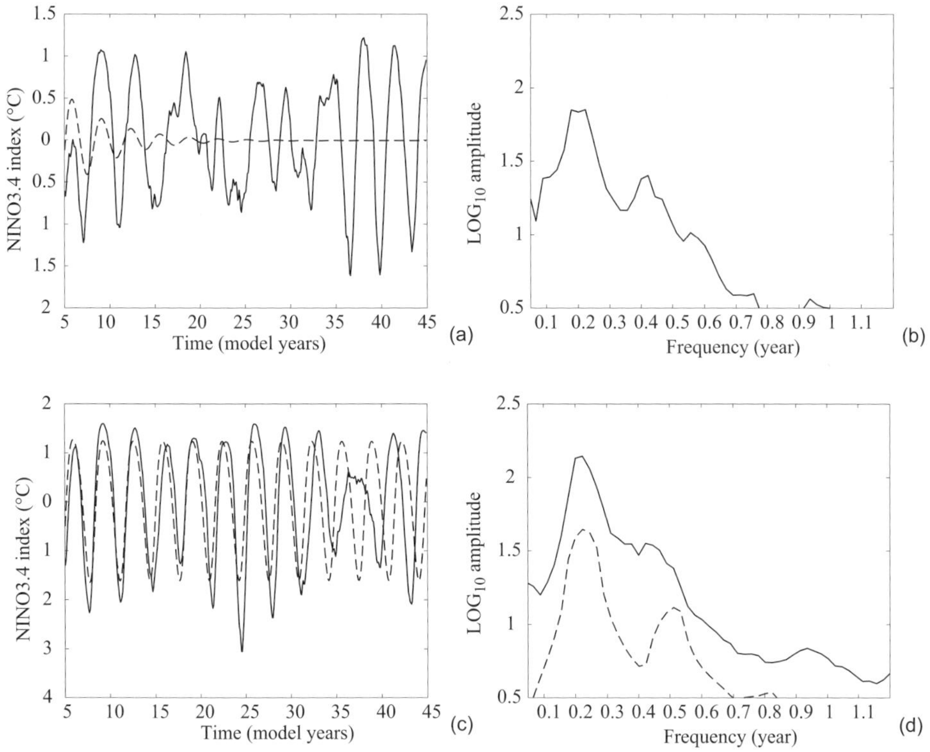

Figure 5.2 The response of the Zebiak–Cane (ZC) model. Dashed curves are for the deterministic model and solid ones are for the stochastic model with red-noise wind-stress forcing. (a) The NINO3.4 index at a coupling strength $\mu = 2.7$; (b) the amplitude spectrum for (a); (c) Same as (a) but at $\mu = 3.02$; (d) the amplitude spectrum for (c). Source: Feng and Dijkstra (2017).

model aimed at modeling the El Niño/Southern Oscillation phenomenon (ENSO). It is a pseudo-spectral model with a number of nodes equal to $N_x \times N_y$ collocation points. We will present results of `pyunicorn` here from 45-year simulations with different numbers of nodes $N = N_x \times N_y$, standard values being $N_x = 30$ and $N_y = 31$. The strength of the ocean–atmospheric feedbacks in this model is measured by a dimensionless coupling parameter μ. See additional details in Feng and Dijkstra (2017).

In the deterministic ZC model, a Hopf bifurcation occurs at the coupling parameter value $\mu = \mu_c = 3.0$ for the standard values of the other parameters (Feng and Dijkstra, 2017). When the coupling strength is smaller than μ_c, for example $\mu = 2.7$, the system is in the subcritical regime in which variables approach a stable fixed point if no noise is present. The response to an initial perturbation is a damped ENSO oscillation. This is shown in the behavior of the NINO3.4 index in Fig. 5.2a (dashed line). When the coupling strength is increased to just above the critical value (e.g., $\mu = 3.02 > \mu_c$) the system enters the supercritical regime where the NINO3.4 index displays an interannual oscillation (dashed line in Fig. 5.2c) and the spectrum shows a main peak at a frequency corresponding to about four years (dashed line in Fig. 5.2d).

In Feng and Dijkstra (2017), an additive red-noise product was used for the zonal wind-stress forcing (Roulston and Neelin, 2000). First, a linear model relating the Florida State University pseudo-wind-stress data (Legler and O'Brien, 1988) over the Pacific Ocean for the period 1978–2004 to the extended reconstructed sea surface temperature (ERSST) (Smith and Reynolds, 2003) for the same period is used to extract a deterministic part of the zonal wind-stress. The zonal wind-stress residual is interpreted as atmospheric noise and decomposed into its EOFs. The first and second EOF and their principal components (PCs) are shown in Fig. 5.3. The first EOF (Fig. 5.3a) captures the residual zonal wind-stress response in the eastern Pacific. The second EOF (Fig. 5.3b) captures the pattern of westerly wind bursts which are generally located west of the date line (Lian et al., 2014). For example, the second PC (PC2) shows (Fig. 5.3d) a strong westerly wind event in March 1997, in agreement with other analyses (Lian et al., 2014; Menkes et al., 2014).

The stochastic component of the zonal wind-stress forcing τ_n^x is finally approximated as (Feng and Dijkstra, 2017)

$$\tau_n^x(x, y) = a_1 E_1(x, y) + a_2 E_2(x, y), \tag{5.1}$$

where $E_i, i = 1, 2$ are the patterns of the first and second EOF and a_i is determined from the fit of an autoregressive model (AR1) $x_{i,t}$ to the PC$_i$ time series using

$$x_{i,t+1} = a_i x_{i,t} + \sigma_i \epsilon_{i,t}. \tag{5.2}$$

Here, each a_i is the lag-1 auto-correlation of PC$_i$ and the term $\sigma_i \epsilon_{i,t}$ represents white noise with a variance σ_i.

This stochastic component (5.1) is added to the deterministic zonal wind-stress and used to force the ZC model with additive noise. The model response is shown as the solid curves in Figs. 5.2a, b (subcritical) and 5.2c, d (supercritical). In the subcritical regime, the red-noise forcing is necessary to excite the ENSO variability, while in the supercritical regime the red-noise forcing simply causes a higher amplitude of ENSO variability; these results are similar to those provided by Roulston and Neelin (2000).

The data to test the performance of `pyunicorn` were obtained from ZC-model outputs of 11 simulations, 540 months (45 years) each, at different values of the coupling strength μ (2.70, 2.80, 2.90, 2.95, 2.98, 3.00, 3.02, 3.10, 3.15, 3.25, and 3.40). They were used to build a PCCN for each value of μ, by using the computed SST anomalies. A standard network has 30 (longitude) $\times$ 31 (latitude) $=$ 930 nodes

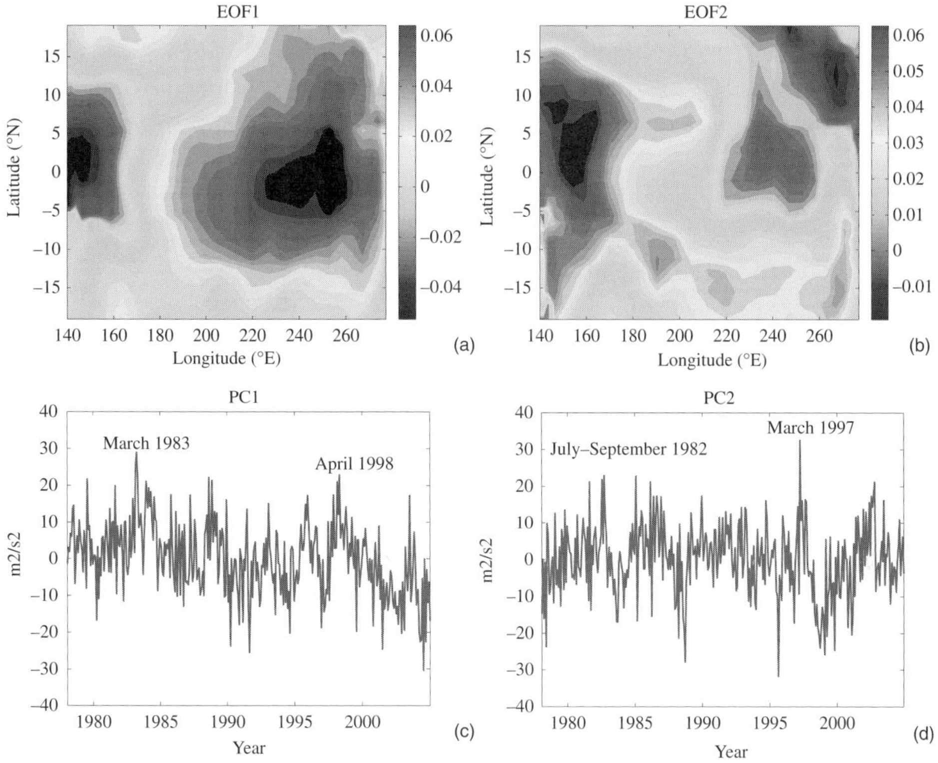

Figure 5.3 (a) The first empirical orthogonal function (EOF) of the zonal wind-stress residual, accounting for 11.9% of the variance. (b) The same as (a) but the second EOF, accounting for 11.6% of the variance. (c) The first principal component (PC1) of the wind-stress residual. (d) The same as (c), but for the second principal component (PC2). Source: Feng and Dijkstra (2017). (A black and white version of this figure appears in some formats. For the color version, please refer to the plate section.)

within the domain (140°E, 280°E) × (20°S, 20°N). Links are assigned to pairs of nodes with correlation higher than a threshold value $\epsilon_C = 0.5$, which guarantees that these links are based on significant ($p < 0.05$) correlations. Also different network sizes (by varying N_x) and threshold values were considered (see Table 5.1).

CPU times used for the network reconstruction and analysis computations are presented in Table 5.1 for the case $\mu = 2.7$. The computation time of the degree field is in all cases negligible with respect to the reconstruction effort. The betweenness computation is the most demanding effort in these networks but it can be computed with `pyunicorn` in reasonable times for these network sizes.

The degree fields of the PCCNs reconstructed using SST for four values of μ are shown in Figs. 5.4a–d. When the coupling strength μ is increased from the

Table 5.1 *Performance of* `pyunicorn` *for the ZC model SST data. Different network sizes and threshold values ϵ_C used in the reconstruction of Pearson correlation networks from the $\mu = 2.7$ dataset and corresponding number of network vertices and edges. The computations were performed on a bullx workstation.*

Vertices	ϵ_C	*Edges*	Reconstruction (s)	Degree (s)	Betweenness (s)
930	0.5	80 162	0.132	9.39×10^{-6}	1.258
930	0.75	27 460	4.88×10^{-2}	8.53×10^{-6}	0.489
930	0.9	8490	2.30×10^{-2}	9.39×10^{-6}	0.121
620	0.5	35 294	5.97×10^{-2}	8.96×10^{-6}	0.392
620	0.75	14 161	2.42×10^{-2}	9.81×10^{-6}	0.167
620	0.9	5674	1.36×10^{-2}	7.68×10^{-6}	0.037
310	0.5	9193	1.37×10^{-2}	8.53×10^{-6}	0.051
310	0.75	4486	7.84×10^{-3}	7.68×10^{-6}	0.016
310	0.9	2161	4.50×10^{-3}	8.11×10^{-6}	0.005

subcritical $\mu < \mu_c$ to the supercritical regime $\mu > \mu_c$, more nodes in the region between 220°E and 280°E get a higher degree. When the critical boundary μ_c is approached, a large-scale coherence appears in the network degree field. Another distinct change of the degree fields (Figs. 5.4a–d) is that the patterns of equatorially symmetric Rossby waves are becoming more prominent when the background state enters the supercritical regime.

Histograms of the degree fields (the degree distributions) for two different values of μ are plotted in Figs. 5.5a, b. For $\mu = 2.7 < \mu_c$ (Fig. 5.5a), the degree distribution is bimodal, where the first peak represents the low-degree nodes in Fig. 5.4a and the second peak (located near a degree of 250) represents the high-degree nodes. When μ is increased (Fig. 5.5b) a peak at a degree of 400 occurs, representing the higher degree nodes in Fig. 5.4c. These arise the ENSO oscillation becomes more dominant in the SST anomalies when the background climate moves into the supercritical regime (Feng et al., 2014; van der Mheen et al., 2013). Further network characteristics are discussed by Feng and Dijkstra (2017).

5.3 Parallel Tools: Par@Graph

Par@Graph is composed of a set of coupled parallel tools designed to leverage the inherited hybrid parallelism in distributed-memory clusters of multi-core (SMP) machines using MPI/OpenMP standards (Ihshaish et al., 2015). The provided tools

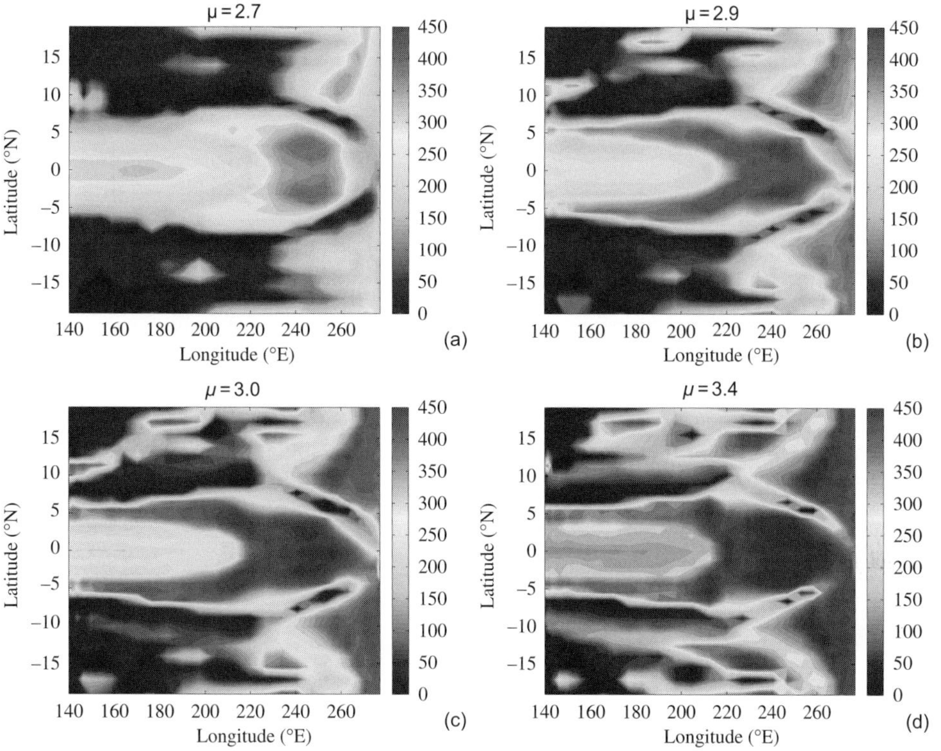

Figure 5.4 (a) Degree field of the PCCN using a threshold $\epsilon_C = 0.5$ reconstructed from the ZC model SST data at a coupling strength $\mu = 2.7$. (b) Same as (a) but at $\mu = 2.9$. (c) Same as (a) but at $\mu = 3.0$. (d) Same as (a) but for $\mu = 3.4$. Source: Feng and Dijkstra (2017). (A black and white version of this figure appears in some formats. For the color version, please refer to the plate section.)

are classified into two major software modules, which we refer to as the network constructor and the analysis engine, together with additional interfacing tools and wrappers.

5.3.1 Description

The network constructor module computes the correlation matrix C from time series at different locations (the network nodes). It also applies a user-defined threshold τ to generate the corresponding network adjacency matrix A. Then it proceeds to the transformation of A into a network data structure which will later be analyzed by the analysis module. The design of the constructor follows a master–worker parallel computing paradigm for distributed-memory parallel clusters of SMPs. The calculation of the correlations between time series is distributed over

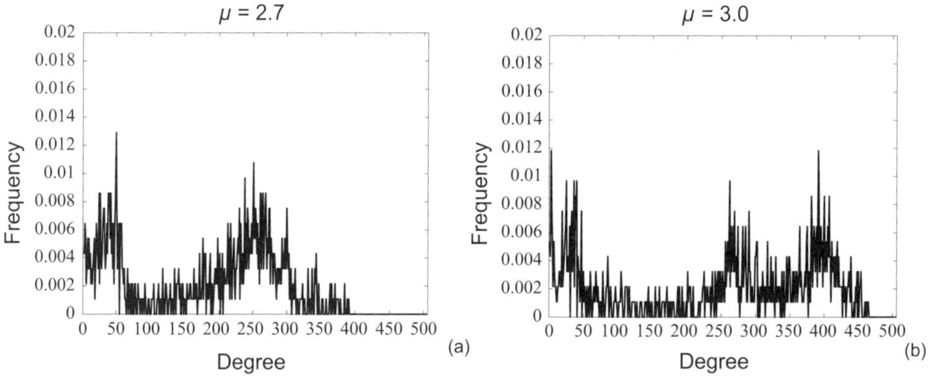

Figure 5.5 (a) Degree distribution of the PCCN using a threshold $\epsilon_C = 0.5$ reconstructed from the ZC model SST data at the coupling strength $\mu = 2.7$. (b) Same as (a) but at $\mu = 3.0$. Source: Feng and Dijkstra (2017).

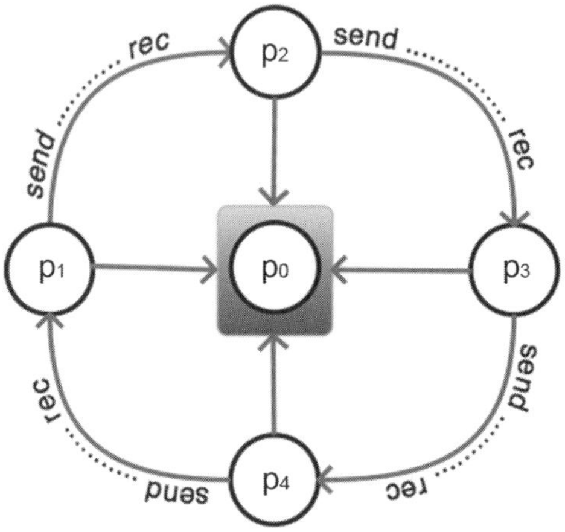

Figure 5.6 Provided a parallel machine of p processors, $p - 1$ processes are initialized and assigned with equal blocks of nodes. Each block's set of time series are correlated, and then these blocks are exchanged $(p - 1)/2$ times (half round of the ring) between processes to complete the all-to-all correlations between the whole set of a node's time series. Conversely, p_0 (the master computing element) is initialized as a master process to gather the resulting calculations and perform the analysis tasks on the resulting network. Source: Ihshaish et al. (2015).

the computing elements (workers), forming a ring topology of processes, shown in Fig. 5.6, which communicate with each other using MPI standards. As soon as a process finds $C_{ij} \geq \tau$, the pair (i, j) is copied to a local process buffer of a user-configurable size, and sends the iteratively filled buffer to the master p_0, where the network is to be analyzed. Note that if the network is weighted, the value of C_{ij} itself is also copied and sent to the master side by side with its pair of nodes i and

j (if required, this is also done for time-lag values). Note that only a subset $\bar{C}$ of C, such that $\forall \bar{C}_{ij} \in \bar{C}, \bar{C}_{ij} \geq \tau$, is sent progressively to the master computing element. This indeed means that the under-threshold values of C are discarded directly at each ring process. This reduces both the amount of data sent to the master element and the memory required there for the construction of the network.

The process of constructing the network itself is performed progressively in the events that the master (p_0) receives edges' coordinates (and attributes, e.g., weights/lags) from any ring process. Initially p_0, having the number of data set grid points, constructs a completely unconnected network, i.e., no edges between graph vertices. As soon as ring processes start sending edge coordinates to p_0, these edges are added to the network. In the long run, constructing the network following this approach results in saving time, since the master is idle (except when receiving data from workers) during the ring-processing iterations. More importantly, because the coming edges are added directly to the graph data structure, memory usage is optimized at the master as data redundancy is markedly minimized. Once correlations and their coordinates are available at the master machine, it consecutively runs graph algorithms to analyze the constructed network. The parallel algorithms developed for network analysis are based on those in `igraph`. This design (coupled with the network constructor) achieves three primary goals: (1) to construct the network rapidly, (2) to enable efficient and safe multithreading of the core library algorithms, and (3) to reduce memory usage for network representation.

With respect to the analyzing algorithms, a set of 20 of the core algorithms of `igraph` have been parallelized in Par@Graph using POSIX threads and OpenMP directives. For instance, in a global transitivity routine, by which the network's average clustering coefficient is obtained, the value *result* is a scalar average value, so that parallelism is straightforward and safe multithreading is achieved by applying *reduction* binary operators over its parallelizable loop. Although this may be approachable in similar cases, unfortunately in most routines *result*'s value does not depend linearly on the iteration variable but in some arbitrary way (depending on the algorithm). This is added to the synchronization overhead, which could be imposed in algorithms where dependent iterative operations are found. This requires which need careful consideration to prevent conflicts commonly caused by the concurrent access to shared memory spaces.

5.3.2 Performance: POP Model Time Series

To illustrate the performance of Par@Graph, we present an analysis (Ihshaish et al., 2015) using data from a high-resolution ocean model (the Parallel Ocean Program, POP), developed at Los Alamos National Laboratory (Dukowicz and

Table 5.2 *Different threshold values τ used in the reconstruction of Pearson correlation networks from the 0.4° and 0.1° POP data sets and corresponding number of network vertices and edges.*

Network	POP	τ	*Vertices*	*Edges*
1	0.4°	0.7	3.0×10^5	3.2×10^8
2	0.4°	0.6	3.0×10^5	1.5×10^9
3	0.4°	0.5	3.0×10^5	2.7×10^9
4	0.1°	0.4	4.7×10^6	1.4×10^{12}

Smith, 1994). This model has a horizontal grid of 3600 × 2400 and hence a nominal horizontal resolution of 0.1°. Control simulation data (Weijer et al., 2012) were used, where POP is forced with a repeated annual cycle from the (normal-year) Coordinated Ocean Reference Experiment (CORE; see www.clivar.org/clivar-panels/omdp/core-2) forcing dataset (Large and Yeager, 2004), with the six-hourly forcing averaged to monthly. Correlation networks were built from one year (year 136 of the control simulation) of the global daily sea surface height (SSH) data. The seasonal cycle was removed by subtracting for each day of the year its five-day running mean averaged over years 131–141.

Two data sets have been used for network reconstruction, one with the actual 0.1° horizontal resolution of the model, resulting in about 4.7×10^6 ocean grid points, and an interpolated one with a lower 0.4° horizontal resolution resulting in about 3×10^5 grid points. The results were computed on a bullx supercomputer (see https://surfsara.nl/systems/cartesius) composed of multiple "fat" computing nodes of four-socket bullx R428 E3 each, having eight-core 2.7 GHz Intel Xeon E5-4650 (Sandy Bridge) CPUs, with a shared Intel smart cache of 20 MB at each socket, resulting in SMP nodes of 32 cores which share 256 GB of memory. The interconnection between those "fat" nodes is built on InfiniBand technology providing 56 Gbits/s of internode bandwidth. The same technology is used to connect the nodes to a Lustre parallel file system of 48 OSTs (object storage targets) each with multiple disks.

To test the performance of Par@aGraph for reconstruction, three weighted correlation networks from the 0.4° POP grid were used. Different edge densities (see Table 5.2) were obtained as a result of applying different threshold values τ for the link definition. The parallel speed-up of Par@Graph and the corresponding computational time are plotted in Fig. 5.7. The execution time falls weakly superlinearly with the number of processors up to 100. Moreover, the performance

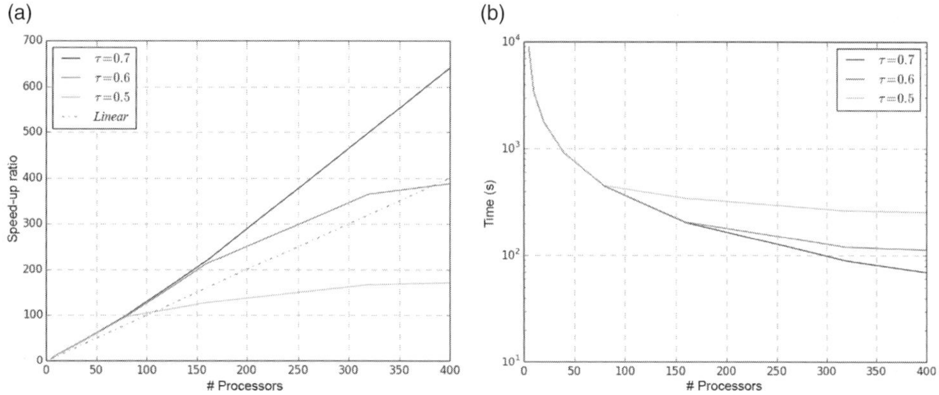

Figure 5.7 (a) Speed-up ratio (the ratio between the time to run the task in a single processor over the time using multiple processors) for the parallel construction of SSH climate networks from the POP model data having 0.4° spatial resolution. The shown speed-up also includes the parallel reading and reordering of the input time series. (b) The corresponding execution times (in seconds) over different sizes of computing processors starting from five processors upwards. Source: Ihshaish et al. (2015).

becomes strongly superlinear for $\tau > 0.5$ as the number of processors increases. This superlinearity is due to a reduction in cache misses at each processor's cache (note that the 20 MB cache is shared among the eight cores) as fewer time series are needed to fit in those shared caches when more cores are implied. In a further analysis it was also observed (Ihshaish et al., 2015) that the reordering of input time series did improve the performance of Par@Graph mainly when the number of processors was less than 100.

Results of the performance tests to compute six standard network properties were shown by Ihshaish et al. (2015). Although there are some differences in the performance gain in each of the algorithms, a general improvement is achieved by the fine-grained parallel implementation over sequential `igraph` algorithms. Similar performance results were obtained for tests using the much larger correlation networks from the 0.1° POP grid, resulting in networks of 4.7×10^6 nodes and edges ranging from 1.5×10^{10} to 1.4×10^{12} for thresholds from 0.8 to 0.4, respectively (although computations of the betweenness centrality and clustering coefficient algorithms for the network of 1.4×10^{12} have not yet been performed).

One of the important questions in physical oceanography deals with the coherence of the global ocean circulation. In low-resolution (non-eddying) ocean models, the flows appear quite coherent with near-steady currents filling the ocean basins. However, as soon as eddies are represented when the spatial resolution is smaller than the internal Rossby radius of deformation, a fast decorrelation is seen in the

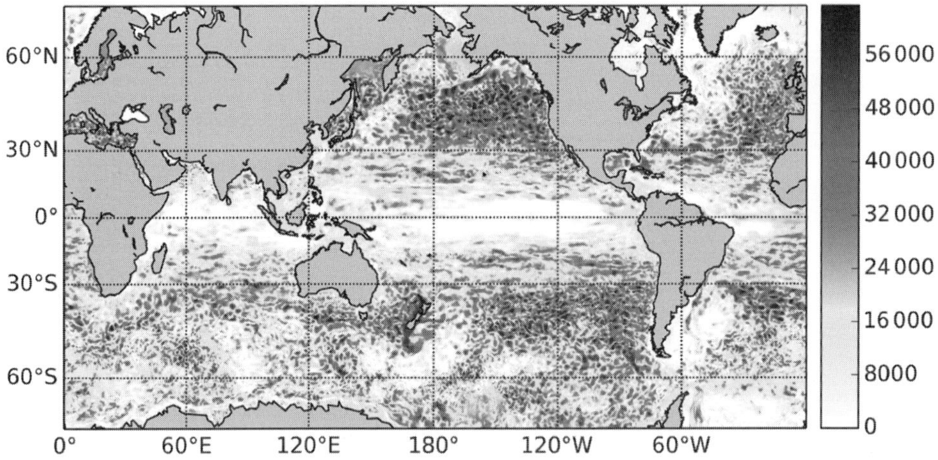

(a)

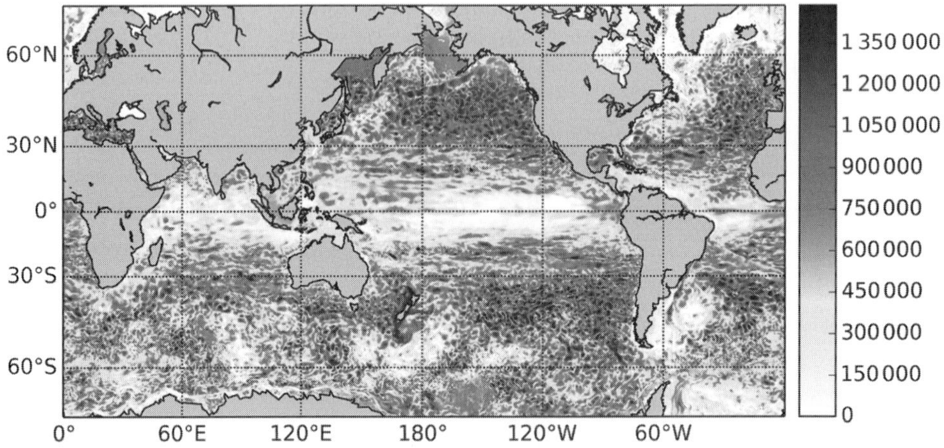

(b)

Figure 5.8 (a) Degree for the SSH POP data interpolated on the 0.4° grid and a threshold of $\tau = 0.5$. (b) Degree field for the 0.1° grid and a threshold $\tau = 0.4$; here, the reconstructed network has 4.7×10^6 nodes and 1.4×10^{12} edges. Source: Ihshaish et al. (2015). (A black and white version of this figure appears in some formats. For the color version, please refer to the plate section.)

flow field. The issue of coherence has, for example, been tackled by looking at the eigenvalues of the transfer matrix (Dellnitz et al., 2009; Froyland et al., 2014) but also complex networks are very suited to address this question (Tantet and Dijkstra, 2014). Preliminary results (Ihshaish et al. 2015) on the degree distribution for the SSH derived networks of the 0.4° and 0.1° POP simulation are shown in

Figs. 5.8(a)–(b). In both cases, a weighted, undirected network was constructed by using the Pearson correlation with zero lag and threshold values $\tau = 0.5$ and $\tau = 0.4$, respectively.

In summary, in this chapter we have provided the background of two software packages, `pyunicorn` and Par@Graph, suitable for climate network analysis. Data for the two benchmark problems, from the Zebiak–Cane model and from the POP model, were used to illustrate the performance and capabilities of both methods. The availability of Par@Graph will allow one to solve a new set of questions in climate research, one of which, the coherence of the ocean circulation at different scales, was briefly discussed here. Apart from higher-resolution data sets of one observable, it will also be possible to deal with data sets of several variables and to more efficiently reconstruct and analyze large (and time-dependent) networks (Berezin et al., 2012). However, apart from climate research, both pyunicorn and Par@Graph will also be very useful for all fields of science where very large-scale networks are relevant.

6

Applications to Atmospheric Variability

In this chapter we apply the network methodology to advance the understanding of atmospheric teleconnection processes at different timescales. We present results obtained by using the methods for constructing climate networks that were summarized in Chapter 4: functional networks built from cross-correlations (zero-lag and lagged), mutual information, event synchronization, and flow networks. A variety of atmospheric phenomena are discussed here, starting from the characterization of connectivity associated with ENSO (Section 6.1) and phenomena with longer timescales (Section 6.2), the identification of the component of the atmospheric variability that is forced by the ocean (Section 6.3), followed by the identification of Rossby waves as the main carriers of teleconnections worldwide (Section 6.4), the description of transport pathways during a blocking event (Section 6.5), and concluding with the spatiotemporal organization of rain and moisture transport in the Indian and South American monsoon systems (Sections 6.6 and 6.7, respectively).

6.1 Network Analysis of ENSO Phases

As discussed in Section 2.4, El Niño–Southern Oscillation (ENSO) has a worldwide impact on the atmospheric circulation. Arizmendi and Barreiro (2017) performed a systematic study to address how the global atmospheric connectivity, in the different seasons of the year, is modulated by ENSO. The study focused on understanding atmospheric connectivity and, in particular, the non-zonal response of the Southern Hemisphere (SH) extratropics to tropical heating anomalies. They analyzed the eddy geopotential height (monthly mean geopotential height at 200 hPa from the NCEP/NCAR CDAS1 reanalysis; see Section 3.1), which was calculated by removing the zonal mean of each latitudinal band for every month. Anomalies were calculated using monthly mean data for each season of the year.

Climate networks were built based on the Pearson correlation (see Section 4.2), and two nodes were linked if their correlation value was above a significance threshold: as in Arizmendi et al. (2014), the 99% quantile value of the distribution of correlation values calculated with surrogate time series (random shuffle) was used.

The area-weighted connectivity (AWC; see (4.8)) was calculated for seasons June–July–August (JJA), September–October–November (SON), December–January–February (DJF), and March–April–May (MAM). In each case, the maps of AWC were differentiated considering only El Niño (EN) years, La Niña (LN), years and Neutral (N) years.[1]

ENSO years (either El Niño or La Niña) were considered from June of year 0 to May of year +1, since normally ENSO starts developing in the boreal summer, peaks in December–January, and decays in boreal fall. The period analyzed (1949–2014) contains 21 EN years and 22 years for both LN and N. Thus, the length of the time series to construct the AWC for each season is 63 months for EN and 66 for LN and N years.

Figures 6.1 and 6.2 show the AWC maps for the different seasons and ENSO phases. In the SH winter, JJA, which usually marks the initial stage of development of El Niño and La Niña, the AWC shows largest values in the tropical band, with maxima over the western Pacific and Maritime continent, as well as in the eastern Pacific (Fig. 6.1). Moreover, connectivity is largest in the SH compared to the NH, as expected, given that during winter the subtropical jet moves equatorward in the Pacific sector, strengthening the Rossby wave sources. In EN years the tropical areas of largest connectivity are similar to those of N years, but values are larger over the western Pacific. In the SH extratropics, the connectivity is largest during N years. During LN years the AWC is significantly larger in the central Pacific. In the SH extratropics the centers of maximum connectivity during LN are in approximate quadrature to those during EN and N years. The plot of the difference AWC (EN–LN, last rows) stresses these differences and also shows the larger connectivity in the tropical Atlantic and Indian Oceans during LN events.

In SON, the AWC patterns are very different for EN, LN, and N years, (Fig. 6.1). During N years the AWC is largest in the tropical band, but shows no well-defined structure. During EN years the AWC is largest in the western tropical Pacific and Maritime continent extending toward the Indian ocean. Moreover, it is possible to detect a clear wave train that emanates from the southeastern Indian ocean and arches toward the south Pacific and then northeastward into the south Atlantic. During LN the maximum AWC is in the central equatorial Pacific and the map

[1] ENSO years were classified according to the Oceanic Niño Index (ONI) of the Climate Prediction Center of NOAA:www.cpc.ncep.noaa.gov/products/analysis_monitoring/ensostuff/ensoyears.shtml.

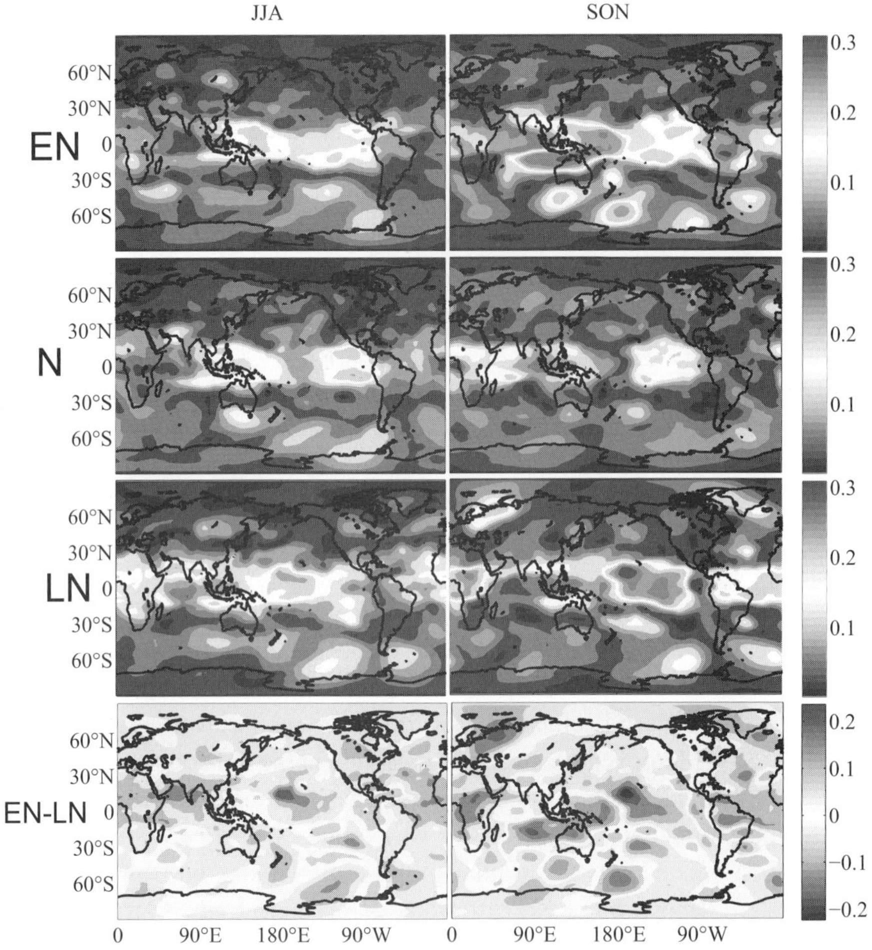

Figure 6.1 AWC during El Niño (first row), neutral (second row), and La Niña (third row) years. The last row shows the difference between El Niño and La Niña years. Left column: June–July–August (JJA) season. Right column: September–October–November (SON) season. Source: Arizmendi and Barreiro (2017). (A black and white version of this figure appears in some formats. For the color version, please refer to the plate section.)

suggests a wave train emanating from there and propagating into the south Pacific and then into the southwestern Atlantic. The difference map EN–LN clearly shows the shift in the maximum of tropical AWC, as well as the fact that connectivity is largest in the tropical Atlantic during LN. These two characteristics are also seen, but to a lesser extent, in JJA. Moreover, the difference also shows the existence of maximum AWC in the shape of a wave train in the extratropical SH during EN years. This wave train of connectivity seen during EN has a pattern similar to

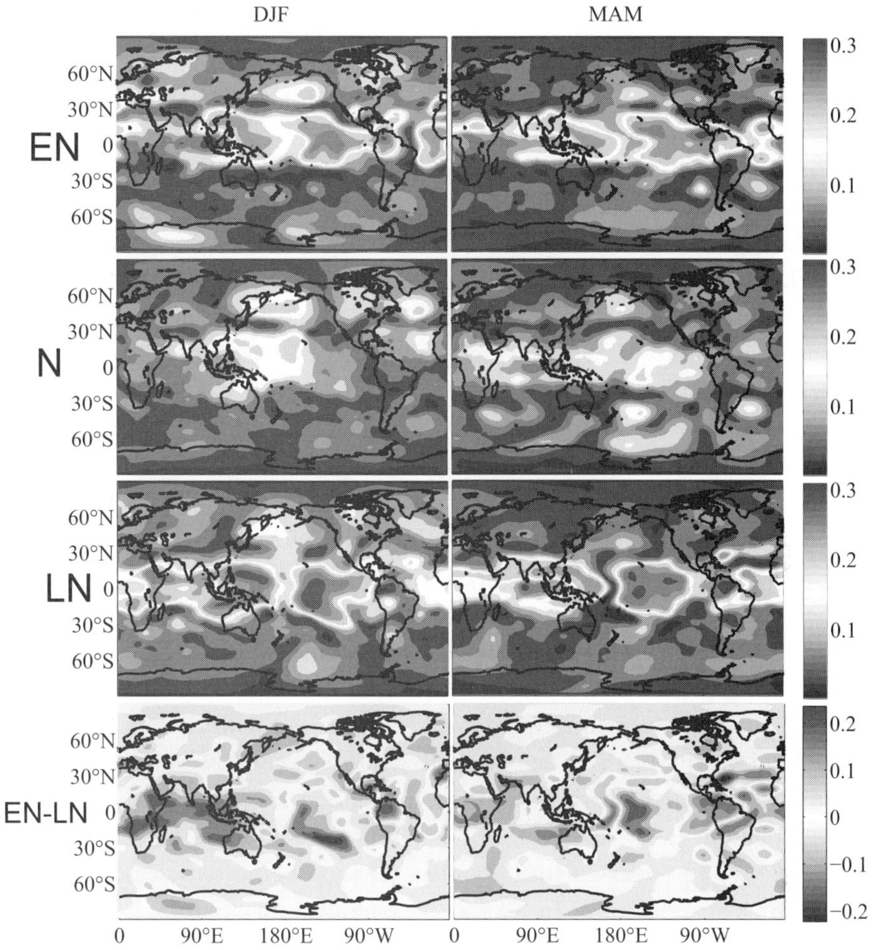

Figure 6.2 Same as Fig. 6.1 but for the December–January–February (DJF) season (left column) and the March–April–May (MAM) season (right column). Source: Arizmendi and Barreiro (2017). (A black and white version of this figure appears in some formats. For the color version, please refer to the plate section.)

the *Pacific South American pattern 2* (PSA 2), while the shape of the connectivity seen during LN in the SH reminds us of the *Pacific South American pattern 1* (PSA 1). The PSA patterns are common modes of atmospheric variability in the SH on several timescales. The PSA 1 mode has been associated with enhanced convection in the western Pacific and suppressed convection over the Indian Ocean, while the PSA 2 mode is linked to tropical heating anomalies in the central Pacific and suppressed convection in the western Pacific. Thus, different from JJA, during SON the SH extratropics is most connected during EN, and both phases of ENSO are more connected than N years.

Regarding the AWC map for DJF, EN, and LN have similar patterns of connectivity in the tropical Pacific, with two areas of large values to the east and west of the dateline, with LN ones being somewhat stronger (Fig. 6.2). LN also shows larger values of AWC in the Indian sector. Connections in the NH extratropics become larger than in the SH, particularly over the north Pacific and western Atlantic oceans, suggesting the presence of the *Pacific North American (PNA) pattern*. Some continental areas become also much more connected during EN years, including the Mediterranean region and northern South America. As a result, tropical South America shows a clear dipole of connectivity in the EN–LN map, such that during EN northern (central) South America is more (less) connected than during LN years. In the extratropical SH, connectivity is low and there is no clear presence of wave trains as during SON, probably because of the climatological southern shift of the subtropical jet stream during summertime. It is worth noting that during N years the tropical connectivity is much lower than during EN and LN years, but is of similar magnitude in the NH extratropics.

During MAM, the tropical connectivity is similar to that seen during DJF, although with smaller values, probably because the ENSO events are in the decaying phase (Fig. 6.2). In particular, the western Pacific becomes much more disconnected, while the central-east Pacific tends to maintain its connectivity. The tropical Atlantic is well connected during EN and LN, with centers of maximum values shifted longitudinally. The main difference between LN and EN is the strong connectivity seen in northern South America during EN, which is absent during LN, as is found in DJF. No significant connectivity is seen in the SH extratropics for either EN or LN years, with the only exception being two centers straddling South America at about 35°S. In N years the overall connectivity is small and there are no clear spatial structures in the AWC.

By performing a composite analysis, the connectivity wave train seen in the AWC for SON was found to be physically consistent with the anomalies present during EN events.

Summarizing, the results of Arizmendi and Barreiro (2017) indicate that during neutral years the tropical band is the most connected region of the world, and the Pacific Ocean is the main hub. Neutral years show a large seasonal dependence: SON is the season with overall least connectivity and has the maximum in the central-east Pacific; DJF (JJA) is the season of maximum connectivity in the NH (SH) probably because the strong winter subtropical jet stream acts as a source of tropically induced perturbations as well as due to the presence of well-known intrinsic modes of variability such as the NAO and Pacific North American pattern in the NH that are more active during wintertime.

Except in the JJA season, which is typically the beginning of the ENSO phenomenon, during the rest of the year the atmospheric connectivity is much larger during El Niño and La Niña years than during neutral years, in particular in the tropical region.

6.2 Evolution of Atmospheric Connectivity in the Twentieth Century

In the previous section we have considered atmospheric connectivity at the ENSO interannual timescale. Arizmendi et al. (2014) have also analyzed the evolution of the connectivity during the twentieth century at different time scales.

The data analyzed were anomalies in the 200 hPa eddy geopotential height of NCEP/NCAR CDAS1 reanalysis during the 1949–2011 period, and also from NOAA twentieth-century reanalysis spanning the period 1901–2009. Climate networks were constructed by using Pearson correlation (PCorr, as in the previous section) and also by using mutual information computed from histograms of anomaly values (MIH) and from ordinal patterns. As discussed in Section 3.3.5 (see Fig. 3.8), ordinal patterns allow tuning to particular timescales, and intraseasonal ($\Delta t = 1$ month), intraannual ($\Delta t = 4$ months), and interannual $\Delta t = 12$ months) values of Δt (the interval at which the time series is sampled to construct the ordinal patterns) were considered.

Figure 6.3 displays the AWC computed from the NCEP/NCAR CDAS1 reanalysis using PCorr and the ordinal patterns M_I at several temporal scales. In Fig. 6.3a we see that the most connected areas are located in the tropics, particularly in the central Pacific and to a lesser extent over the Atlantic Ocean, South Asia, and over Indonesia between the Indian Ocean and western Pacific. This larger connectivity in the tropics is fully consistent with the results presented in the previous section. Some extratropical regions are also relevant for their number of connections, such as the south and north Pacific, which reflect the atmospheric teleconnection patterns in both hemispheres. Also, comparing the AWC in Figs. 6.3b–d it is clear that outside the tropical region there is relatively larger connectivity in the northern Pacific on intraannual timescales, a weakly connected central Pacific on intraseasonal timesscales, and a strongly connected south Pacific on interannual timescales. Moreover, there is a weaker indication of a wave train from the Indian Ocean on all timescales.

Figure 6.4 displays how the connectivity in the western Pacific and central Pacific regions (see the precise definitions in Arizmendi et al. (2014)) has evolved in time. Here, the AWC was calculated from NOAA twentieth-century reanalysis. The temporal evolution of the AWC for both regions depends on the methodology used to construct the network. The overall evolution of connectivity of PCorr, MIH, and interannual networks is very similar, but intraannual and intraseasonal networks behave differently. In the western Pacific (Fig. 6.4a) the mean connectivity presents

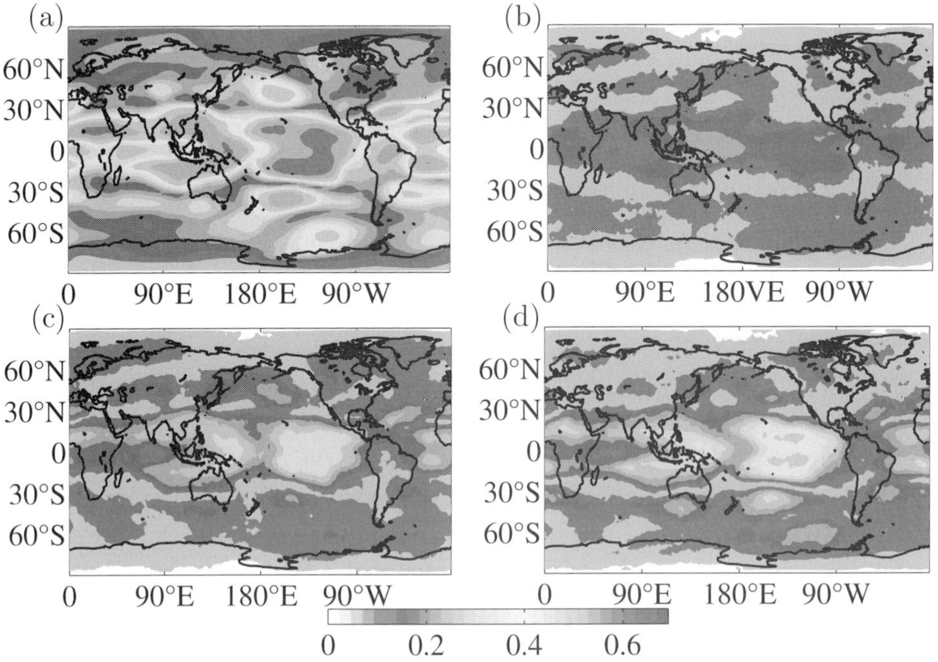

Figure 6.3 Area-weighted connectivity maps obtained from the 200 hPa eddy geopotential height NCEP/NCAR reanalysis data (1949–2011). The statistical similarity measures used are: Pearson correlation (a), mutual information with the ordinal patterns methodology: intraseasonal $\Delta t = 1$ month (b), intraannual $\Delta t = 4$ months, (c) and interannual $\Delta t = 12$ months (d). Source: Arizmendi et al. (2014). (A black and white version of this figure appears in some formats. For the color version, please refer to the plate section.)

a general decrease for the PCorr, MIH, and interannual networks from the beginning of the century up to approximately 1940, when the connectivity starts an increasing trend until the end of the century. The connectivity of the intraannual network is quite constant during the first decades, and from 1956 onward it starts increasing up to the end of the record. On the other hand, the AWC of the intraseasonal networks has a small but positive trend during the entire period considered.

In the central Pacific, PCorr, MIH, and interannual networks show similar temporal variation in connectivity (Fig. 6.4b), starting with a negative trend until 1940 and then increasing the connectivity toward the end of the period. On the other hand, the intraannual and intraseasonal networks maintain the negative trend for a few more years, starting the increasing period in 1956, as in the western Pacific box for the intraannual case. Except in the intraseasonal case, the connectivity values of the central Pacific are larger than those in the western Pacific.

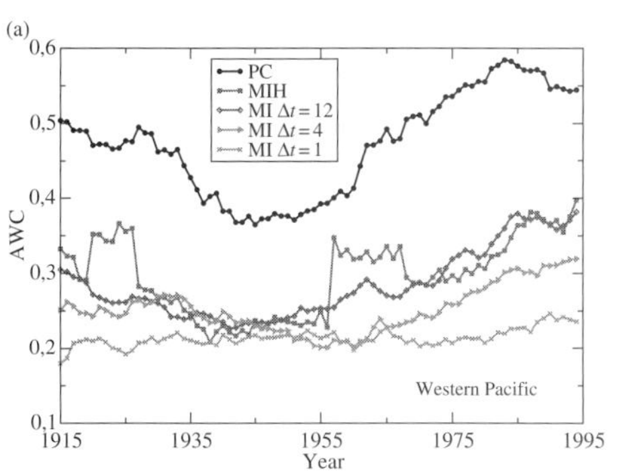

Figure 6.4 Temporal evolution of the AWC in the western Pacific (a) and in the central Pacific (b). The networks are built by evaluating 30-year windows of the 200 hPa eddy geopotential height field (NOAA reanalysis) with the similarity measures indicated in the legend. For each mean AWC value the time period corresponding to 15 years before and 15 years after were considered. Source: Arizmendi et al. (2014).

Figure 6.5 shows the statistically significant Pearson correlation values of the eddy geopotential at each location, with the series averaged inside the central Pacific and the western Pacific boxes, for 1915, 1940, and 1995 (NOAA reanalysis). In agreement with the time evolution shown in Fig. 6.4, it is clearly seen that in the 30 years centered in 1940 the connectivity of these regions is smaller. For the western Pacific it is evident that the increase in global connectivity seen in the last decades of the century is particularly due to stronger connections with the tropical Atlantic. These maps also show that the central Pacific and western Pacific are not independent of each other, which is expected because phenomena like El Niño connect both regions through changes in the Walker circulation.

6.3 Forced and Internal Atmospheric Variability

In order to advance the understanding of the physical mechanisms underlying the links uncovered by the network analysis, it is important to distinguish between links that are mainly due to internal atmospheric variability, and those that uncover the influence of the oceans.

In this section we focus on climate networks that represent internal atmospheric variability and those that represent variability forced by oceanic conditions. The results presented here summarize the work in Arizmendi et al. (2014) and Deza et al. (2014).

Atmospheric variability can be considered, to first order, as a superposition of an internal part due to intrinsic dynamics, and an external part due to the variations

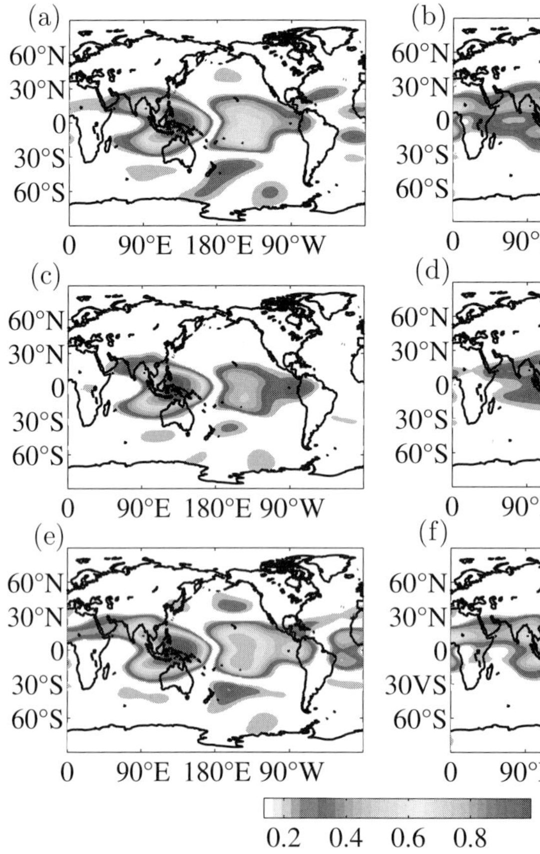

Figure 6.5 Maps of the cross-correlation between the eddy geopotential time series at each location and the time series averaged inside the western Pacific (left column) and in the central Pacific (right column), using the NOAA reanalysis data. Correlations are computed by taking 30-year windows centered at 1915 ((a) and (b)), 1940 ((c) and (d)), and 1995 ((e) and (f)). Only statistically significant values are shown in color. Source: Arizmendi et al. (2014). (A black and white version of this figure appears in some formats. For the color version, please refer to the plate section.)

of the boundary conditions, primarily driven by the sea surface temperature (SST) forcing (James, 1994; Trenberth et al., 1998). These two components can be separated by using the output of an ensemble of runs from an atmospheric general circulation model (AGCM, see Section 1.3) forced with prescribed historical SSTs (Barreiro et al., 2002; Bracco et al., 2004; Molteni, 2003; Straus and Shukla, 2010).

Separating forced from internal atmospheric variability might allow for improvements in seasonal climate prediction. In many geographical regions, the atmosphere is strongly influenced by local or remote SST variations that force persistent regional anomalies (Shukla, 1998). Because the evolution of the tropical oceans and soil moisture presents some predictability at timescales longer than the atmosphere, prediction of atmospheric variables beyond the synaptic timescale of 7–10 days can be possible, provided that the atmospheric dynamics are forced by the boundary conditions (Shukla, 1998).

The usual modeling strategy to separate atmospheric components of variability consists of forcing AGCMs with idealized or observed SST. The experiment involves the generation of an ensemble of runs initialized differently but forced with the same SST as boundary conditions. Then, the simulated time series of anomalies of a climatic field (e.g., SAT anomalies [SATA]) is considered as a linear combination of internal and forced variability, e.g., $x = x_{for} + x_{int}$. Thus, for each run i it results in

$$x^i = x^i_{for} + x^i_{int} = x_{for} + x^i_{int}, \tag{6.1}$$

as x_{for} is considered to be independent on the initial conditions.

Averaging over N runs this yields

$$\bar{x} = x_{for} + \frac{1}{N}\sum_i x^i_{int}. \tag{6.2}$$

Thus, the ensemble mean is a biased estimate of the forced component. If N is large enough, the second term will be small as each model run will have different and not very correlated values. Thus, to first order $\bar{x} \approx x_{for}$.

It is then possible to identify the internal variability as

$$x^i_{int} \approx x^i - \bar{x}, \tag{6.3}$$

yielding N different internal variability realizations, which are processed separately, producing N different climate networks. Thus, this method allows us to construct two types of networks, one in which the links represent similarities in forced atmospheric variability (the *forced variability* network), and the other a set of networks displaying similarities in internal atmospheric variability (referred as *internal variability* networks), which are calculated for each set of initial conditions, and then averaged over the ensemble of N runs.

As in the previous section, Arizmendi et al. (2014) constructed the climate networks using the mutual information computed with ordinal patterns, which allowed focusing the analysis at specific timescales (intraseasonal, intraannual, or interannual). To separate internal and forced variability they used the AGCM from the International Centre for Theoretical Physics (ICTP-AGCM). This consists of a full atmospheric model with simplified physics and a horizontal resolution of T30 ($3.75^\circ \times 3.75^\circ$, which results in $N = 608$ grid points or network nodes) with eight vertical levels (Molteni, 2003). The model is forced with historical global

SST (ERSSTv.2) (Smith and Reynolds, 2004). To separate forced and internal atmospheric variability, an ensemble of nine runs forced with the same boundary (SST) conditions but initialized with slightly different atmospheric conditions was constructed.

In the ensemble of nine runs, SST is taken as a boundary condition and it is not changed by the atmospheric flow. In the climate system, however, there is a two-way interaction between the ocean and the atmosphere. This limitation is especially important in the extratropics, where the SST evolution strongly depends on the atmospheric forcing (Barsugli and Battisti, 1998; Frankignoul and Hasselmann, 1977).

The forced-component AWC maps, obtained with the mutual information methodology based on ordinal patterns, are shown in Figs. 6.6a, c, e. All timescales considered present similar spatial structures with maxima in the tropics and wavy patterns in the extratropics. The main difference is the intensity of the connectivity. On intraannual timescales Northern Hemisphere regions are more connected than on intraseasonal timescales and are larger than in the Southern Hemisphere. On interannual timescales the connectivity is similar in the northern and southern extratropics, and the model recovers the highly connected region in the south Pacific located at about 35°S, 150°W.

Comparison with the AWC maps presented in Section 6.2 reveals very similar structures, suggesting that most of the connectivity seen in Fig. 6.3 is due to oceanically forced variability. The largest differences occur on intraseasonal timescales in the Indo-Pacific region, where the model shows weaker connectivity.

The AWC maps of internal atmospheric variability are shown in of Figs. 6.6b, d, f. Networks constructed with Pearson correlation and ordinal patterns-based M_I with $\Delta t = 1$ and 4 months (intraseasonal and intraannual scales, respectively), have mostly the same structure but differ on where global maxima are located. On intraseasonal timescales the regions of maximum connectivity are in the western Pacific and in the southern extratropics. There are clear connections in the Southern Hemisphere between the south Pacific, south America, and the south Atlantic. This result, together with the AWC maps of the forced atmospheric variability, suggest that the changes in connectivity observed in the reanalyses during the twentieth century (Section 6.2) are due to changes in both forced and intrinsic components. On intraannual timescales, on the other hand, the AWC maps present the same well-connected spots as the intraseasonal case, but also additional regions, mainly in the north Pacific and north Atlantic. The wavy patterns suggests the existence of Rossby wave propagation all over the globe. The Pearson correlation network has different link density, but the structure is quite similar to the intraannual case.

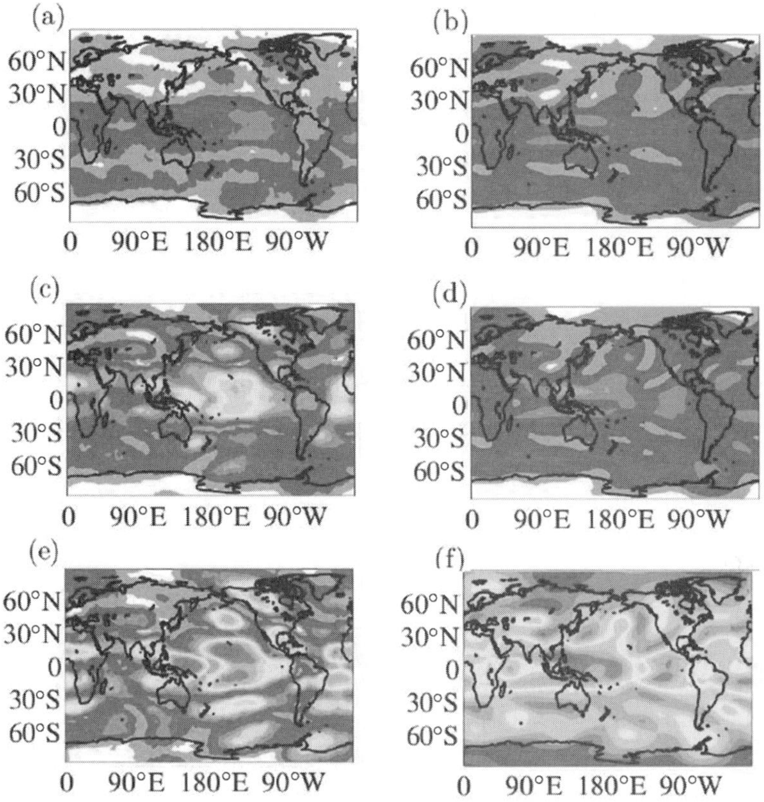

Figure 6.6 Area-weighted connectivity maps computed from 200 hPa eddy geopotential height in the period 1954–2006. In the left column the maps of the ICTP-AGCM oceanically forced component are computed using the mutual information from ordinal patterns for intraseasonal ((a), $\Delta t = 1$ month), intraannual ((c), $\Delta t = 4$ months), and interannual ((e), $\Delta t = 12$ months) timescales. In the right column the AWC maps of internal variability (averaged over the nine runs) are computed using the mutual information with intraseasonal ((b), $\Delta t = 1$ month) and intraannual ((d), $\Delta t = 4$ months) ordinal patterns. Panel (f) presents the AWC map obtained with Pearson correlation. Source: Arizmendi et al. (2014). (A black and white version of this figure appears in some formats. For the color version, please refer to the plate section.)

A similar analysis to separate forced from internal variability was performed by Deza et al. (2014) using monthly averaged SAT. Strong connectivity in the tropics and the Pacific, Indian, and Atlantic basins was observed. It is worth noting that while tropical connectivity is relatively symmetrical about the equator for the Pacific and Indian Oceans, the north Atlantic is significantly more connected than the ocean south of the equator. The presence of highly connected spots was observed in the extratropics, especially in the Pacific basin but also in the Indian and Atlantic Oceans. Deza et al. (2014) found that most of those links were strongly correlated to ENSO.

6.4 Atmospheric Rossby Waves

As we have seen, the climate network approach allows us to uncover physical mechanisms that provide connections between atmospheric locations. In the previous sections we have seen several times the importance of Rossby waves (see Section 2.2) in setting such connections. By distinguishing positive and negative correlations in a ground-temperature lagged-correlations climate network, Wang et al. (2013) found that the links associated with atmospheric Rossby waves are dominant in the Southern Hemisphere.

They used the anomalies from the daily near-surface (1000 hPa) atmospheric temperature from the NCEP/NCAR reanalysis spanning the years 1948–2010. These data are mapped on 726 nodes over the globe, and climate networks are constructed for each year as described in Section 4.2. In this approach one computes, for each pair of nodes, the maximum and the minimum lagged cross-correlations and, from them, one assigns positive or negative seasonal weighted links to the pair, which are normalized measures of the maximum and minimum values of the lagged correlations found when varying the lag time (see Section 4.2). Then, Wang et al. (2013) analyzed the negative links by focusing on two properties: the negative link weight W_{ij} and the time delay τ_{ij}^* at which the most negative correlation is attained.

Figure 6.7a depicts the negative link-weight statistics for the real and for shuffled data during boreal winter (November to February). Large negative mean-link-weight values (mean is over years) exist in the probability density function (PDF) of the real data but not in the shuffled data, and therefore they are not likely to occur by chance. Moreover, low dispersion (std) of the time delays associated to the most negative links (see Fig. 6.7b) indicates that these links have well-defined properties and have statistical and physical significance.

Next, the world is divided into three geographical zones, the Southern Hemisphere (SH, from 22.5°S to the South Pole), the Northern Hemisphere (NH, from 22.5°N to the North Pole), and the equator (between 22.5°S and 22.5°N). The (geographical and temporal) mean link weight $\langle W_{s_1,s_2} \rangle_d$ between sites at a fixed geographical distance d is shown in Fig. 6.8a as a function of d. It is clear that in the SH there is a special distance of ~3500 km and another one of ~10 000 km. In the NH region a similar but weaker dependence is observed, while in the equatorial region no special distances are appreciated. Between the negative peaks there is a maximum at 7000 km. These distances may be associated with atmospheric

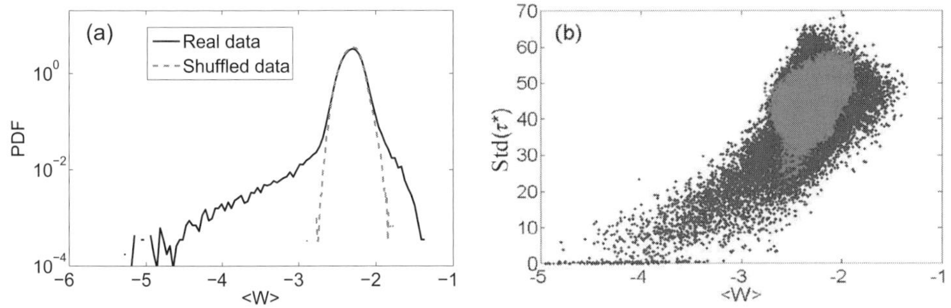

Figure 6.7 (a) The probability density function of the mean weight of negative links in the globe during November–February (the mean is over years), in both real (solid line) and shuffled (dashed line) data. (b) The dependence of the standard deviation of time delay τ^* on the mean weight $\langle W \rangle$ for all possible links, for real (black) and shuffled (grey) data. Note that large negative values of $\langle W \rangle$ have small standard deviation of τ^*, suggesting that they are real climate links, not appearing by chance. Source: Wang et al. (2013).

Rossby waves (see Section 2.2) with a wavelength of ~7000 km, consistent with the expected wavelength range of 5000 km to 8000 km (Chang, 1999). The negative peaks at 3500 km and 10 000 km most probably represent a 1/2 wavelength and a 3/2 wavelength of the Rossby wavelength. For comparison, the mean of the traditional Pearson correlation coefficient, at the same lag time determined for W, as a function of d is also shown in Fig. 6.8b. It shows a much weaker peak at ~3500 km than the link weight W, and it does not show any preferred peak at around ~ 10 000 km. This highlights the power of the methodology based on the separation of positive and negative links.

To further consolidate the association of the observed pattern in the climate network with Rossby waves, Wang et al. (2013) also compared the seasonality of this pattern with the known seasonal characteristics of Rossby waves. The negative and positive weights of all SH links, for the winter and summer months separately, are shown in Figs. 6.9a–d. Each point represents an average (over years) link weight $\langle W \rangle$ and it is plotted as a function of the distance d between the nodes it connects. The negative weights have a pronounced enhanced distribution of large weights for d ~3500 km during both summer and winter months, while for the SH summer months (November to February) there is an additional preferred distance of ~10 000 km (Fig. 6.9b), both in accordance with the 1/2 and 3/2 wavelengths of atmospheric Rossby waves. Around the full wavelength distance (i.e., $d \approx 7000$ km), an enhanced distribution of large positive weights is observed, as expected (Figs. 6.9c, d). However, a larger abundance of strong waves (represented by links at 1/2, 1, and 3/2 wavelengths) during summer, in comparison to winter

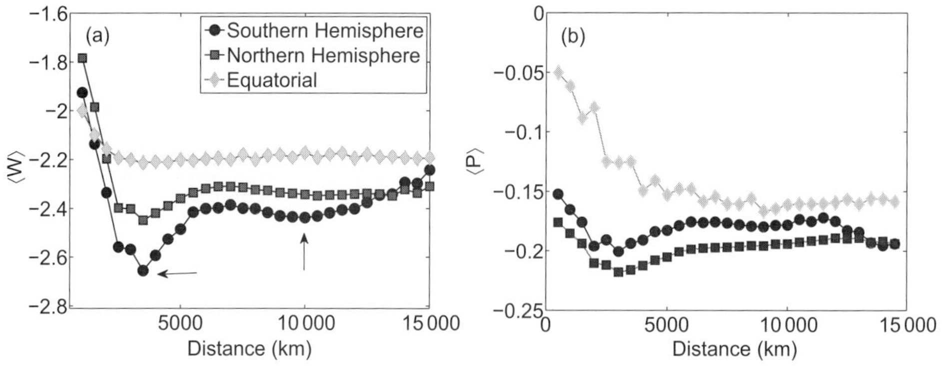

Figure 6.8 (a) The mean (over years, and over nodes at distance d) negative link weight $\langle W \rangle$ and (b) the mean Pearson correlation coefficient $\langle P \rangle$ (at the same lag time identified for the negative link weights) as a function of distance d, for the SH (circles), NH (squares), and the equator (diamonds) regions. The figure is computed from the globe surface (1000 hPa) temperatures during the November to February months. Source: Wang et al. (2013).

months, is clearly seen in Figs. 6.9a–d, in agreement with the clearer pattern of Rossby waves found during the SH summer (Chang, 1993, 1999; Chang and Yu, 1999). In the distribution of positive weights (Figs. 6.9c, d), one also sees the links that emerge around $d < 2000$ km due to the proximity effect.

Atmospheric Rossby waves have a characteristic group velocity that can be approximately estimated based on the climate network results shown above. It can be calculated by dividing d by τ^* for each link, under the assumption that τ^* is a good estimate for the underlying dynamical delay between the two sites (nodes). Since τ^* is only meaningful for links with weights above the background noise level (see Fig. 6.7b), the analysis only considers real links with weights $| < W > | > 2.8$. Furthermore, to avoid links that are prone to the proximity effect (as such links cannot easily fit in the current Rossby wave interpretation), the positive links with $d < 5000$ km are neglected when considering positively correlated links. Also, a distinction is made between links with $d \in [2000, 5000]$ km which are called "near negative links," and "far negative links" with $d > 8000$ km. The PDFs of time delay τ^* of the near negative links, the far positive links, and the far negative links are shown in Figs. 6.9e, f. By convention, positive τ^* means an eastward energy flow, which is the typical case for most observed links. The results point to a time lag of $\tau \approx 1$ day for the near negative links (i.e., links in the first peak in Fig. 6.9a), $\tau \approx 2$–3 days for the far (beyond the proximity effect) positive links, and $\tau \approx 3$–4 days for the far negative links (i.e., belonging to the second peak in the scatter plot of Figs. 6.9a, b). Based on these values, our estimated group velocities are in the

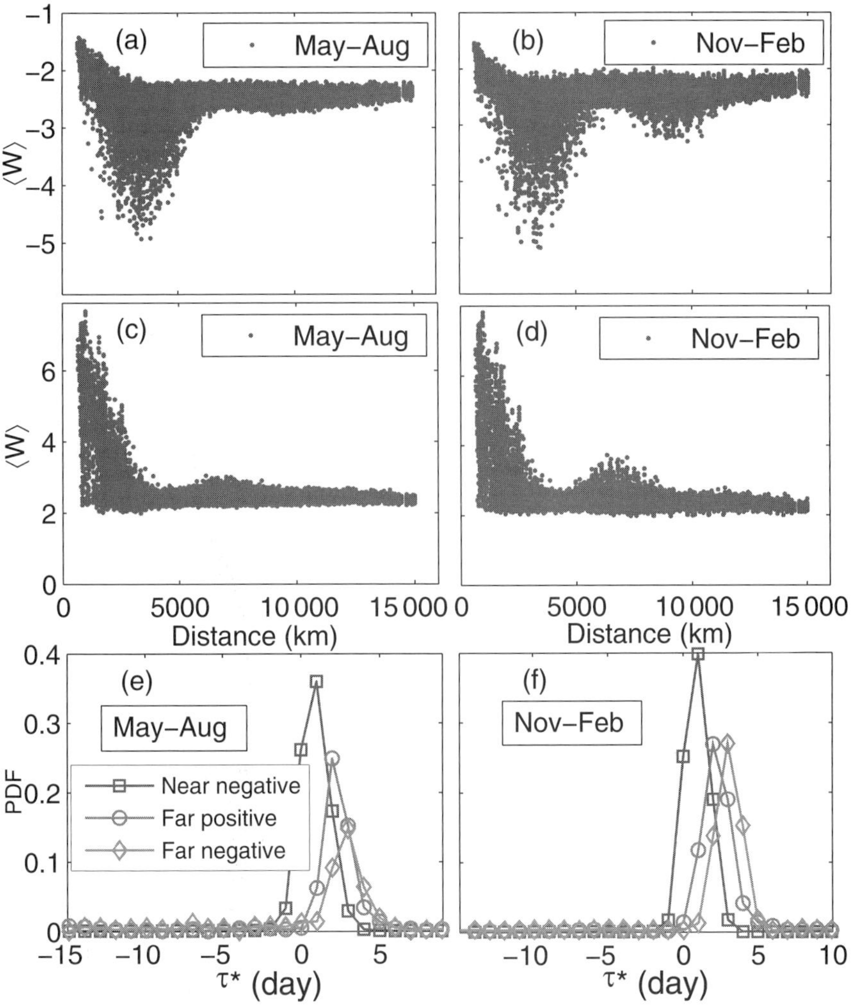

Figure 6.9 The dependence of the mean (over years) weight of negative links on the distance d in the SH during (a) winter and (b) summer months. (c),(d) Same as (a),(b) for positive links. The PDF of time delays τ^* of near negative (squares, $d \approx$ 3500 km), far positive (circles, $d \approx$ 7000 km), and far negative (diamonds, $d \approx$ 10 000 km) weighted links, in the SH during (e) winter and (f) summer. Source: Wang et al. (2013).

range $[20-35]\mathrm{m\,s^{-1}}$, consistent with the range $[23-32]\mathrm{m\,s^{-1}}$ reported in previous studies (Berbery and Vera, 1996; Chang and Yu, 1999).

The climate network has a unique geographical structure that can be compared with the geographical distribution of occurrence of Rossby waves. Since the network is directed – positive τ indicates eastward flow while negative τ

indicates westward flow – the node degree is distinguished between a link that is pointing toward a node (where the number of links pointing to a specific node is referred to here as "in-degree"), or away from the node (referred to here as "out-degree"). Physically, since the weighted degree represents the coherence of the waves, the area with high out- and in- weighted degree could represent places along which wave propagation is seen to be most coherent. Figure 6.10 depicts the mean (over years) in- and out-degrees of each node, excluding the equatorial region (that conforms with a pattern that is not related to the current discussion). The observed structure is consistent with the structure of Rossby wave activity (Chang and Yu, 1999). First, the wave band in the SH from May to August (SH winter, Figs. 6.10c, d) is broader than that of the SH summer (November to February, Figs. 6.10a, b). Second, the atmospheric Rossby wave structure in the NH summer is less pronounced. Third, the wave structure in the SH summer (November to February) lies on a band centered near 50°S. Finally, the transient heat flux (usually used to map storm tracks which are influenced by Rossby waves (Ashkenazy et al., 2008)) and the network pattern of Fig. 6.10 are qualitatively similar. All the above characteristics are consistent with properties of Rossby wave propagation.

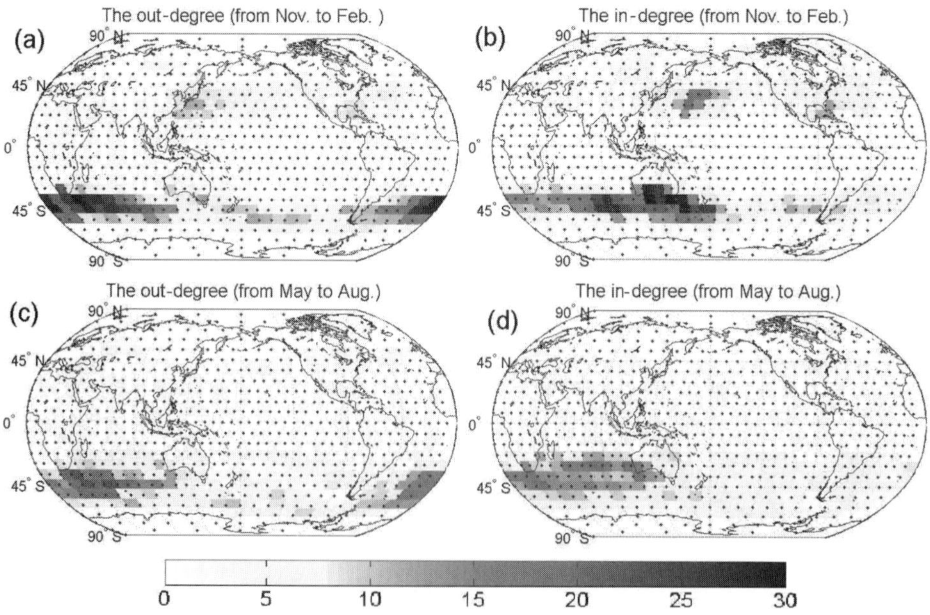

Figure 6.10 The (a) out- and (b) in-degree of the climate network structure from November to February (SH summer). The (c) out- and (d) in-degree of the climate network structure from May to August (SH winter). The specific region with enhanced presence of outgoing and incoming links indicates eastward propagation, similar to Rossby waves traveling on top of a jet. Source: Wang et al. (2013).

6.5 Atmospheric Blocking Events

The low-frequency Rossby waves that characterize the climate networks of the previous sections transport *energy* between remote locations. In this section we consider *mass* transport, and address shorter timescales, of the order of days or weeks.

The methodology developed by Ser-Giacomi et al. (2015a,c) to extract optimal paths of transport between different parts of the globe was described in Section 4.4.4. It was developed in terms of flow networks (Section 4.4), and optimal paths between two sites is the name we give to paths, or sequential visits to network nodes, that fluid is most likely to follow when connecting the two particular nodes. Thus, they are most probable paths (MPPs), although a more intuitive characterization is to say that they are the paths transporting more fluid between two specified locations. Using that methodology we present here optimal transport paths for the atmospheric circulation during a particularly important blocking event occurring in summer 2010 (in particular we focus our study on the period July 20–30) over Eastern Europe and Russia (Ser-Giacomi et al., 2015a). These regions experienced a strong, unpredicted, heatwave during the summer of 2010. Extreme temperatures resulted in over 50 000 deaths and inflicting large economic losses upon Russia (Matsueda, 2011).

Physically, the origins of this heatwave were in an episode of atmospheric blocking that produced anomalously stable anticyclonic conditions, redirecting the trajectories of migrating cyclones. Atmospheric blockings can remain in place for several days (sometimes even weeks) and are of large scale (typically larger than 2000 km). In particular, the Russian blocking of summer 2010 was morphologically of the type known as *Omega block* that consists of a combination of low–high–low pressure fields in the longitudinal direction with geopotential lines resembling the Greek letter Ω (see Fig.6.11a). Omega blockings bring warmer and drier conditions to the areas that they impact, and colder, wetter conditions in the upstream and downstream (Black et al., 2004). Ser-Giacomi et al. (2015a) studied the specific period extended from July 20 to July 30.

Atmospheric data were provided by the National Centers for Environmental Prediction (NCEP) Climate Forecast System Reanalysis (CFSR) through the Global Forecast System (GFS) (Saha et al., 2010). The temporal resolution is one hour and the spatial horizontal resolution $0.5^\circ \times 0.5^\circ$. The spatial coverage contains a range of longitudes of 0°E to 359.5°E and latitudes of 90°S to 90°N.

The variables needed as input to the Lagrangian dispersion model described below include dew-point temperature, geopotential height, land cover, planetary boundary layer height, pressure and pressure reduced to mean sea level, relative humidity, temperature, zonal and meridional component of the wind, vertical

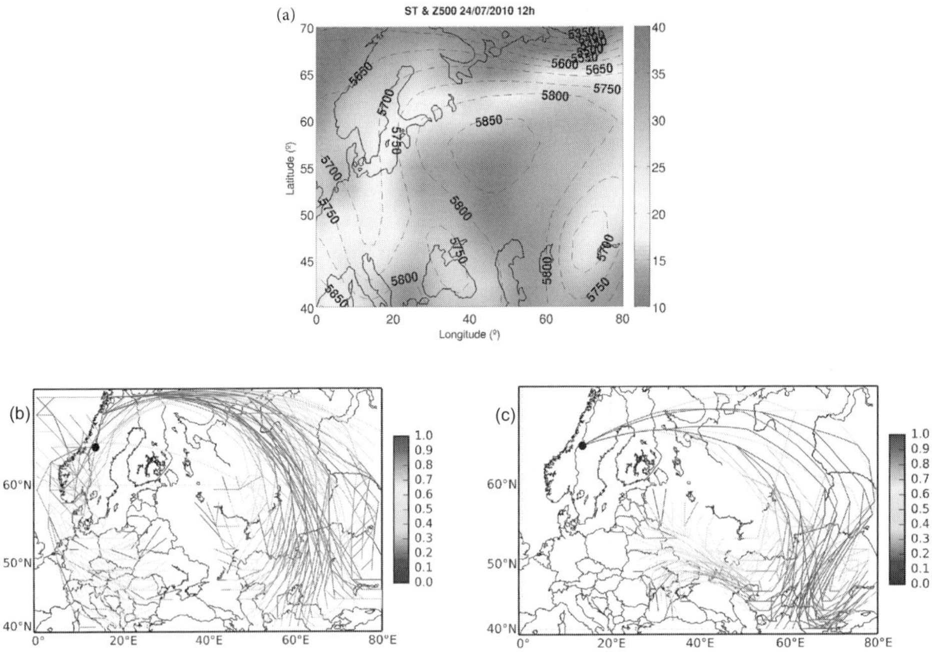

Figure 6.11 (a): Geopotential height at 500 hPa (contours, in m) and temperature (color code, in degrees C) over the region of interest, on July 24, 12:00 UTC. (b) and (c) Paths of $M = 9$ steps of $\tau = 12$ hours in the flow network with starting date July 25 2010 (b) and July 20 2010 (c), represented as straight segments (in fact, maximal arcs on the Earth sphere) joining the path nodes. MPPs originating from a single node (black circle) and ending in all accessible nodes. Color gives the P_{IJ}^{M} value of the paths in a normalized log scale between the minimum value (deep blue) and the maximum (dark red). In (b) the probabilities range from 10^{-3} to 10^{-14}; in (c), from 10^{-3} to 10^{-15}. (A black and white version of this figure appears in some formats. For the color version, please refer to the plate section.)

velocity, and water equivalent to accumulated snow depth. All these fields are provided by CFSR data on 26 pressure levels.

The numerical model used to integrate particle velocities and obtain trajectories is the Lagrangian particle dispersion model FLEXPART version 8.2 (Stohl et al., 2005, 2011). FLEXPART simulates the long-range and mesoscale transport, diffusion, dry and wet deposition, and radioactive decay of tracers released from point, line, area, or volume sources. It most commonly uses meteorological input fields from the numerical weather prediction model of the European Centre for Medium-Range Weather Forecasts (ECMWF), as well as the GFS from NCEP (the one used by Ser-Giacomi et al. (2015a)). Trajectories are produced by integrating the equation (the input velocity data are interpolated on the present particle position):

$$\frac{d\mathbf{X}}{dt} = \mathbf{v}(\mathbf{X}(t)), \tag{6.4}$$

with t being time, $\mathbf{X}$ the vector position of the air particle, and $\mathbf{v} = \bar{\mathbf{v}} + \mathbf{v}^{\mathbf{t}} + \mathbf{v}^{\mathbf{m}}$ the wind vector. FLEXPART takes the grid scale wind $\bar{\mathbf{v}}$ from the CFSR, but complements it with stochastic components $\mathbf{v}^{\mathbf{t}}$ and $\mathbf{v}^{\mathbf{m}}$ to better simulate the unresolved turbulent processes occurring at small scales. The turbulent wind fluctuations $\mathbf{v}^{\mathbf{t}}$ are parameterized by assuming a Markov process via a Langevin equation, and the mesoscale wind fluctuations $\mathbf{v}^{\mathbf{m}}$ are implemented also via an independent Langevin equation by assuming that the variance of the wind at the grid scale provides information on the subgrid variance. Variables entering the parameterizations are obtained from the meteorological CFSR fields (Stohl et al., 2005, 2011).

Ser-Giacomi et al. (2015a) focused on the domain between 0°E–80°E and 40°N–70°N. In order to define the nodes of the network this region was discretized in 626 equal-area boxes using a sinusoidal projection. This means that the discretization produced a square grid in the variables x and y, where $x = \phi \cos\theta$ and $y = \theta$, with ϕ and θ the standard longitude and latitude coordinates, respectively. In this way, latitudinal extension of each node-box was $\Delta y = \Delta\theta = 1.5°$, but the longitudinal one varied depending on the latitude, so that each node box has an area of 27 722 km^2. Typical horizontal box sizes are of the order of 166.5 km. Networks were constructed at time intervals of $\tau = 12$ hours, which is enough to follow the dynamics of the blocking event. It has been shown in an oceanic flow network (Ser-Giacomi et al., 2015c) that the value of τ has a minor influence on optimal paths. The total time-interval considered Mτ, with M the number of steps, is more important. Each node was uniformly filled with 800 ideal fluid particles that were released at 5000 m of height, a representative level in the middle troposphere. FLEXPART trajectories are fully three-dimensional, but initializing at each time step particles in a single layer effectively neglects the vertical dispersion (which is of the order of 800 m in the $\tau = 12$ h time step), putting the focus on the pathways of large-scale horizontal transport.

The flow network was constructed from the trajectory integrations with the formalism of Section 4.4, and the relevant paths followed by the air masses were characterized with the tools described in Section 4.4.4.

Figure 6.11b shows all the optimal paths leaving a node in the Scandinavian Peninsula at July 25 and arriving to all nodes which are reached in $M = 9$ steps (i.e., 4.5 days). The graphical representation joins with maximal arcs the center of the grid boxes identified as pertaining to the MPP. The actual particle trajectories between two consecutive boxes are not necessarily such arcs. The paths are colored according to their probability value P_{IJ}^{M}. The MPPs with highest probability (reddish colors) follow a dominant anticyclonic (i.e., clockwise) route bordering the high-pressure region (see Fig. 6.11a, but note that this is at a particular time, whereas the trajectory plots span a range of dates of more than four days) without penetrating it. There is also a branch of MPPs with much smaller probabilities (yellow and bluish colors) that are entrained southward by a cyclonic circulation.

Despite the persistent character of the Eulerian block configuration, sets of Lagrangian trajectories become highly variable in time. See, for example, the set of MPPs starting from the same initial location but five days earlier (Fig. 6.11c). The southward cyclonic branch is now absent, all MPPs following initially the anticyclonic gyre. Remarkably, the set of trajectories bifurcates into two branches when approaching what seems to be a strong hyperbolic structure close to 40°N 60°E. A hint of the presence of second hyperbolic structure is visible at the end of the westward branch, close to 50°N 30°E.

Figure 6.12 displays additional MPPs starting also at July 20, but initialized inside the main anticyclonic region of the blocking, and in two low-pressure regions flanking it. Figure 6.12a clearly shows the main anticyclonic circulation, highlighting also the escape routes from the high-pressure zone, associated with the hyperbolic regions described above. The other two panels show the cyclonic circulations

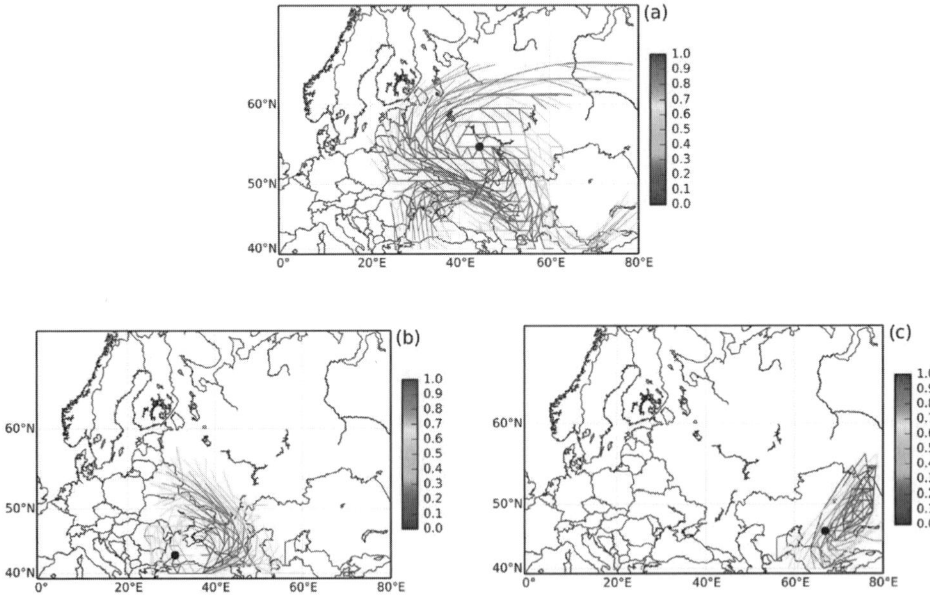

Figure 6.12 Optimal paths of nine steps of $\tau = 12$ hours with starting date July 20 2010, entrained in the high- and in the two low-pressure areas of the blocking. Same coloring scheme as in Fig. 6.11. (a) Probabilities ranging from 10^{-3} to 10^{-16}. (b) Probabilities ranging from 10^{-2} to 10^{-16}. (c) Probabilities ranging from 10^{-3} to 10^{-13}. (A black and white version of this figure appears in some formats. For the color version, please refer to the plate section.)

at each side of the high, in a characteristic Omega-blocking configuration. The compactness of the trajectories inside the eastern low-pressure area, which form a very localized and coherent set with practically no escape in the 4.5-day time interval displayed, is remarkable.

The range of colors in Figs. 6.11 and 6.12 indicates that, given an initial box, not all MPPs leading to different locations are equally probable. This is quantified by the probability P_{IJ}^{M}, which gives a weight to each MPP (see Section 4.4.4). Indeed P_{IJ}^{M} takes a very large range of values. Very low probability values arise because of the exponential explosion of the number of paths between two nodes with increasing M. Given these low values of P_{IJ}^{M} except for the smallest values of M, one should ask how representative the MPPs are for the full set of paths. This was addressed (Ser-Giacomi et al., 2015a) by analyzing the distributions of the parameter $\lambda_{IJ}^{M}(r, \epsilon)$ giving the relative importance (see Section 4.4.4) of the different types of paths obtained for successive values of r, the parameter controlling how exhaustively we search into the set of highly probable paths (HPPs) (i.e., the paths with probability larger than ϵP_{IJ}^{M}). The analysis reveals that λ-values are small when considering only the MPPs ($r = 0$), but the distributions shift toward higher values for richer path-sets obtained for increasing r. The mean values of the λ distributions decrease with M, reflecting the lack of representativeness of the smallest sets of paths for large M. However, already for $r = 1$ the set of HPPs has a mean value of λ higher than 0.5, indicating high relevance in an important range of time steps.

Thus, for the values of M and ϵ discussed here, the set of HPPs with $r = 1$ seems to be rich enough to represent the transport pathways. But how different is the geometry of the different paths in this HPP set? And how different is it from the MPPs? An examination of the HPPs with $r = 1$ and $\epsilon = 0$ reveals (Ser-Giacomi et al., 2015a) that in all the cases the sets remain coherent and narrow tubes of trajectories define roughly the same pathway as the MPP. It can also be shown that although the tube width increases with M, it remains always below the typical linear box size of approximately 166.5 km, indicating that the tubes remain narrow. Thus, Ser-Giacomi et al. (2015a) concluded that, despite the decreasing probability of the MPPs for increasing M, they remain good indicators of the dominant pathways in the transport network.

6.6 Indian Monsoon

The previous sections addressed how the use of complex networks can help in understanding atmospheric teleconnections and mass transport. This section and the next section are devoted to the study of monsoon phenomena.

As explained in Section 2.5, the Indian Summer Monsoon (ISM) is a large-scale atmospheric phenomenon in the modern climate and one of the active components of the global climate system in the tropics. The ISM plays a crucial role in the daily life of the Indian population, and is important for other parts of the globe as well, because of the monsoon's coupling with climate drivers such as ENSO, the Indian Ocean Dipole, and the Equatorial Indian Ocean Oscillation (Achuthavarier et al., 2012; Gadgil, 2004; Sabeerali et al., 2011; Sankar et al., 2011; Wu et al., 2003, 2012).

Climate networks have been used to study the monsoon variability in various timescales (from the Late Holocene to the present day); they been used to study the relationships between the two monsoon systems (ISM and the East Asian Monsoon, EAM); to determine spatial structures and directionalities in ISM extreme rainfall; to reveal the seasonal evolution of the extreme precipitation over the Indian subcontinent; and to identify ISM spatial patterns that can be used as indicators for the coming monsoon season. The following subsections summarize the various ways in which climate networks have contributed to advancing the understanding of the Indian Monsoon.

6.6.1 Paleoclimate Networks

In the work by Rehfeld et al. (2012) on Asian Monsoon dynamics, the network approach was used to study paleorecords, which are geographically heterogeneously distributed. An example of nodes of the paleoclimate network is shown in Fig. 6.13. The irregularity in the sampling of paleoclimate data and the reduced extension of the spatial domain require a challenging reconstruction of the data (Rehfeld et al., 2011, 2012), before the climate network method is applied.

After reconstruction of the paleoclimate data, several similarity measures can be used in order to reveal the connection between dynamics in the nodes of the paleoclimate network, such as Gaussian kernel-based Pearson correlation and mutual information (see Sections 3.2.1 and 3.3.2) (Rehfeld et al., 2011, 2012). Rehfeld et al. (2012) proposed to use several similarity measures for all pairs of nodes and establish statistically significant links for each pair of nodes in the network and for each similarity measure. Then, it was proposed to sum up adjacency matrices of the different measures, and yield a *weighted adjacency matrix*. The nodes i and j of the paleoclimate network described by such weighted adjacency matrix are linked if any $A_{ij} > 0$.

Rehfeld et al. (2012) used the paleoclimate network to reveal a strong influence of the ISM on the East Asian Summer Monsoon (EASM) during the late Holocene period, but with varying strength depending on the epochs. The study focused on

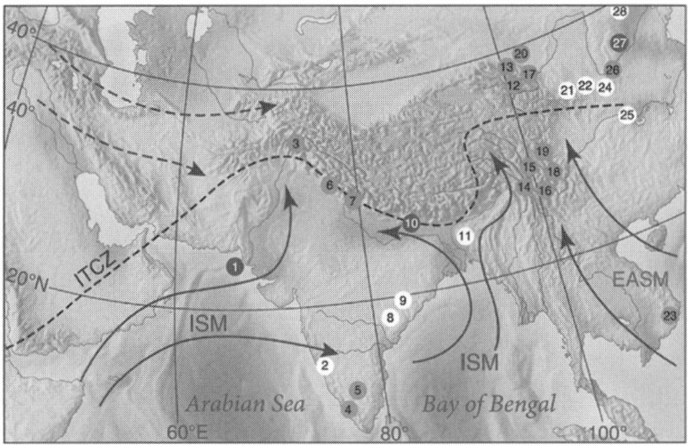

Figure 6.13 Study area with generalized summer wind directions of the Indian Summer Monsoon and East Asian Summer Monsoon (gray arrows), the westerlies (dashed arrows), as well as the spatial coverage of the records considered in the paleoclimate networks. Colors of the dots indicate the type of archive: orange = tree sites, white = stalagmites, purple = other archives (marine sediment, ice core, reconstruction using historic documents, and tree ring data). Source: Rehfeld et al. (2012). (A black and white version of this figure appears in some formats. For the color version, please refer to the plate section.)

the Medieval Warm Period (MWP), the Little Ice Age (LIA), and the recent period of warming (RWP) in East Asia, constructing paleoclimate networks for each time period. For example, during the MWP it was found there was a strong influence of the ISM on the EAM. On the other hand, during the LIA, the ISM circulation was weaker and did not extend as far east (see Fig. 6.14). Analysis of the paleoclimate network for the RWP in East Asia shows a currently ongoing transition phase toward a stronger ISM penetration into China.

6.6.2 Monsoon Extreme Rainfall Events

By using event synchronization networks (ESNs) that combine the event synchronization method with the network approach (see Section 4.5), Malik et al. (2010, 2012) have analyzed the spatial structures, organization, and scales of the extreme rainfall during the present-day ISM (1951–2007).

As seen in Fig. 6.15, network measures such as degree reveal well-defined spatial structures in the monsoonal extreme precipitation. Applying the ESN approach to the analysis of the extreme monsoonal rainfall allows identifying the structure and organization of the extreme rainfall field in terms of its spatial discontinuity. The median link length scale reveals extreme rainfall events are synchronized up to 250 km for most of the region (see Fig. 6.15a).

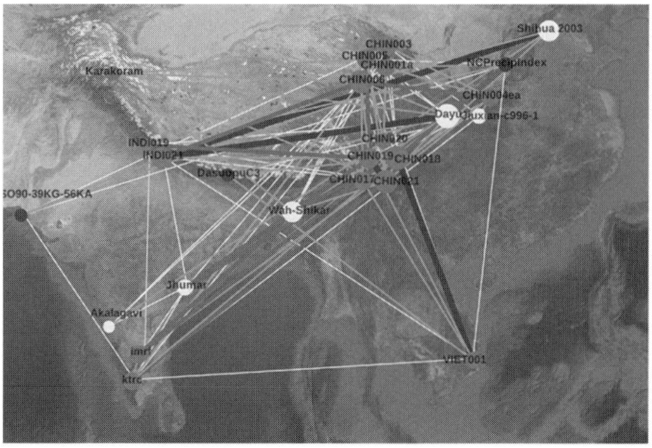

Figure 6.14 Network for the Little Ice Age (LIA): a network embedded in the observation space with true geo-coordinates. The darker and thicker a link, the higher its weight; the size of a node corresponds to its weighted node degree, whereas the node color indicates the type of archive. Source: Rehfeld et al. (2012). (A black and white version of this figure appears in some formats. For the color version, please refer to the plate section.)

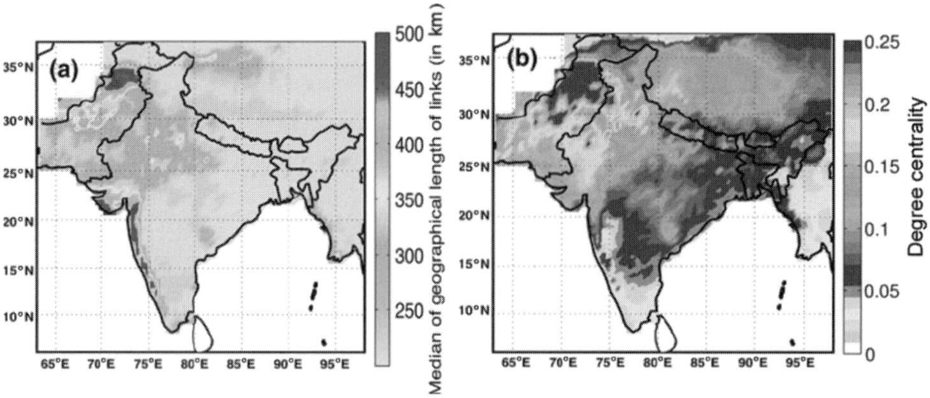

Figure 6.15 Network measures for the extreme rainfall network: median geographical link length (a), and degree (b). Source: adapted from Malik et al. (2012). (A black and white version of this figure appears in some formats. For the color version, please refer to the plate section.)

Figure 6.16 shows the accuracy of prediction based on the synchronization of extreme rainfall events without delay, e.g., which happen on the same day. As can be seen (Malik et al., 2010), for most of the subcontinent the accuracy of prediction ϵ is above 70% and even reaches up to 100% in places.

Stolbova et al. (2014) analyzed the spatial structure and organization of present-day extreme rainfall during the pre-monsoon, monsoon, and post-monsoon periods.

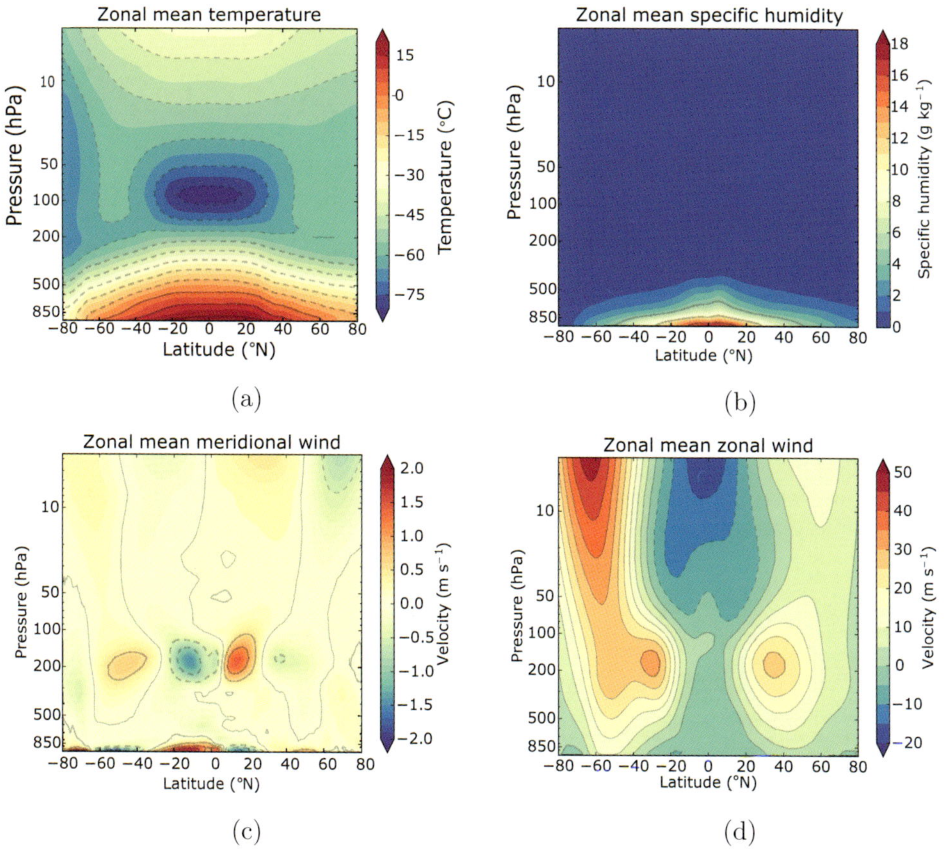

Figure 1.6 (a) Zonal mean temperature, (b) zonal mean moisture, (c) zonal mean meridional velocity, and (d) zonal mean zonal velocity. All fields are obtained by averaging over the CESM model years 170–199.

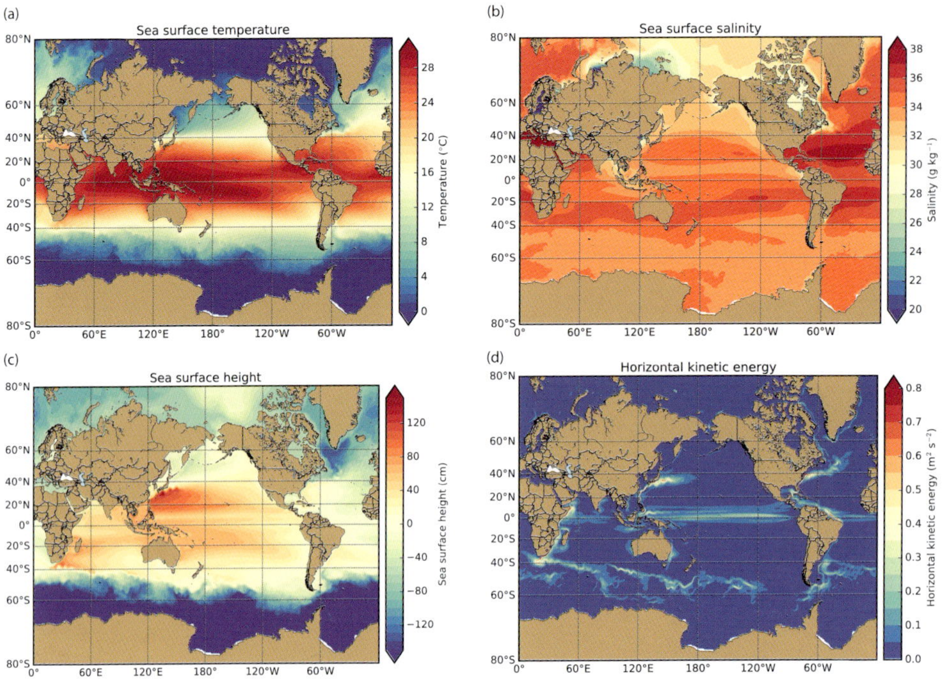

Figure 1.7 (a) Sea surface temperature, (b) sea surface salinity, (c) sea surface height, and (d) surface kinetic energy. All fields are obtained by averaging over the CESM model years 170–199.

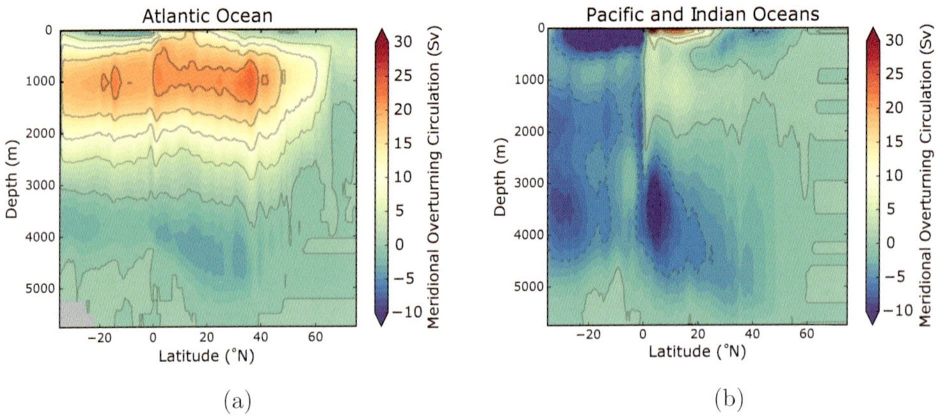

Figure 1.9 (a) Atlantic MOC and (b) Indo-Pacific MOC. All fields are obtained by averaging over the CESM model years 170–199.

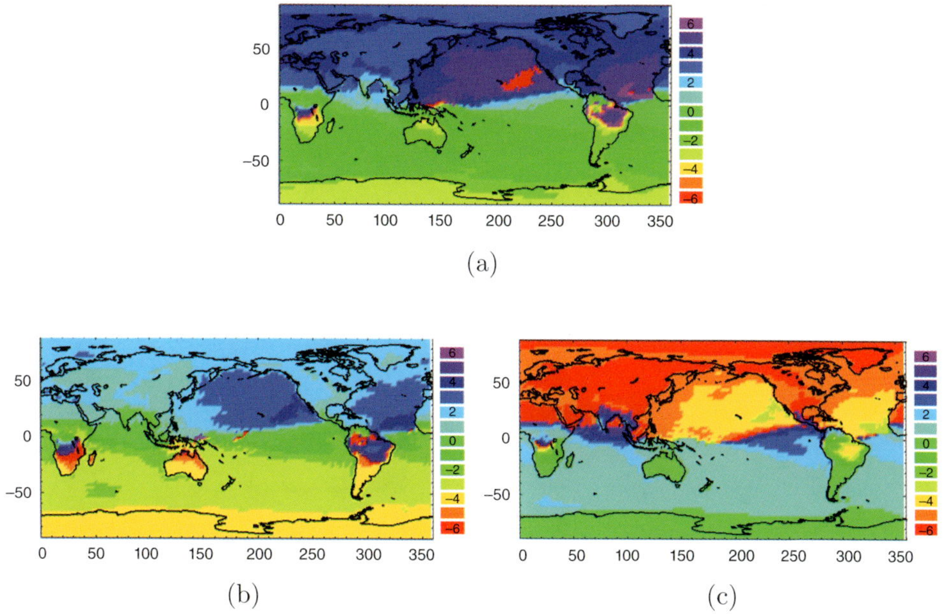

Figure 3.2 Lag times computed from cross-correlation analysis of monthly surface air temperature reanalysis. The lag is defined in the interval [−6,6] months. Panel (a) displays the lag times between the time series at a reference point located in El Niño basin and all the other time series; in panel (b) the reference point is located in the Indian Monsoon area; and in panel (c) it is in central Australia.

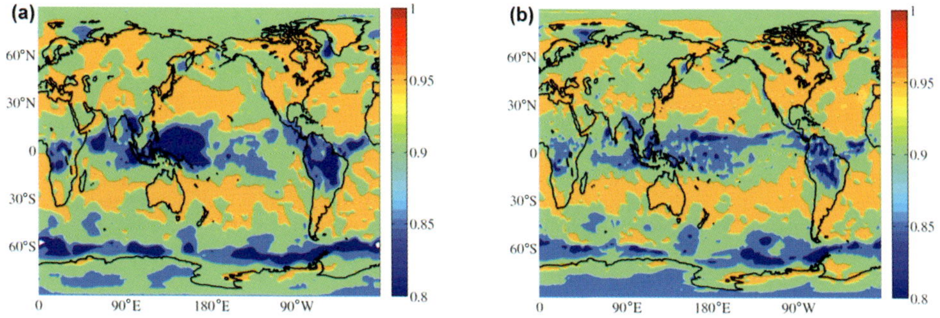

Figure 3.7 Shannon entropy of monthly surface air temperature time series. (a) NCEP CDAS1 reanalysis, (b) ERA Interim reanalysis. The number of bins M is the same for all time series within a reanalysis database, but is adjusted in each database to take into account the different length of the time series: in panel (a) $M = 40$; in panel (b) $M = 20$. Source: Arizmendi et al. (2017).

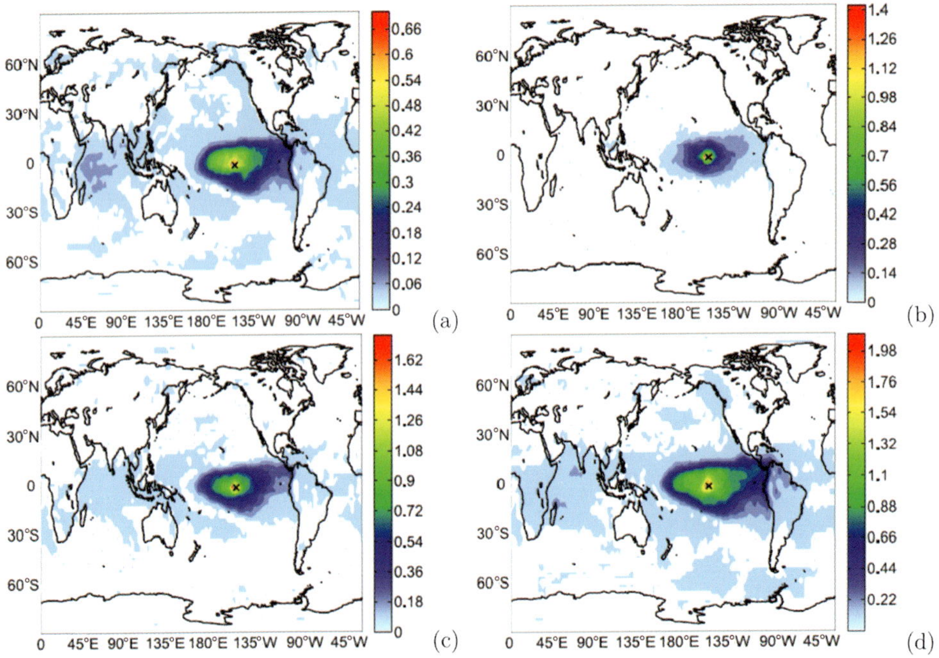

Figure 3.9 Mutual information (M_I) values of monthly SATA from NCEP/NCAR reanalysis. The value of M_I between the time series recorded at a given reference point (indicated with a cross) and all the time series recorded at different points is shown in color code. The probability distributions used in panel (a) are computed from the histograms of anomaly values (using six bins to compute the histogram), while in (b)–(d) OPs are used: in (b), the patterns are formed by three consecutive months, in (c) by three equally spaced months covering a one-year period, and in (d) by three months in consecutive years. Only significant M_I values are shown: a M_I value is considered significant if it is larger than $\mu + 3\sigma$, with μ and σ being the mean value and the standard deviation of the distribution of M_I values computed from surrogates. Source: Deza (2015).

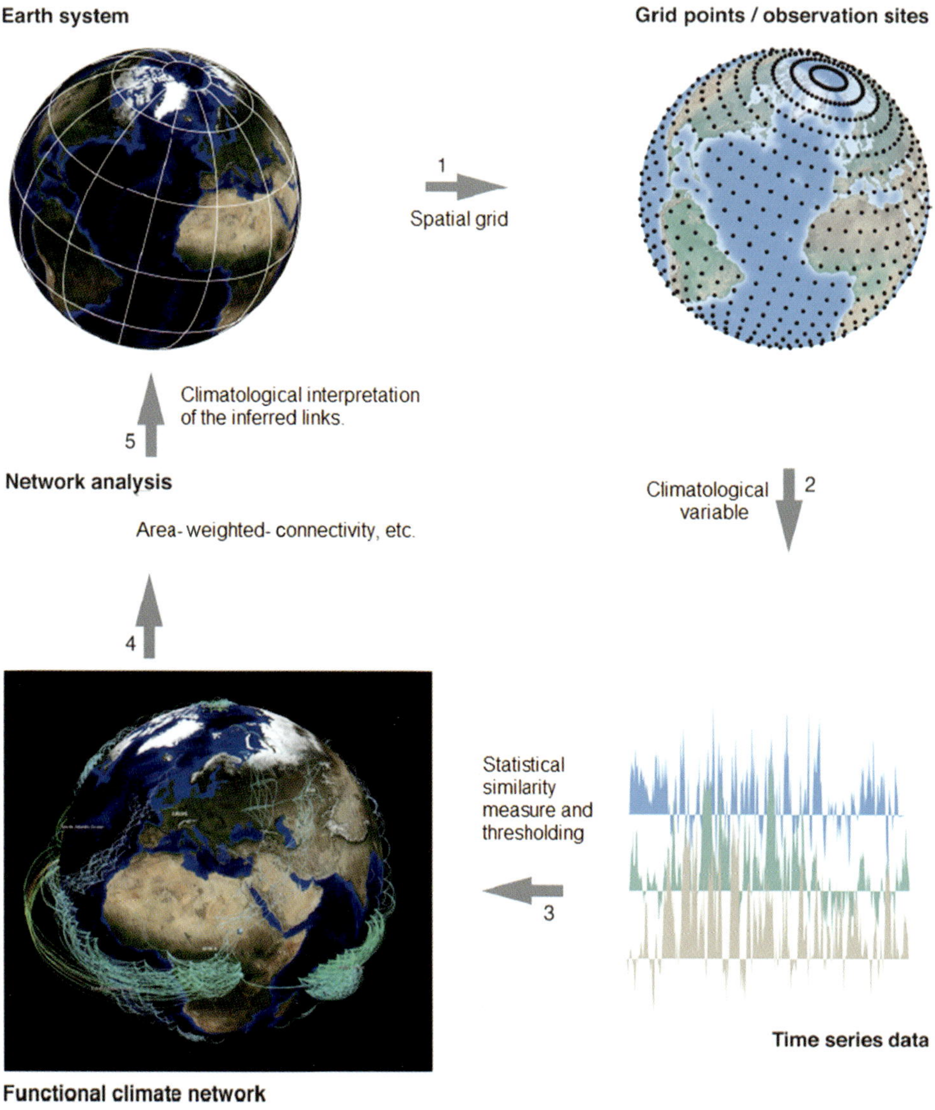

Figure 4.1 Methodology used to extract a climate network from climatic signals. Source: Donges et al. (2015).

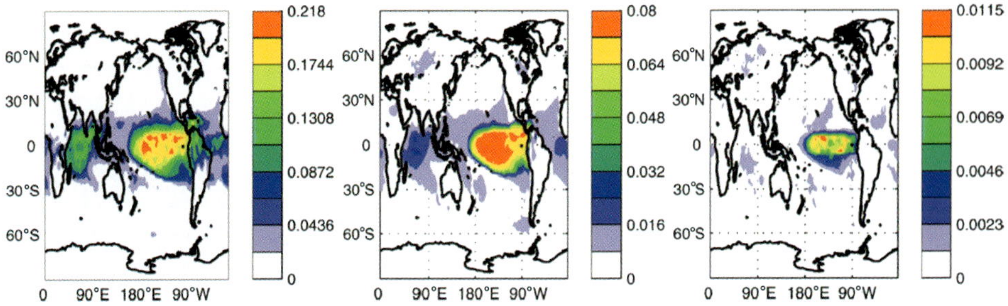

Figure 4.2 Area-weighted connectivity in color code, when the climate network is inferred by using the mutual information computed from the probabilities of ordinal patterns defined by comparing SATA (NCEP/NCAR monthly reanalysis) in four consecutive years. The color code is such that the white (red) regions indicate the geographical areas with zero (greatest) connectivity. The significance threshold increases from left to right. Source: Barreiro et al. (2011).

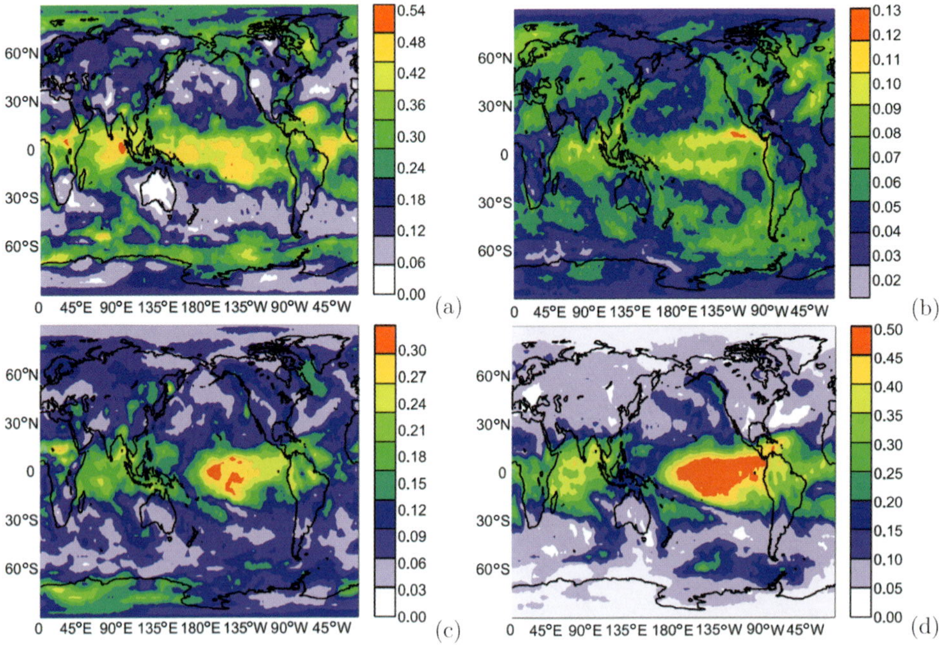

Figure 4.3 Area-weighted connectivity in color code, when the mutual information M_I is computed from (a) the histograms of anomaly values and (b)–(d) from ordinal probabilities: in (b), the are patterns formed with three consecutive months, in (c) by three equally spaced months covering a one-year period, and in (d) by three months in consecutive years. Only significant M_I values are shown: an M_I value is considered significant if it is larger than $\mu + 3\sigma$, with μ and σ being the mean value and the standard deviation of the distribution of M_I values computed from surrogates. The maps of significant M_I values in the ENSO region were shown in Fig. 3.9. Source: Deza (2015).

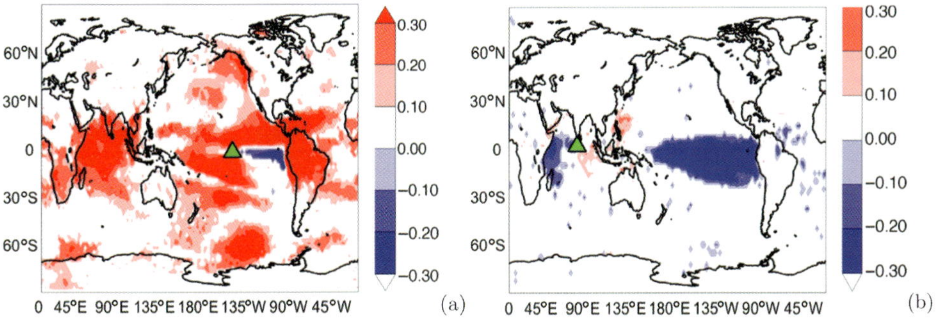

Figure 4.4 Directionality of the links in a node in the central Pacific (a) and in a node in the Indian Ocean (b), indicated with a triangle. The color code indicates the directionality index: outgoing links are shown in red while incoming links are shown in blue. The time scale of information transfer is τ = 30 days. Source: Deza et al. (2015).

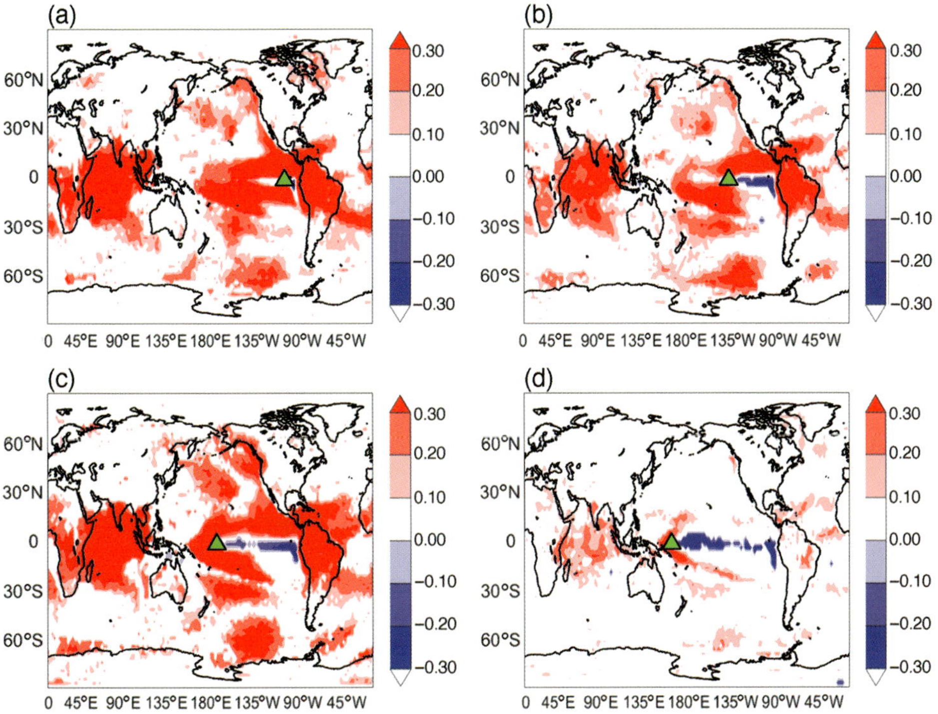

Figure 4.5 Directionality over the equatorial pacific with $\tau = 30$ days. It is observed that most of the regions over the central and eastern Pacific Ocean have strong effects over a large part of the world; especially over the tropical areas and the rest of the Pacific Ocean. Source: Deza et al. (2015).

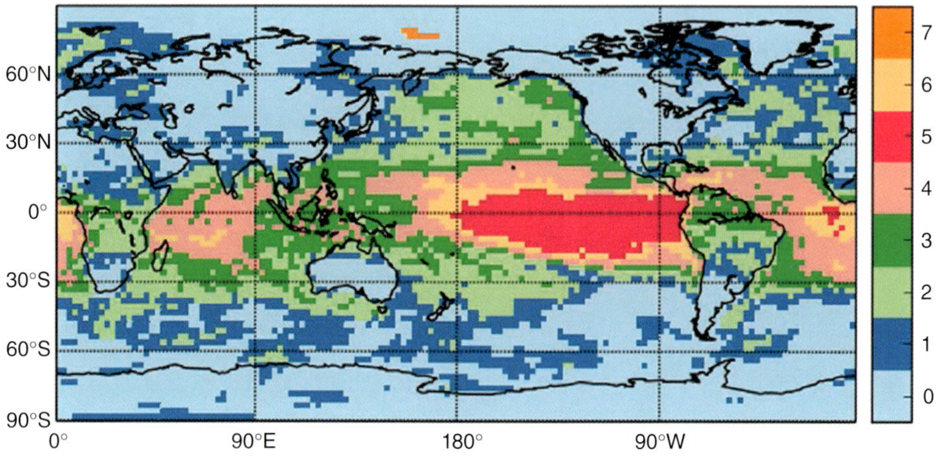

Figure 4.6 Climate communities inferred with the *Infomap* algorithm. The adjacency matrix is obtained from symbolic statistical analysis of SAT anomalies. Regions depicted with the same color belong to the same community. Four macro-communities are identified: extratropical continents and oceans, tropical oceans, and El Niño basin. Source: Tirabassi and Masoller (2016).

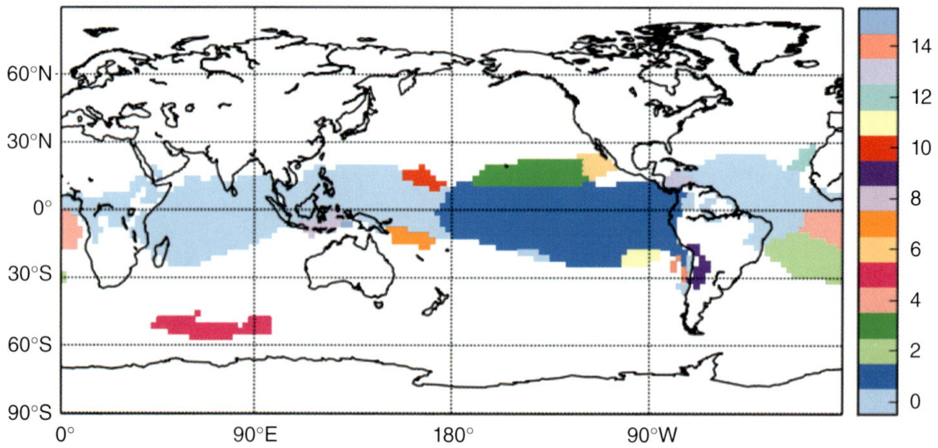

Figure 4.7 Community structure uncovered by *Infomap* algorithm when the network is constructed using the Pearson cross-correlation coefficient as a measure of dynamical similarity. Source: Tirabassi and Masoller (2016).

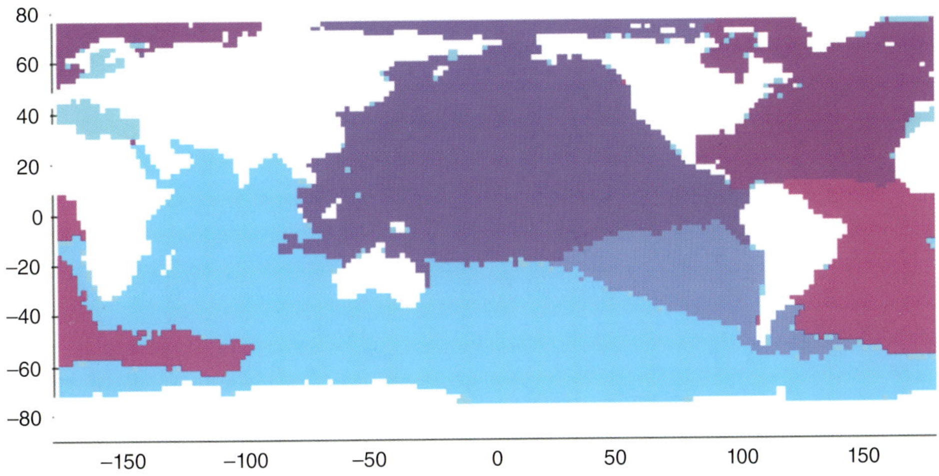

Figure 4.10 A spectral partition of the global ocean into five communities, using four left eigenvectors of the transport matrix describing surface ocean transport during 1000 years. Source: Froyland et al. (2014).

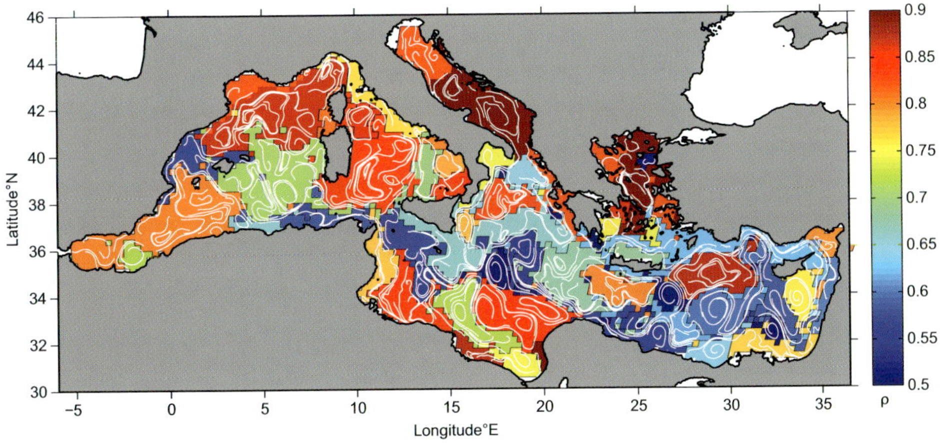

Figure 4.11 A partition of the Mediterranean Sea into 65 communities by Infomap, from the transport matrix describing surface transport for one month starting on January 1 2011. Each community is colored with the value of its coherence ratio ρ (4.27)) White lines are average streamlines during the month considered. Source: Rossi et al. (2014). (A black and white version of this figure appears in some formats. For the color version, please refer to the plate section.)

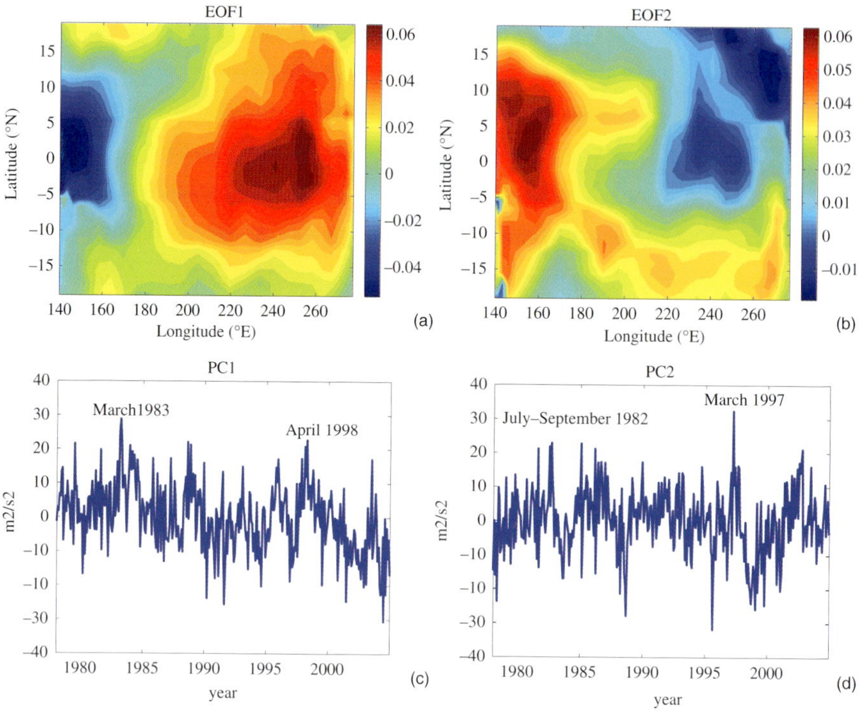

Figure 5.3 (a) The first empirical orthogonal function (EOF) of the zonal wind-stress residual, accounting for 11.9% of the variance. (b) The same as (a) but the second EOF, accounting for 11.6% of the variance. (c) The first principal component (PC1) of the wind-stress residual. (d) The same as (c), but for the second principal component (PC2). Source: Feng and Dijkstra (2017).

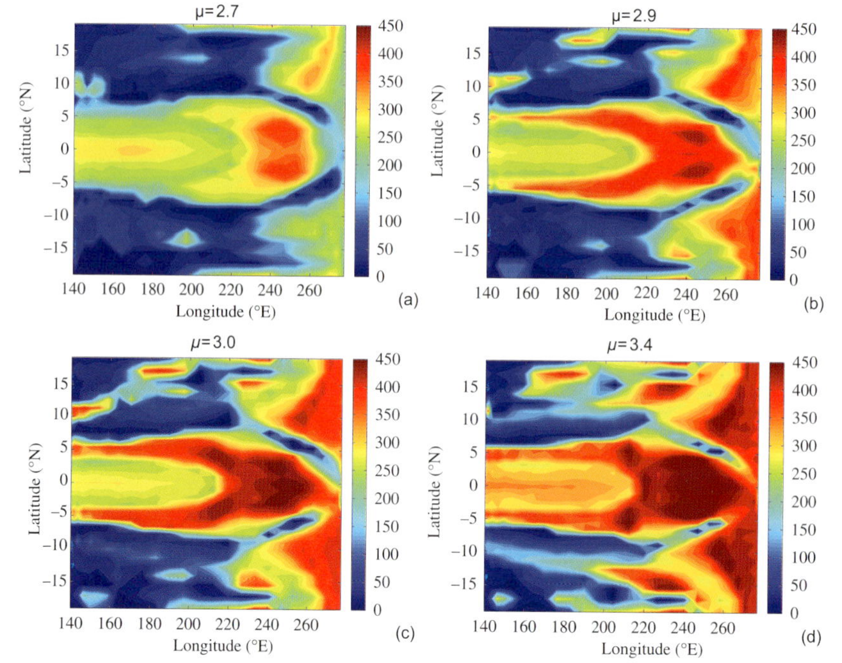

Figure 5.4 (a) Degree field of the PCCN using a threshold $\epsilon_C = 0.5$ reconstructed from the ZC model SST data at a coupling strength $\mu = 2.7$. (b) Same as (a) but at $\mu = 2.9$. (c) Same as (a) but at $\mu = 3.0$. (d) Same as (a) but for $\mu = 3.4$. Source: Feng and Dijkstra (2017).

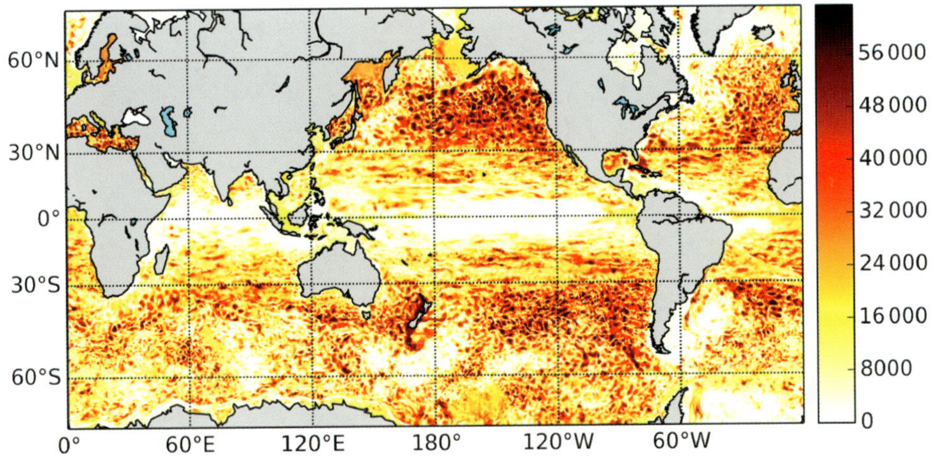

(a)

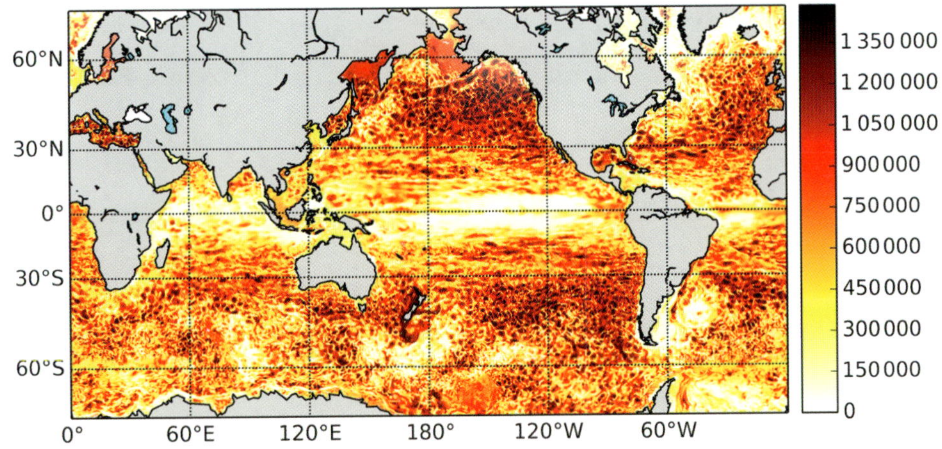

(b)

Figure 5.8 (a) Degree for the SSH POP data interpolated on the 0.4° grid and a threshold of $\tau = 0.5$. (b) Degree field for the 0.1° grid and a threshold $\tau = 0.4$; here, the reconstructed network has 4.7×10^6 nodes and 1.4×10^{12} edges. Source: Ihshaish et al. (2015).

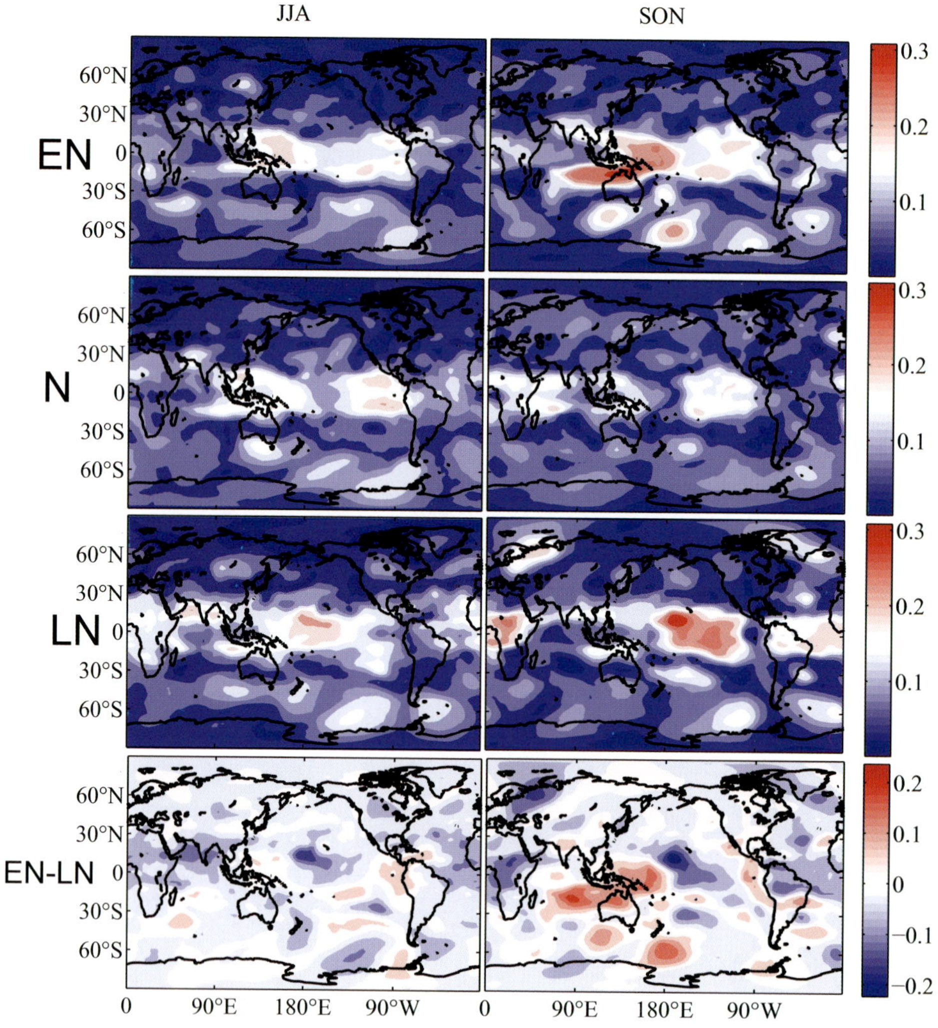

Figure 6.1 AWC during El Niño (first row), neutral (second row), and La Niña (third row) years. The last row shows the difference between El Niño and La Niña years. Left column: June–July–August (JJA) season. Right column: September–October–November (SON) season. Source: Arizmendi and Barreiro (2017).

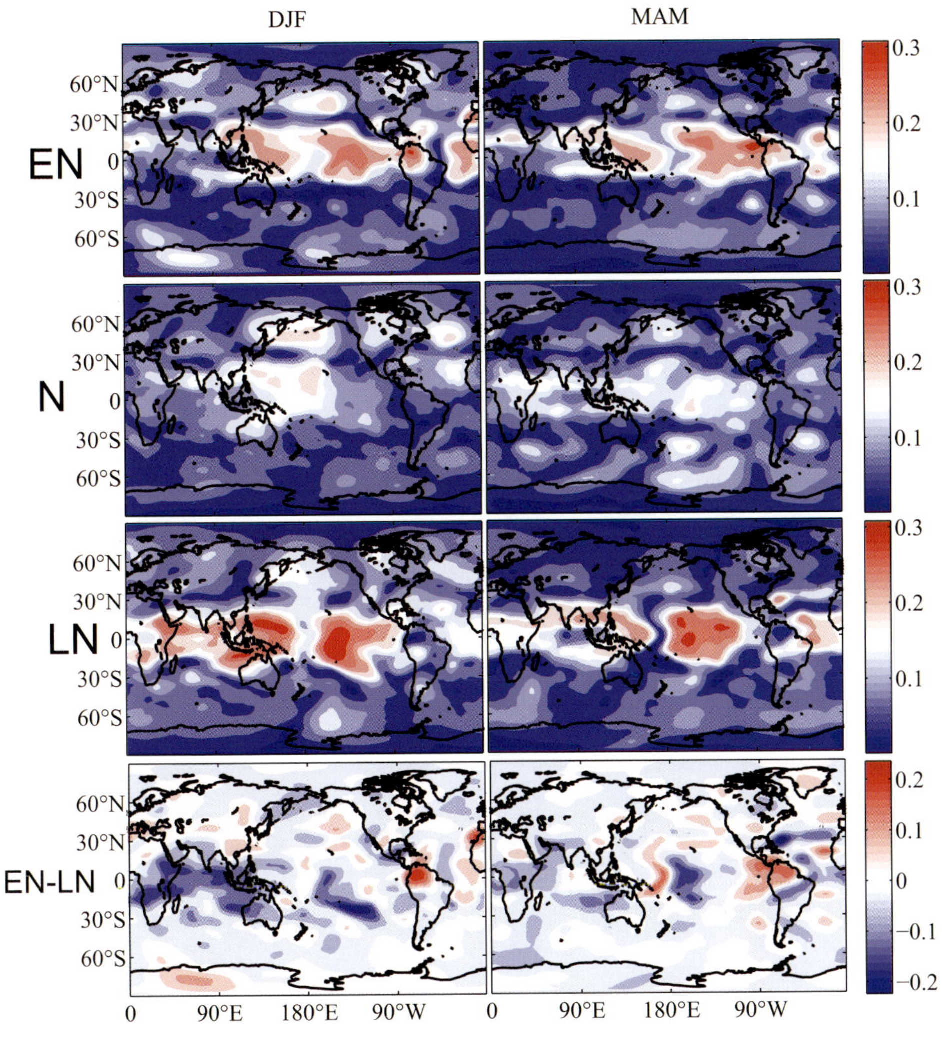

Figure 6.2 Same as Fig. 6.1 but for the December–January–February (DJF) season (left column) and the March–April–May (MAM) season (right column). Source: Arizmendi and Barreiro (2017).

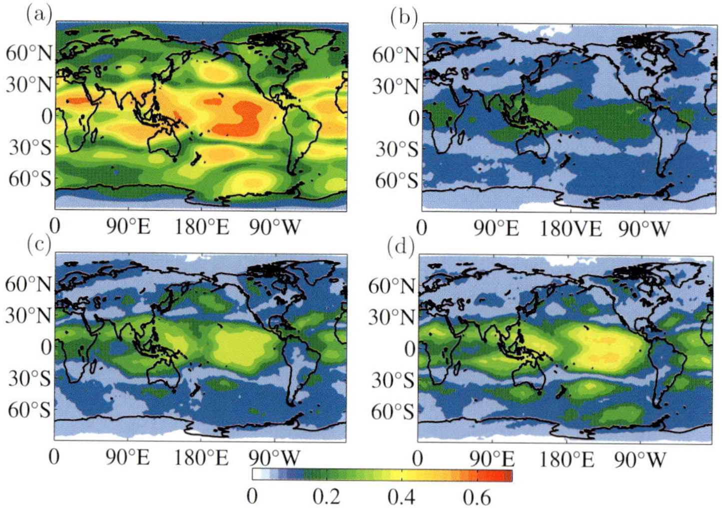

Figure 6.3 Area-weighted connectivity maps obtained from the 200 hPa eddy geopotential height NCEP/NCAR reanalysis data (1949–2011). The statistical similarity measures used are: Pearson correlation (a), mutual information with the ordinal patterns methodology: intraseasonal $\Delta t = 1$ month (b), intraannual $\Delta t = 4$ months, (c) and interannual $\Delta t = 12$ months (d). Source: Arizmendi et al. (2014).

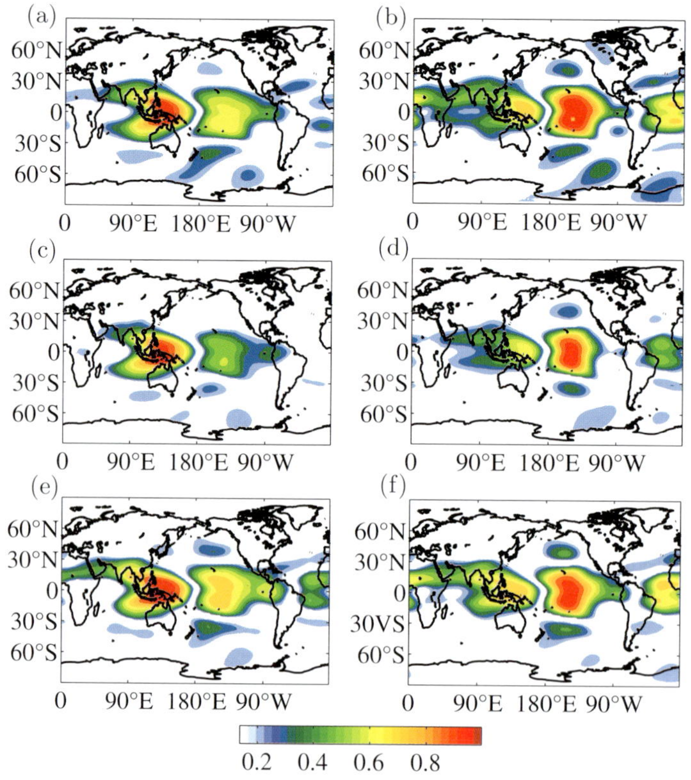

Figure 6.5 Maps of the cross-correlation between the eddy geopotential time series at each location and the time series averaged inside the western Pacific (left column) and in the central Pacific (right column), using the NOAA reanalysis data. Correlations are computed by taking 30-year windows centered at 1915 ((a) and (b)), 1940 ((c) and (d)), and 1995 ((e) and (f)). Only statistically significant values are shown in color. Source: Arizmendi et al. (2014).

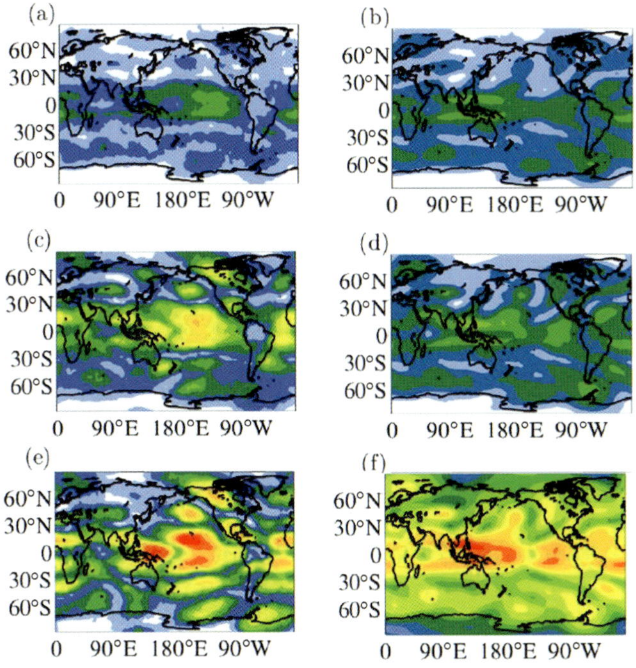

Figure 6.6 Area-weighted connectivity maps computed from 200 hPa eddy geopotential height in the period 1954–2006. In the left column the maps of the ICTP-AGCM oceanically forced component are computed using the mutual information from ordinal patterns for intraseasonal ((a), $\Delta t = 1$ month), intraannual ((c), $\Delta t = 4$ months), and interannual ((e), $\Delta t = 12$ months) timescales. In the right column the AWC maps of internal variability (averaged over the nine runs) are computed using the mutual information with intraseasonal ((b), $\Delta t = 1$ month) and intraannual ((d), $\Delta t = 4$ months) ordinal patterns. Panel (f) presents the AWC map obtained with Pearson correlation. Source: Arizmendi et al. (2014).

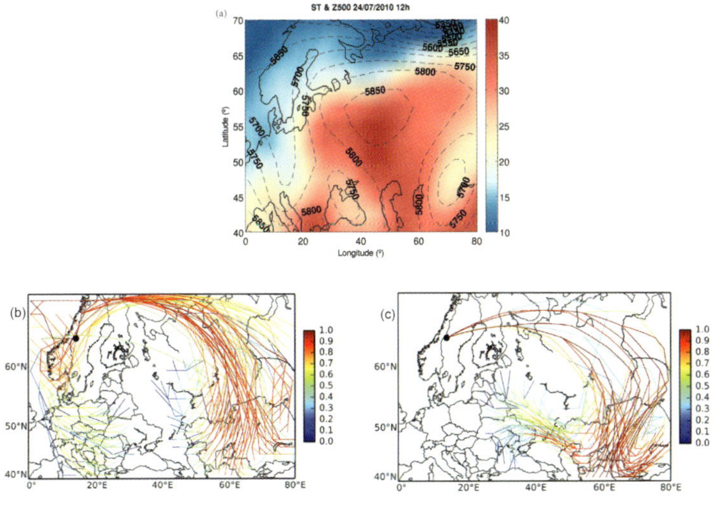

Figure 6.11 (a): Geopotential height at 500 hPa (contours, in m) and temperature (color code, in degrees C) over the region of interest, on July 24, 12:00 UTC. (b) and (c) Paths of $M = 9$ steps of $\tau = 12$ hours in the flow network with starting date July 25 2010 (b) and July 20 2010 (c), represented as straight segments (in fact, maximal arcs on the Earth sphere) joining the path nodes. MPPs originating from a single node (black circle) and ending in all accessible nodes. Color gives the P_{IJ}^{M} value of the paths in a normalized log scale between the minimum value (deep blue) and the maximum (dark red). In (b) the probabilities range from 10^{-3} to 10^{-14}; in (c), from 10^{-3} to 10^{-15}.

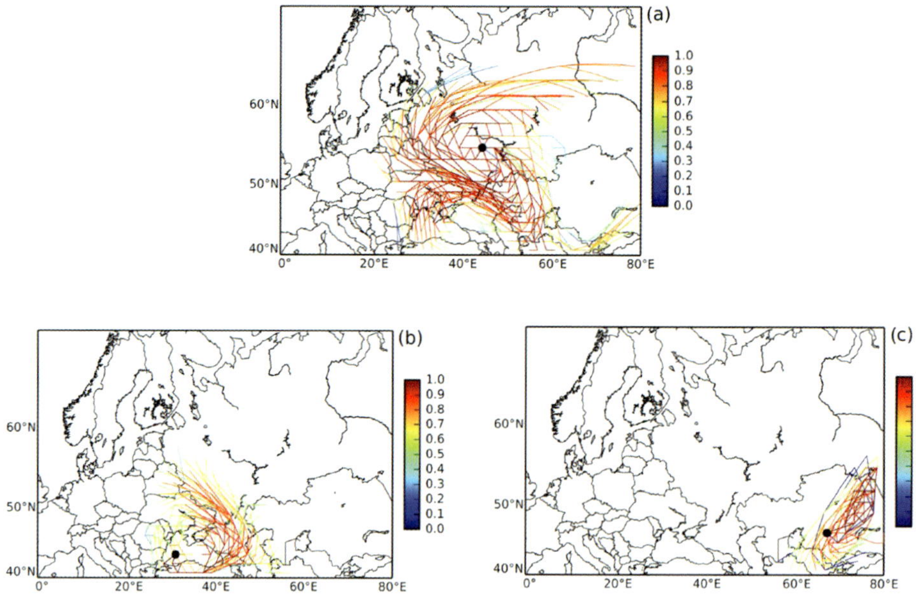

Figure 6.12 Optimal paths of nine steps of $\tau = 12$ hours with starting date July 20 2010, entrained in the high- and in the two low-pressure areas of the blocking. Same coloring scheme as in Fig. 6.11. (a) Probabilities ranging from 10^{-3} to 10^{-16}. (b) Probabilities ranging from 10^{-2} to 10^{-16}. (c) Probabilities ranging from 10^{-3} to 10^{-13}.

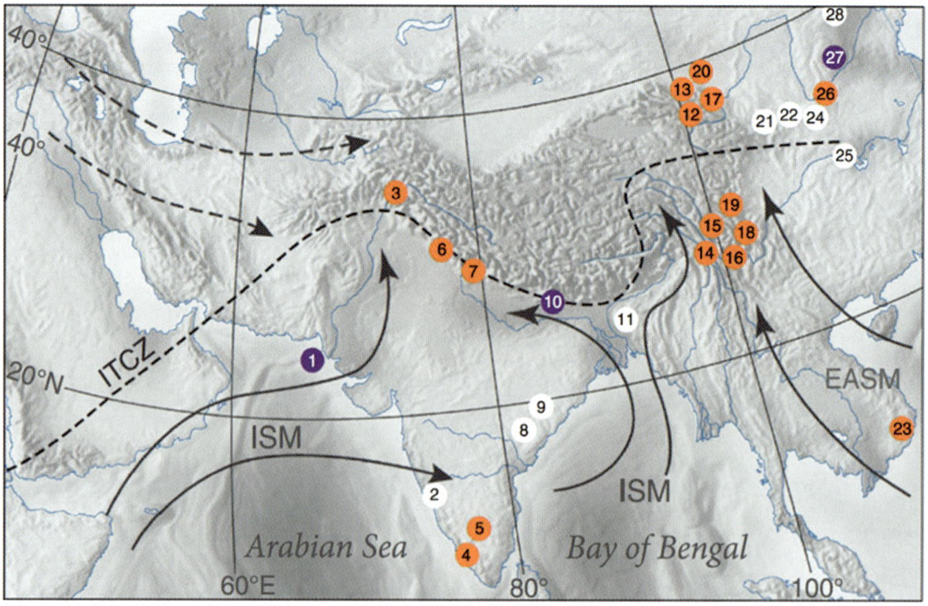

Figure 6.13 Study area with generalized summer wind directions of the Indian Summer Monsoon and East Asian Summer Monsoon (gray arrows), the westerlies (dashed arrows), as well as the spatial coverage of the records considered in the paleoclimate networks. Colors of the dots indicate the type of archive: orange = tree sites, white = stalagmites, purple = other archives (marine sediment, ice core, reconstruction using historic documents, and tree ring data). Source: Rehfeld et al. (2012).

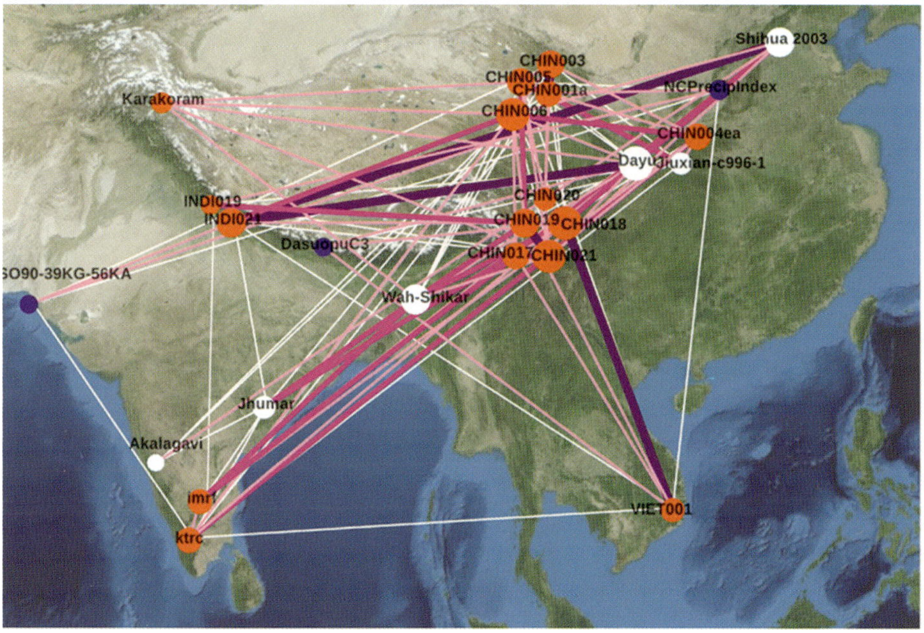

Figure 6.14 Network for the Little Ice Age (LIA): a network embedded in the observation space with true geo-coordinates; a force-weighting algorithm was applied in which linked nodes are attracted and unlinked nodes repelled, providing a complimentary network view independent of the nodes' locations. The darker and thicker a link, the higher its weight; the size of a node corresponds to its weighted node degree, whereas the node color indicates the type of archive. Source: Rehfeld et al. (2012).

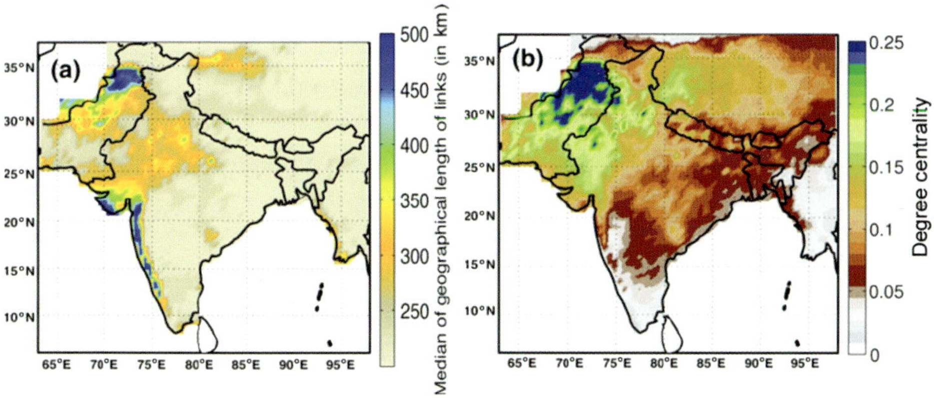

Figure 6.15 Network measures for the extreme rainfall network: median geographical link length (a), and degree (b). Source: adapted from Malik et al. (2012).

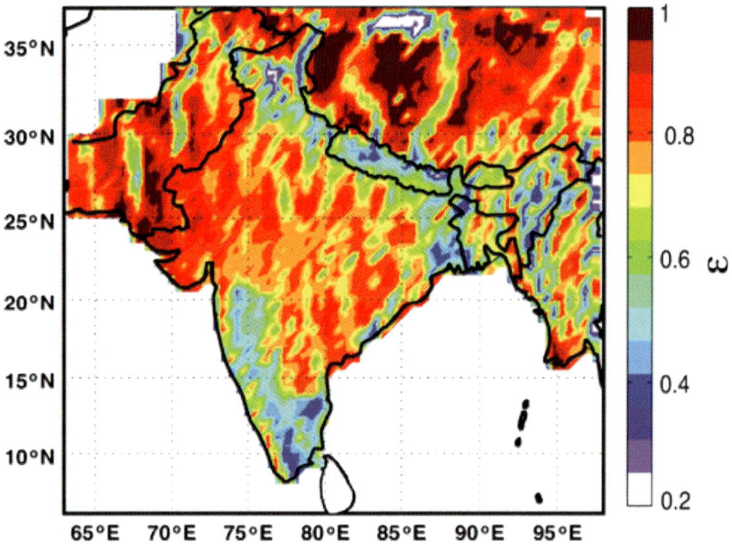

Figure 6.16 The accuracy of prediction ϵ on a scale from 0 to 1. Source: Malik et al. (2010).

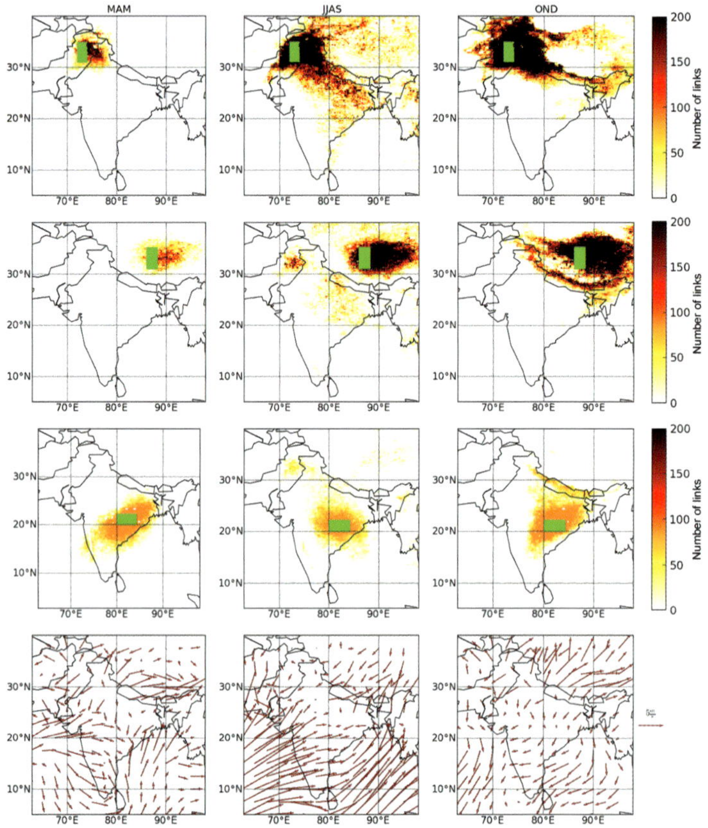

Figure 6.18 Links between a set of 153 reference grid points to other grid points for pre-monsoon (MAM), summer (JJAS), and post-monsoon (OND) seasons. From top to bottom: North Pakistan (NP), Tibetan Plateau (TP), Eastern Ghats (EG) (TRMM). The bottom panels display the mean surface winds between 1982 and 2012 for the different seasons (NCEP/NCAR). Source: Stolbova et al. (2014).

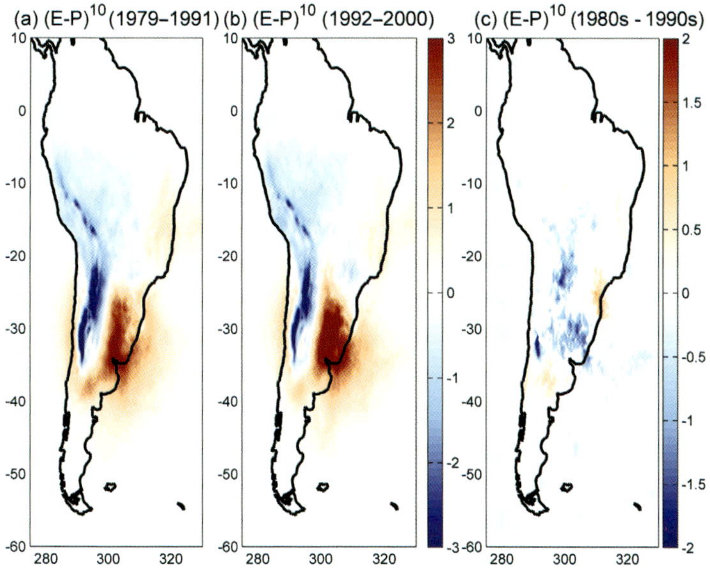

Figure 6.22 Ten-day average of the net budget evaporation minus precipitation, $(E - P)^{10}$, during (a) the 1980s (1979–1991), (b) the 1990s (1992–2000). (c) Difference between the 1980s and 1990s. Only locations with anomalies significant at the 90% confidence level are colored. Units: mm/day. Source: Martin-Gomez et al. (2016).

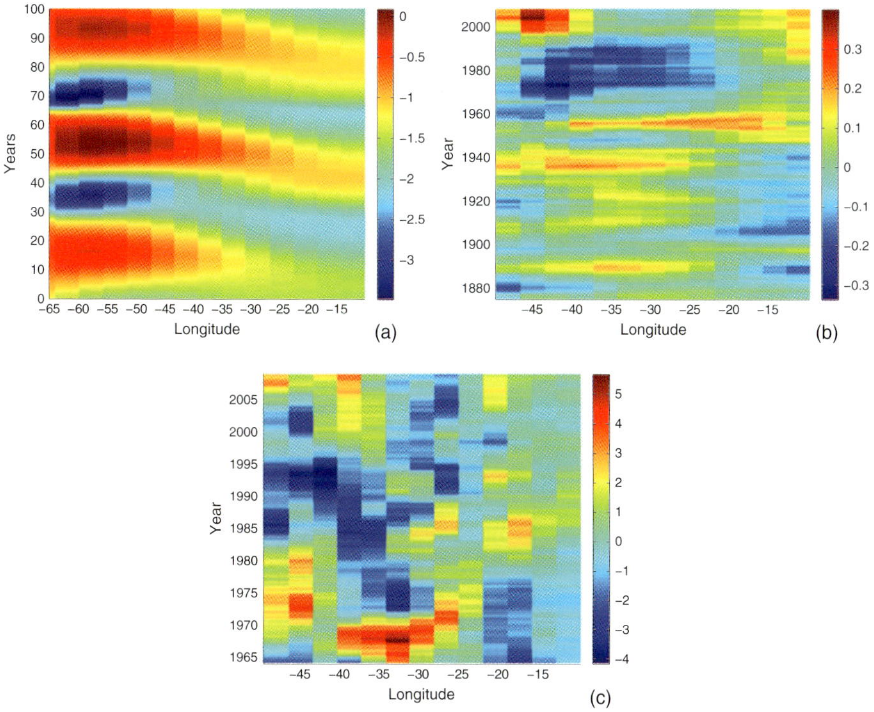

Figure 7.6 Hovmöller diagram of the 30–60°N averaged North Atlantic SST anomalies from (a) the idealized ocean model simulations; (b) eight-year running mean HadISST dataset (1870–2012). (c) the same as (b) but first the surface heat flux contribution (from the ORA-S3 data set, see http://apdrc.soest.hawaii.edu) to SST is subtracted from the HadISST data set (only over the period 1959–2009). Source: Feng and Dijkstra (2014).

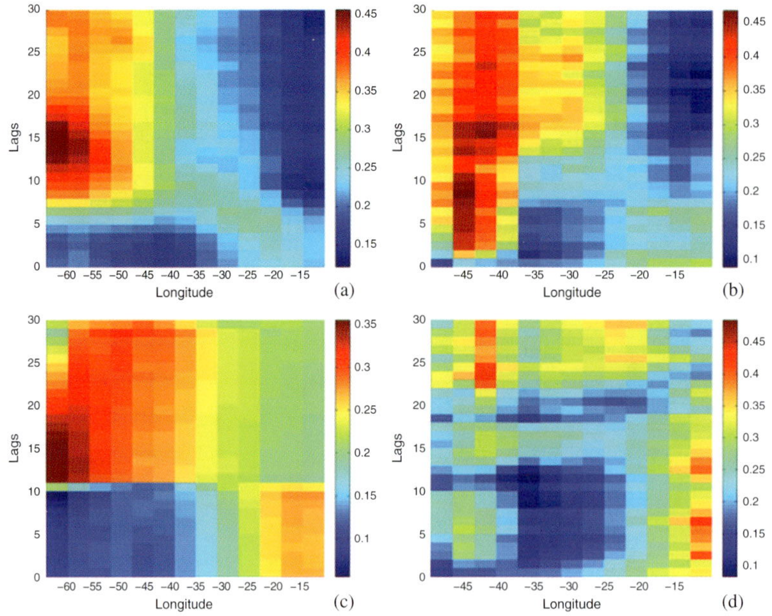

Figure 7.7 Hovmöller diagram of the 30–60°N averaged relative degree fields of (a) the idealized model data, (b) the observations, (c) the idealized model data, and (d) the observations. The networks of (a) and (b) are constructed by using mutual information under a link density of 0.2. The networks of (c) and (d) are constructed by using Pearson correlation also under a link density of 0.2. The time lag τ used to reconstruct the networks is in the vertical axes. Source: Feng and Dijkstra (2014).

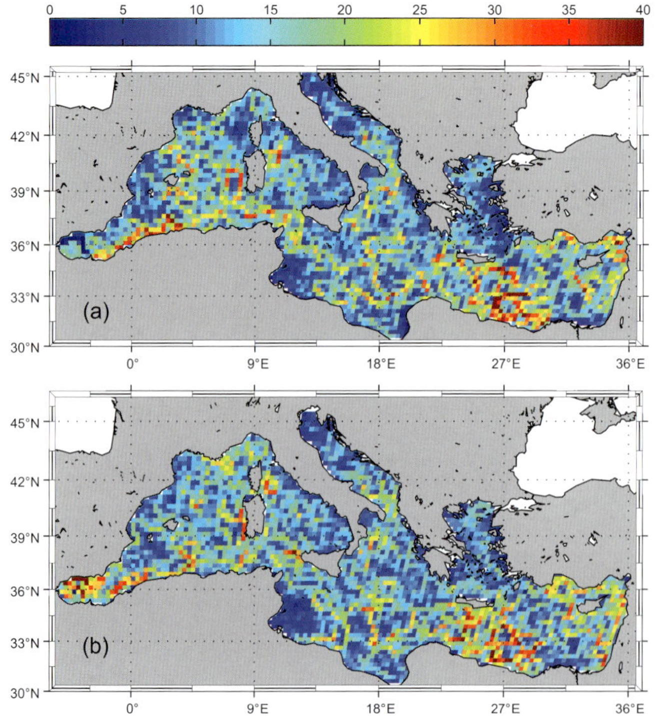

Figure 7.9 Degree of the nodes in the flow network defined by $\mathbf{P}(t_0, \tau)$, for t_0 = July 1, 2011 and $\tau = 15$ days. (a) The in-degree $K_I(i)$. (b) The out-degree $K_O(i)$. Source: Ser-Giacomi et al. (2015b).

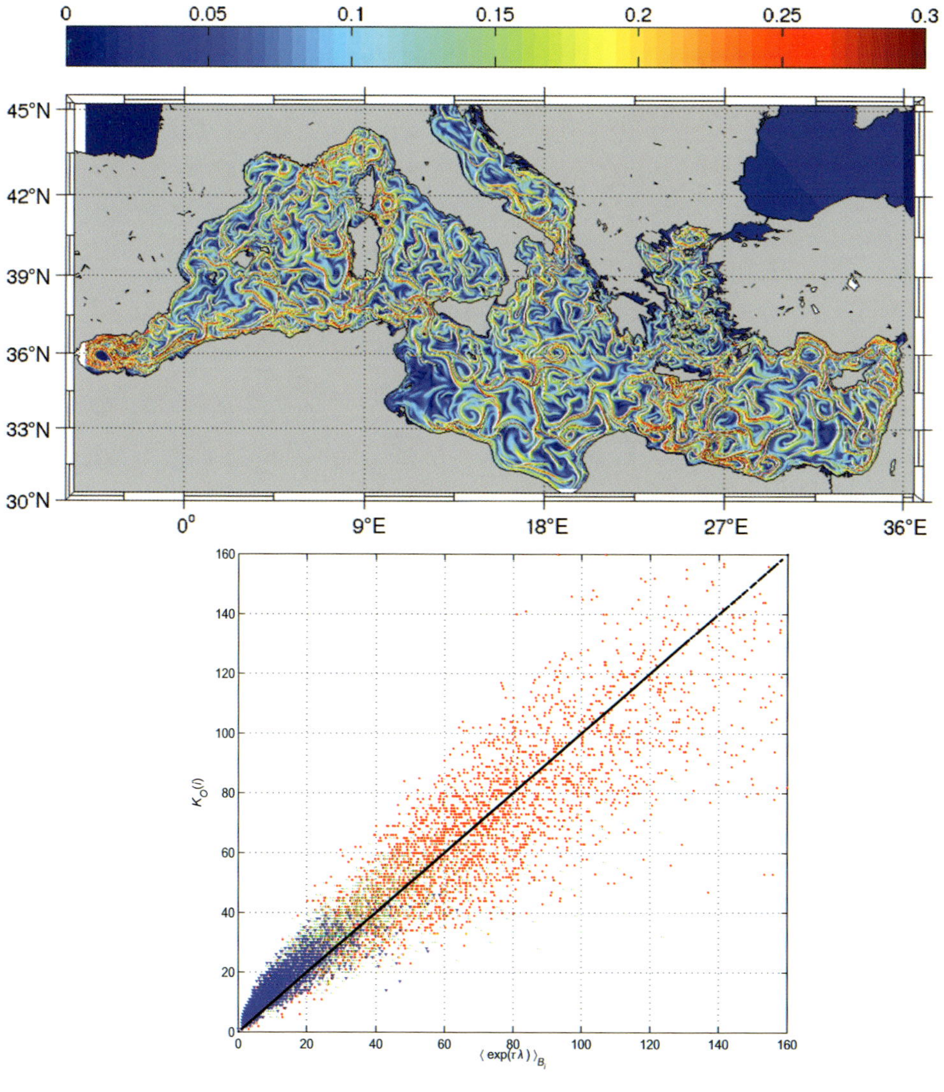

Figure 7.10 (a) An example of forward FTLE field $\lambda(\mathbf{x}_0, t_0, \tau)$ at t_0 = July 1, 2011, and $\tau = 15$ days. Color bar in day^{-1}. (b) Values of the out-degree $K_O(i)$ of each node i versus the average value of the stretching factor $e^{\tau\lambda}$ in that node. t_0 = July 1, 2011. Blue symbols are from $\tau = 15$ days, green from $\tau = 30$ days, and red from $\tau = 60$ days. Black line is the main diagonal. Source: Ser-Giacomi et al. (2015b).

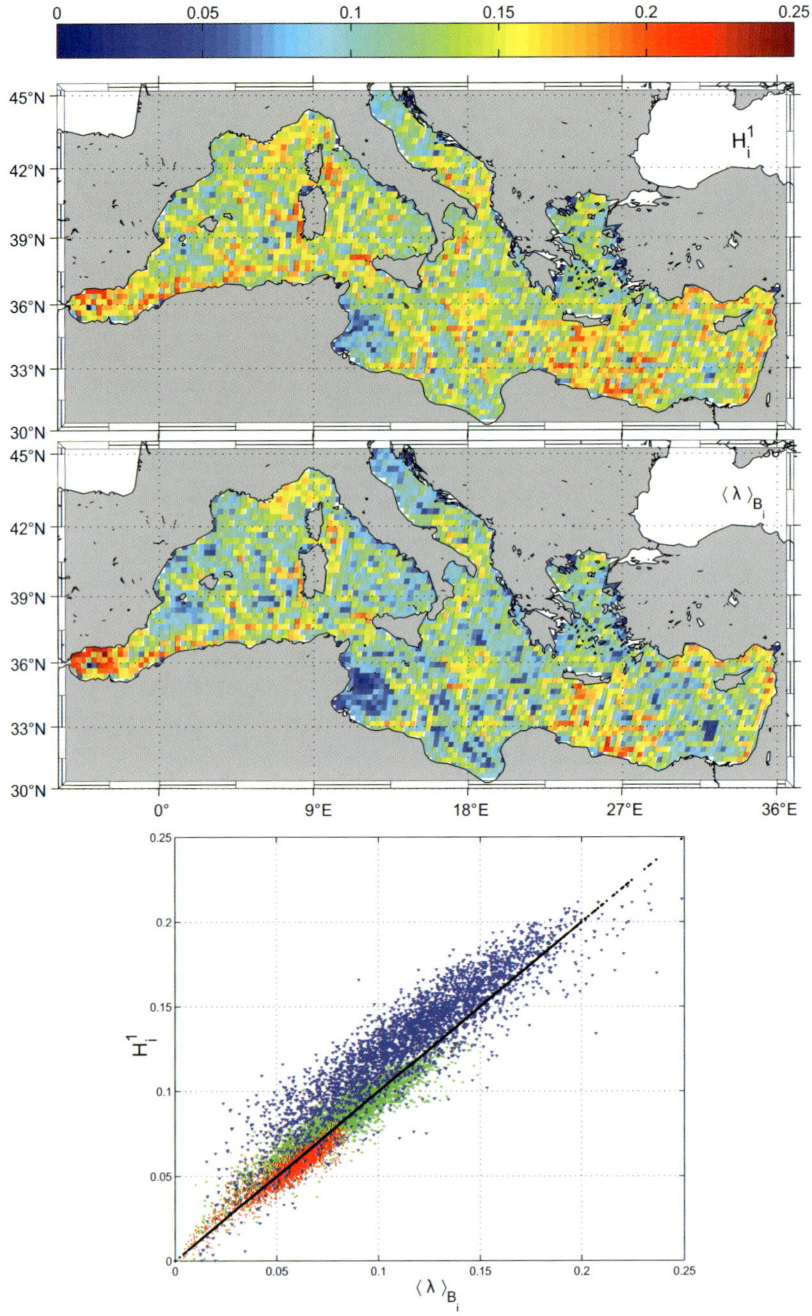

Figure 7.11 (a) The network entropy $H_i^1(t_0, \tau)$. (b) Coarse-graining of the Lyapunov field in Fig. 7.10a into the discretization boxes: $\lambda_i(t_0, \tau) \equiv \langle \lambda(\mathbf{x}_0, t_0, \tau) \rangle_{B_i}$. In both cases, data from the network constructed for t_0 = July 1, 2011, and $\tau = 15$. Color bar in day^{-1}. (c) Values of the network entropy $H_i^1(t_0, \tau)$ of each node i versus the average value of the Lyapunov exponent in that node, $\lambda_i(t_0, \tau)$. t_0 = July 1, 2011. Blue symbols are from $\tau = 15$ days, green from $\tau = 30$ days, and red from $\tau = 60$ days. Black line is the main diagonal. Source: Ser-Giacomi et al. (2015b).

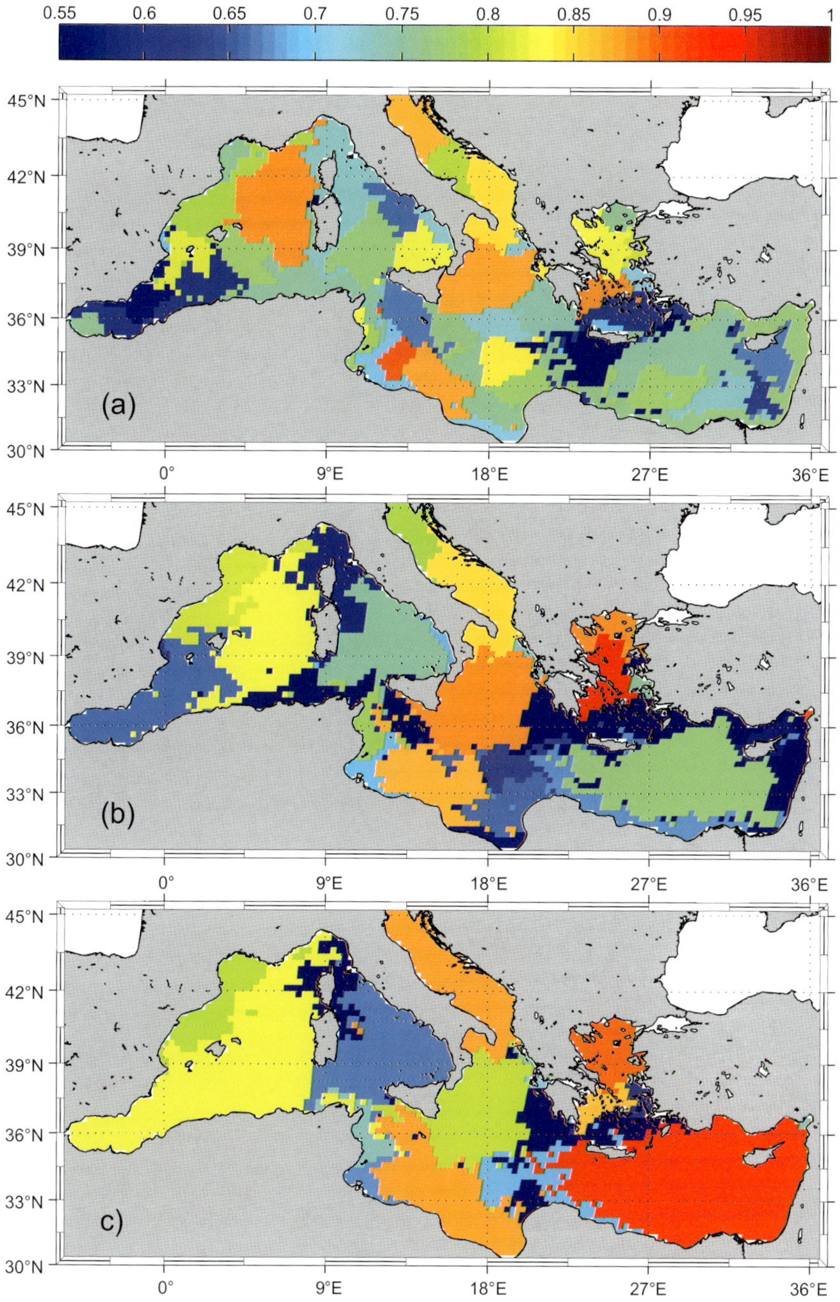

Figure 7.12 *Infomap* partition of flow networks in the Mediterranean Sea, defined by $\mathbf{P}(t_0, \tau)$, into communities or provinces for increasing values of τ. Each province is colored by its coherence ratio value from (4.27), as given in the color bar. In all panels t_0 = July 1, 2011. (a) $\tau = 30$ days; the number of communities is $p = 56$, the global coherence (see Section 4.4.3) $\rho_{t_0}^{\tau}(\mathcal{P}) = 0.76$, and the global mixing $M_{t_0}^{\tau}(\mathcal{P}) = 0.47$. (b) $\tau = 60$ days; $p = 33$, $\rho_{t_0}^{\tau}(\mathcal{P}) = 0.73$, $M_{t_0}^{\tau}(\mathcal{P}) = 0.54$. (c) $\tau = 90$ days; $p = 22$, $\rho_{t_0}^{\tau}(\mathcal{P}) = 0.80$, $M_{t_0}^{\tau}(\mathcal{P}) = 0.59$. Source: Rossi et al. (2014).

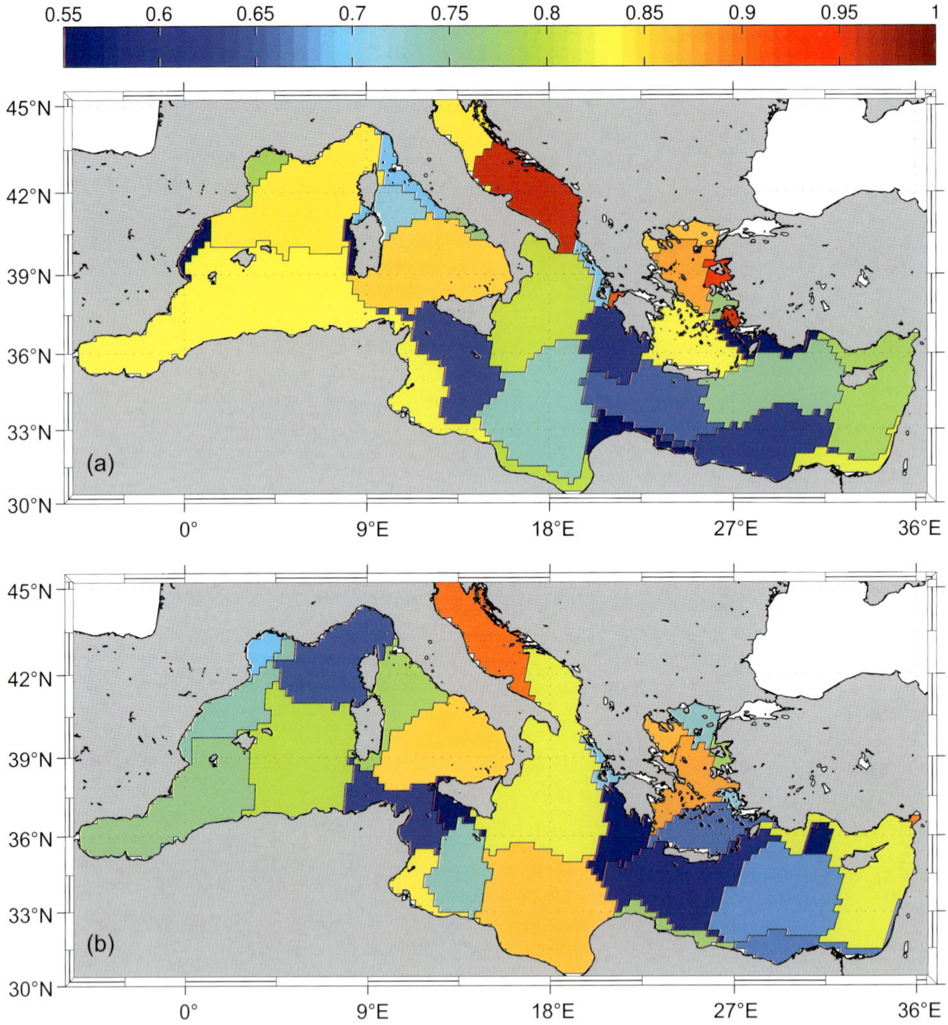

Figure 7.13 *Infomap* communities obtained from the average networks given by $\overline{\mathbf{P}(t_0, \tau)}$, with $\tau = 30$ days. Each community is colored by its coherence ratio. (a) The average is over the ten matrices corresponding to t_0 = January 1 in ten years (2002–2011) of simulation; the number of communities is $p = 34$, the global coherence $\rho^{\tau}_{t_0}(\mathcal{P}) = 0.78$, and the global mixing $M^{\tau}_{t_0}(\mathcal{P}) = 0.68$. (b) The average is over the ten matrices corresponding to t_0 = July 1 in the ten years 2002–2011; $p = 30$, $\rho^{\tau}_{t_0}(\mathcal{P}) = 0.77$, $M^{\tau}_{t_0}(\mathcal{P}) = 0.69$. Source: Rossi et al. (2014).

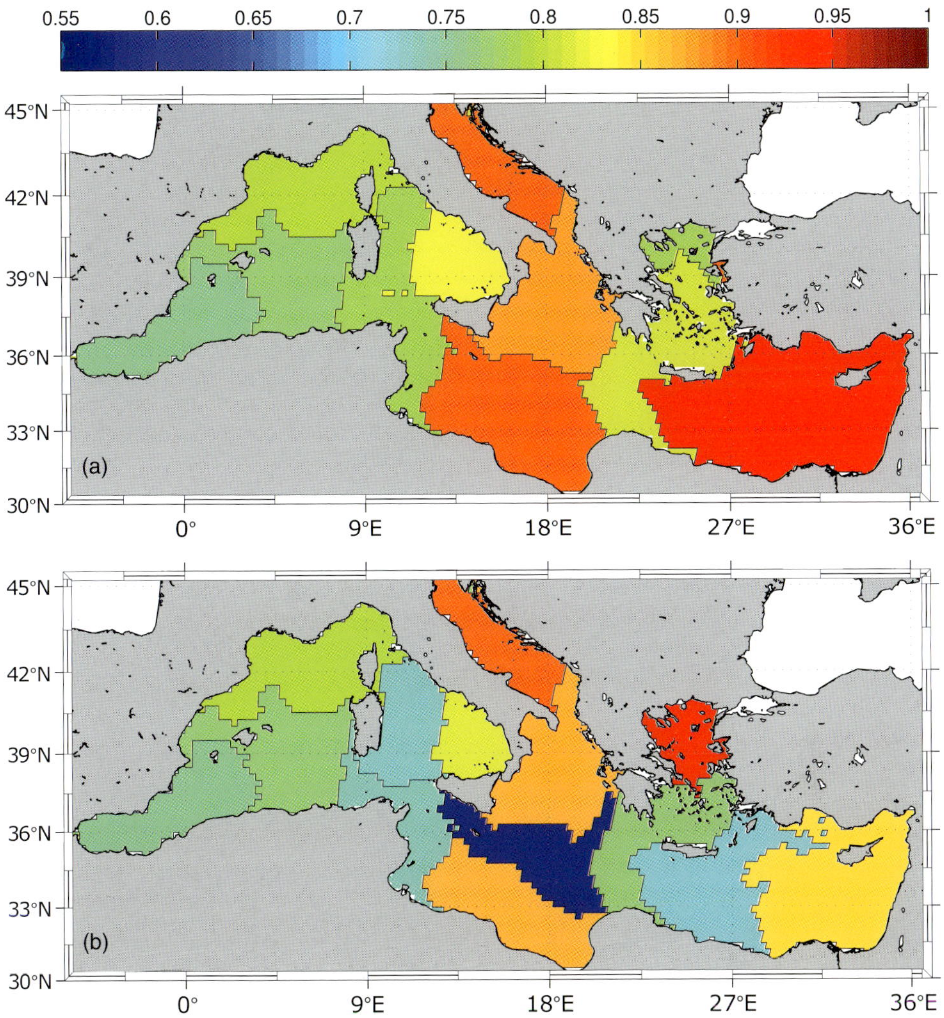

Figure 7.15 Community decomposition by the spectral method with fuzzy c-means clustering described by Froyland and Dellnitz (2003). The matrix used is the same average $\overline{\mathbf{P}(t_0, \tau)}$ as in Fig. 7.13b, i.e., with t_0 = July 1, averaged over the ten years 2002–2011, and τ = 30 days. Ten eigenvalues are used. (a) The number of communities is fixed to be $p = 10$; the global coherence is $\rho_{t_0}^{\tau}(\mathcal{P}) = 0.85$; and the global mixing is $M_{t_0}^{\tau}(\mathcal{P}) = 0.62$. In the Aegean, the southern yellow community is the only independent one: The portions of the Aegean further north are clustered by the c-means algorithm as being part of the same province as areas in the central Mediterranean with the same color. (b) $p = 14$; $\rho_{t_0}^{\tau}(\mathcal{P}) = 0.78$, $M_{t_0}^{\tau}(\mathcal{P}) = 0.64$. Source: Rossi et al. (2014).

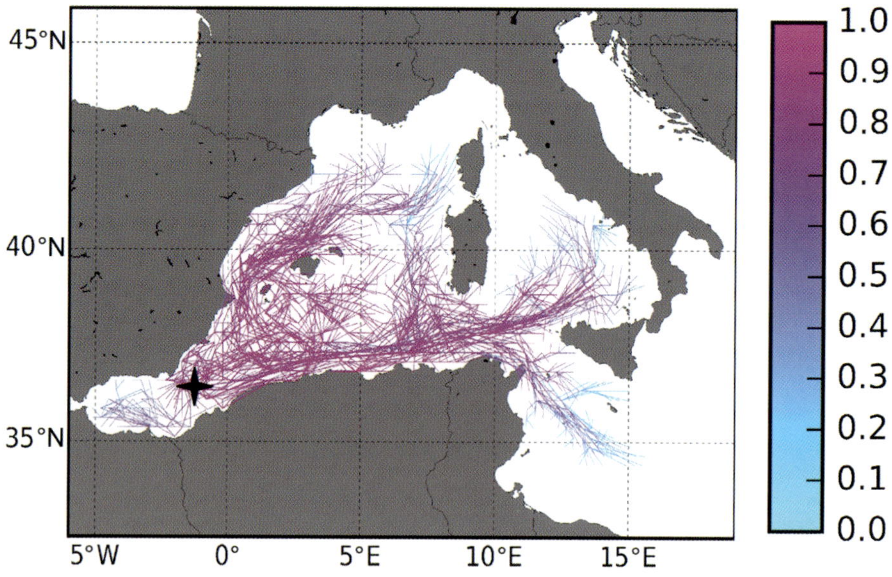

Figure 7.16 Paths of $M = 9$ steps (three months) in the Mediterranean flow network with starting date January 1, 2011, represented as straight segments joining the path nodes. Most probable paths (MPPs) originating from a single node (black star) and ending in all accessible nodes. Color gives the P_{IJ}^{M} value of the paths in a normalized log-scale between the minimum value (10^{-15}, light turquoise) and the maximum (10^{-5}, dark pink).

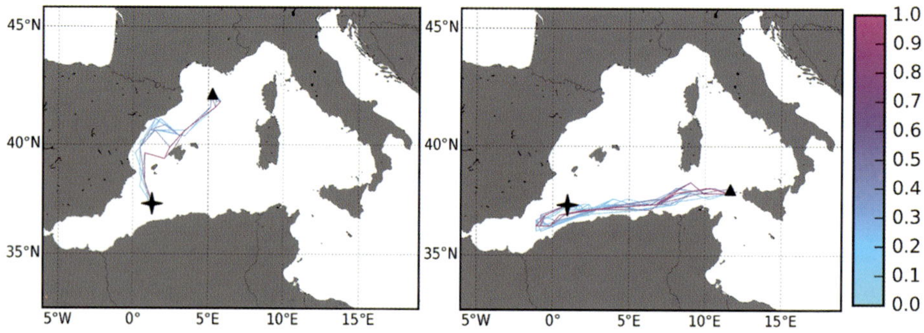

Figure 7.19 All the paths of $M = 9$ steps (three months) in the $\mathcal{K}_{IJ}^{M}(\epsilon)$ set with $\epsilon = 0.1$, represented as straight segments joining the path nodes, initial point marked by a cross and final point marked by a triangle. Starting date January 1, 2011. (a) The 18 HPPs, out of a total of 54/,276 paths between the two sites. The MPP, with $P_{IJ}^{M} = 3 \times 10^{-9}$, is displayed in dark red, whereas the other paths are colored with a normalized logarithmic scale according to their $(p_{IJ}^{M})_{\mu}$ values in $[\epsilon P_{IJ}^{M}, P_{IJ}^{M}]$. (b) shows the 39 HPPs, out of a total of 61×10^{6}, in a similar logarithmic scale normalized in $[\epsilon P_{IJ}^{M}, P_{IJ}^{M}]$ with $P_{IJ}^{M} = 1.4 \times 10^{-6}$. Source: Ser-Giacomi et al. (2015c).

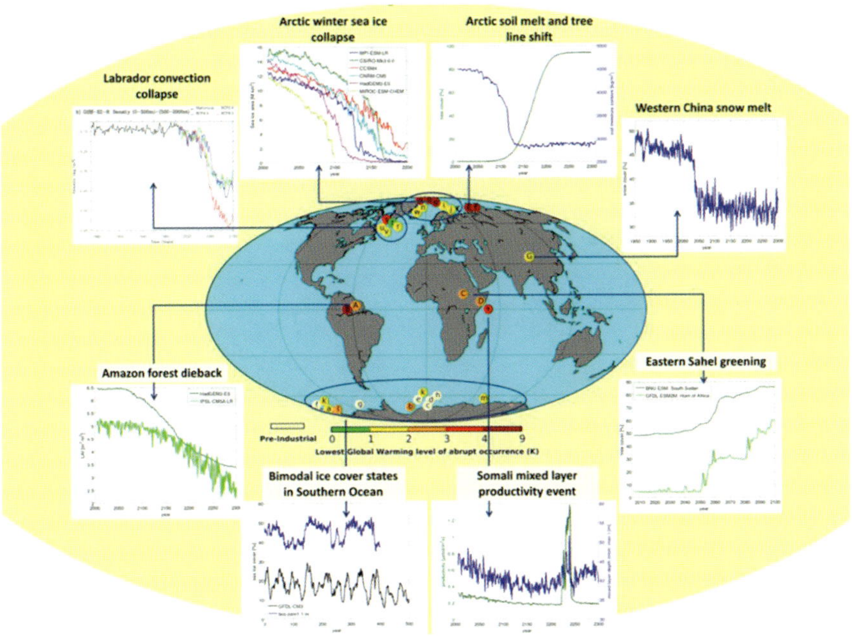

Figure 8.2 Abrupt shifts in complex climate model simulations. Source: Bathiany et al. (2016).

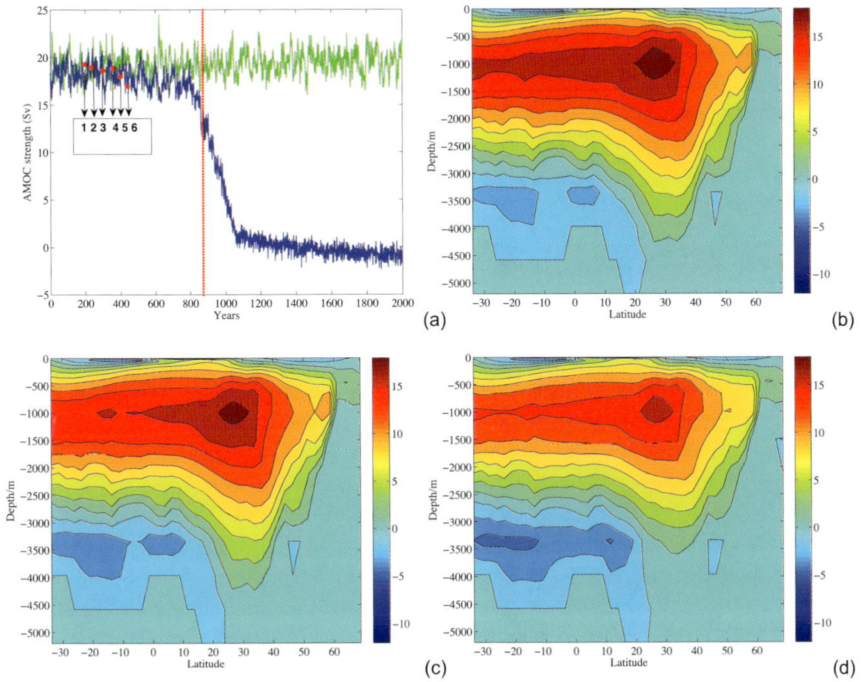

Figure 8.4 (a) Time series of the MOC (in Sv, 1 Sv = 10^6 m^3s^{-1}) at 26°N and 1000 m depth in the Atlantic for the control simulation (green curve) and the freshwater perturbed simulation (blue curve) of the FAMOUS model. The red dots labeled from 1 to 6 show the average values of MOC from the corresponding equilibrium simulations and the red broken line indicates the collapse time $\tau_c = 874$ years. (b) Annual mean MOC streamfunction pattern of equilibrium simulation 1. (c) Same as (b) but of equilibrium simulation 4. (d) Same as (b) but of equilibrium simulation 6. Source: Feng et al. (2014).

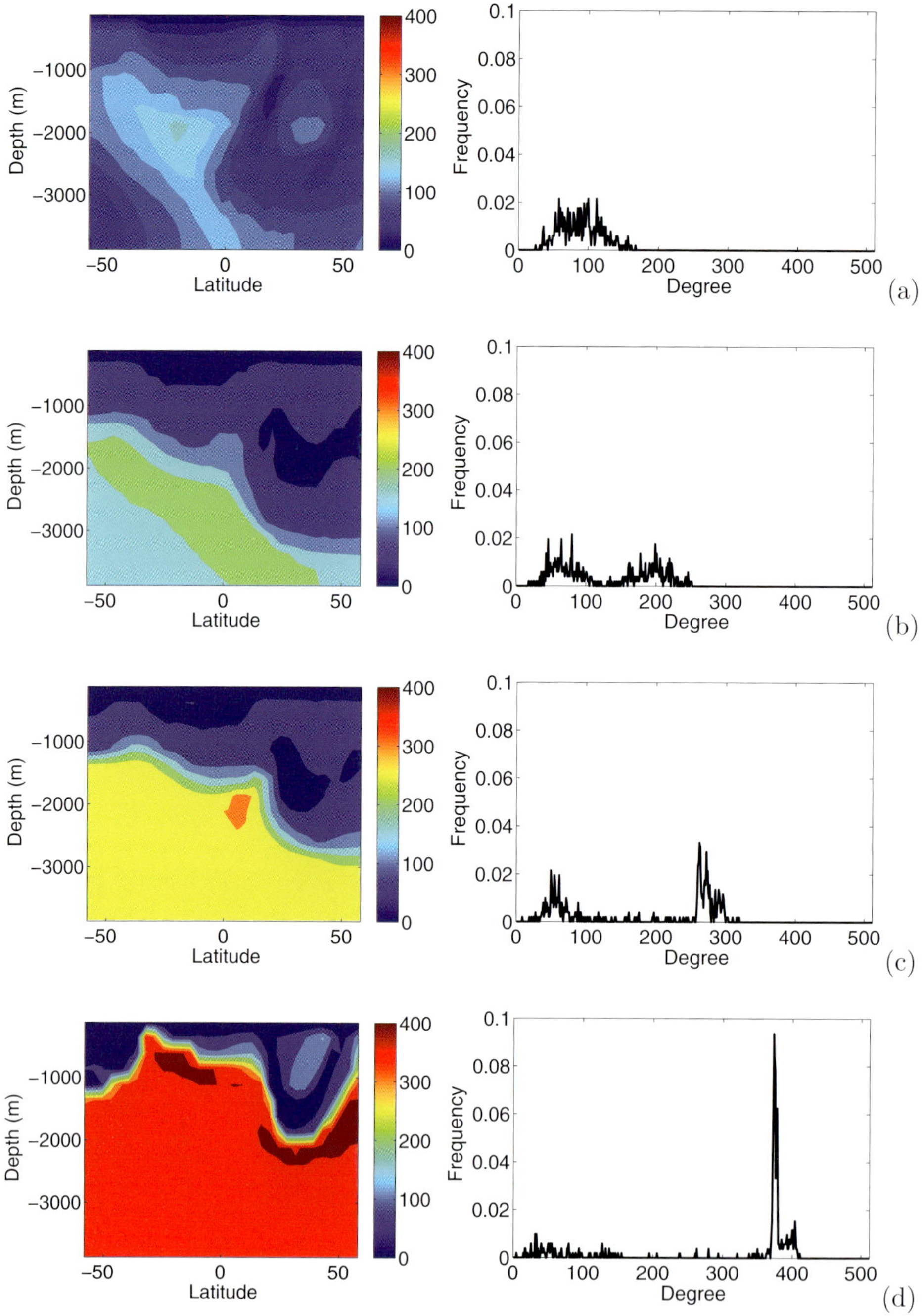

Figure 8.5 Degree fields (left column) and degree distributions (right column) of the PCCNs (with a threshold $\tau = 0.7$) for the temperature field time series obtained at different locations A, B, C, and D, as indicated in Fig. 8.3a under transient noise ($\eta = 0.1$) forcing and $\gamma = 0.3$. (a) Point A: $\beta = 0.05$ m/year; (b) point B: $\beta = 0.139$ m/year; (c) point C: $\beta = 0.154$ m/year; (d) point D: $\beta = 0.166$ m/year. Source: van der Mheen et al. (2013).

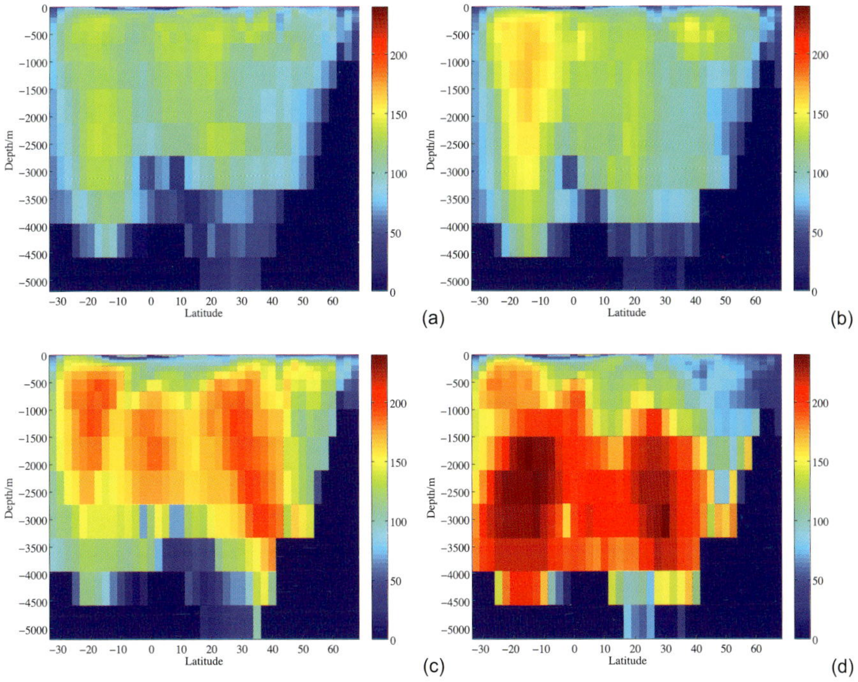

Figure 8.8 (a) Degree field of the PCCN reconstructed from the MOC streamfunction data of equilibrium simulation 1 in Fig. 8.4 using a threshold of 0.5. (b) Same as (a) but of equilibrium simulation 2. (c) Same as (a) but of equilibrium simulation 4. (d) Same as (a) of equilibrium simulation 6. Source: Feng et al. (2014).

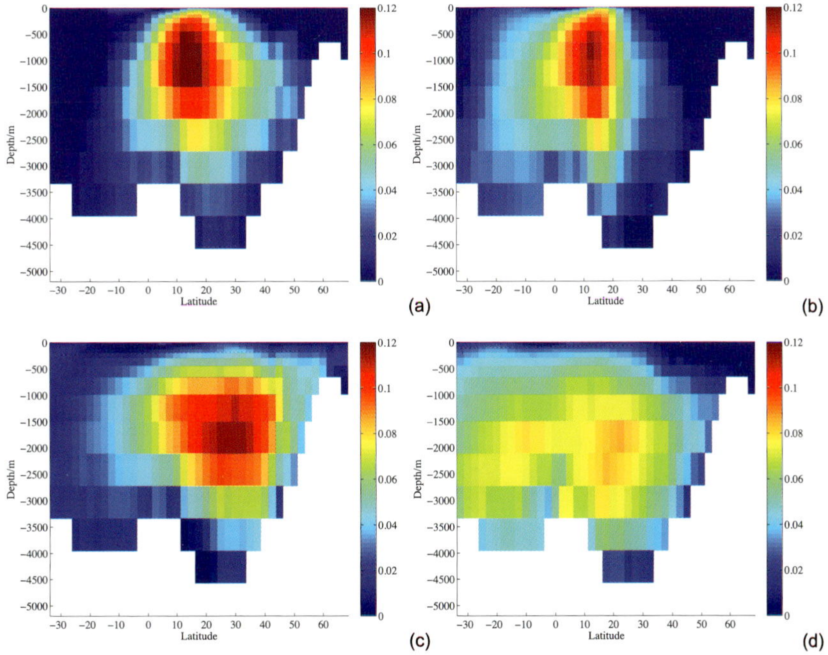

Figure 8.9 (a) The first EOF of the MOC data of equilibrium simulation 1, explaining 26.3% of the variance. (b) Same as (a) but of equilibrium simulation 2, explaining 24.1% of the variance. (c) Same as (a) but of equilibrium simulation 4, explaining 37.1% of the variance. (d) Same as (a) but of equilibrium simulation 6, explaining 37.6% of the variance. Source: Feng et al. (2014).

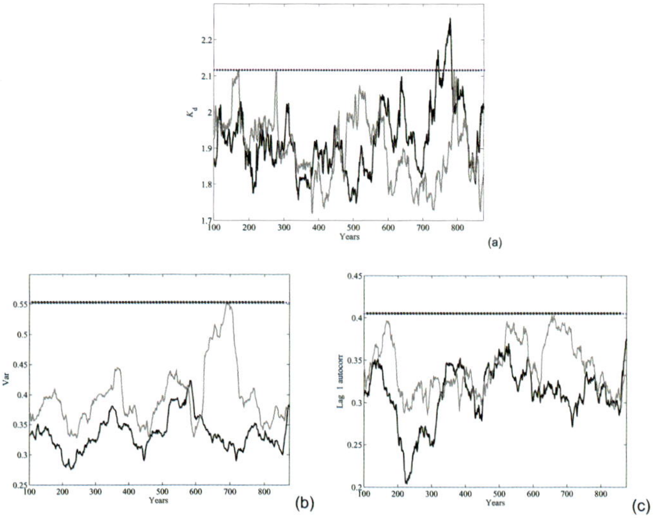

Figure 8.10 Warning indicators for the complete Atlantic MOC field. Grey curves are for the control simulation and black curves are for the hosing simulation. The dashed horizontal lines indicate the corresponding maximum values of the control simulation taken over the total time interval $[0, \tau_c]$. (a) the kurtosis indicator K_d gives the early warning signal at 738 years and lasts for 44 years. (b) The traditional variance indicator Var gives no early warning signal before the collapse time $\tau_c = 874$ years. (c) The traditional lag-1 auto-correlation indicator gives no early warning signal before the collapse time τ_c. Source: Feng et al. (2014).

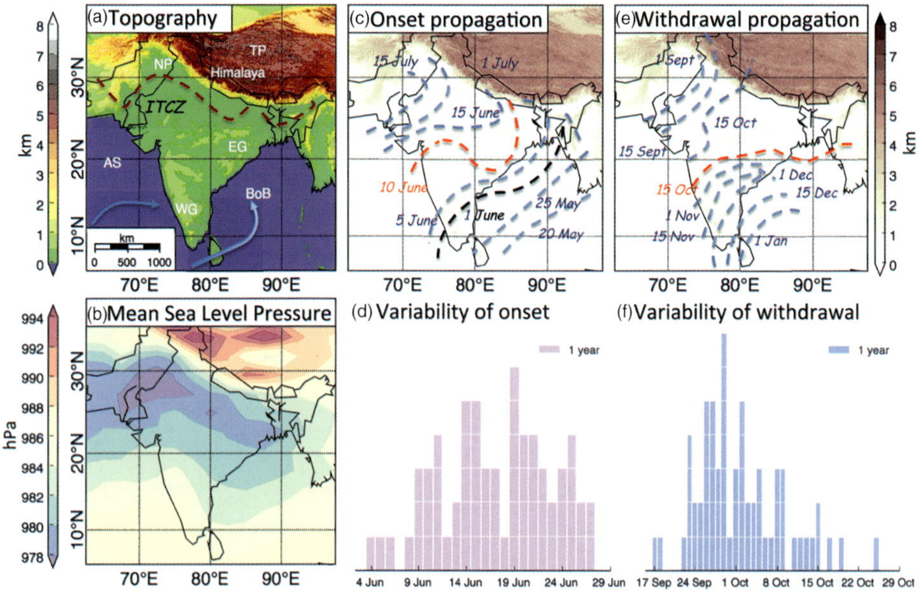

Figure 9.1 (a) Topography of the Indian subcontinent with key features of the Indian Summer Monsoon: Himalaya, Tibetan Plateau (TP), North Pakistan (NP), Eastern Ghats (EG), Western Ghats (WG), Arabian Sea (AS), Bay of Bengal (BoB), Intertropical Convergence Zone (ITCZ). Blue arrows indicate near-sea-level wind direction. (b) Composites of mean sea-level pressure (June to September, 1951–2014 based on NCEP/NCAR data). (c) and (e) Schematic representation of the long-term average propagation (since 1951) of the advance and withdrawal of monsoon over the Indian subcontinent (northern limit of monsoon). Dashed black line shows averaged monsoon onset for the Kerala region forecasted by the IMD, and dashed red line for the Eastern Ghats (the region of main interest in this study). (d) and (f) Histograms of onset and withdrawal dates for the Eastern Ghats region (1951–2014). Source: Stolbova et al. (2016).

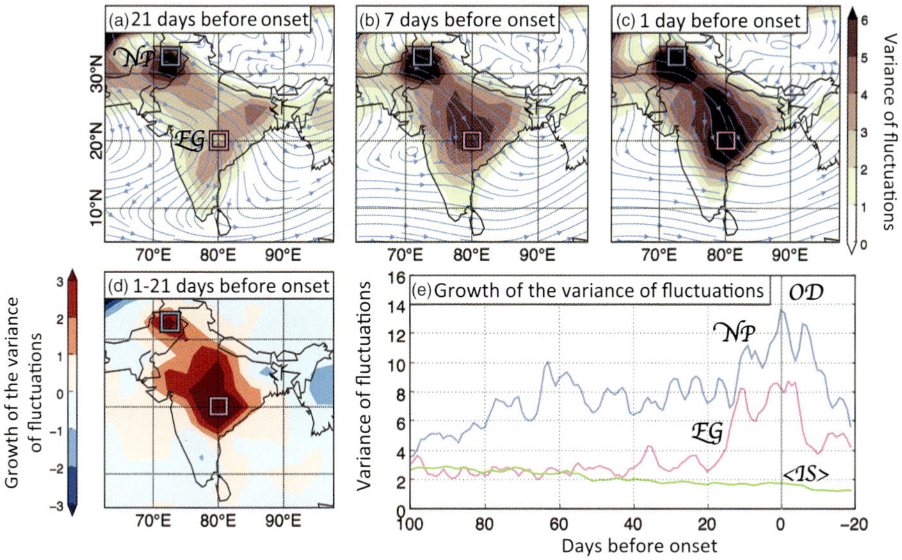

Figure 9.2 Pre-monsoon growth of the variance of fluctuations (σ^2) of the weekly mean values of near-surface air temperature (T) 21 days, 7 days, and 1 day before the monsoon onset at the Eastern Ghats (EG): (a)–(c) (d) = (c) – (a). Composites are for the period 1971–2001 and were calculated from the ERA40 reanalysis data set, 700 hPa winds are indicated by the blue lines. The two boxes refer to RPs: North Pakistan (NP, blue) and EG (pink). E, growth of the variance of fluctuations in NP (blue), EG (pink), and averaged over the Indian subcontinent (<IS>) at approaching OD of the Indian monsoon. OD and WD of the monsoon are taken from Singh and Ranade (2010) and data from the IMD. Source: Stolbova et al. (2016).

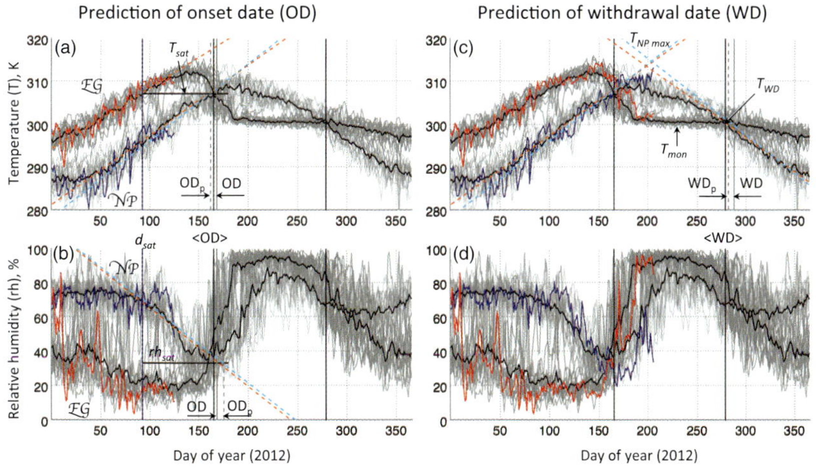

Figure 9.3 Prediction of OD and WD of Indian Monsoon: case study 2012. Left column: prediction of the OD; right column: WD in the Eastern Ghats (EG). (a), (c): air temperature at 1000 hPa; (b), (d) relative humidity at 1000 hPa. Time series from reference points: 14-year mean (black) and 2012 values for North Pakistan (NP) (blue) and the EG (red). Gray lines show time series from the NP and EG for the training period of 14 years. Saturation temperature T_{sat} (a) and saturation humidity rh_{sat} (c) are marked by horizontal black solid lines ($T_{sat} = T_{onset}$, T_{onset} and rh_{sat} calculated as intersection of mean time series for the training period from the EG and NP), and day of the saturation (d_{sat}) (when temperature in the EG in 2012 reaches T_{sat}) with dark blue. The orange line indicate trends to the mean time series in the NP and EG for the training period, light blue is trends for 2012. Black solid lines indicate mean values of the OD (<OD>) and WD (<WD>) for the training period. Dotted gray lines correspond to the predicted onset (OD_p) and withdrawal dates (WD_p), while solid gray lines are actual onset and withdrawal dates for 2012. Source: Stolbova et al. (2016).

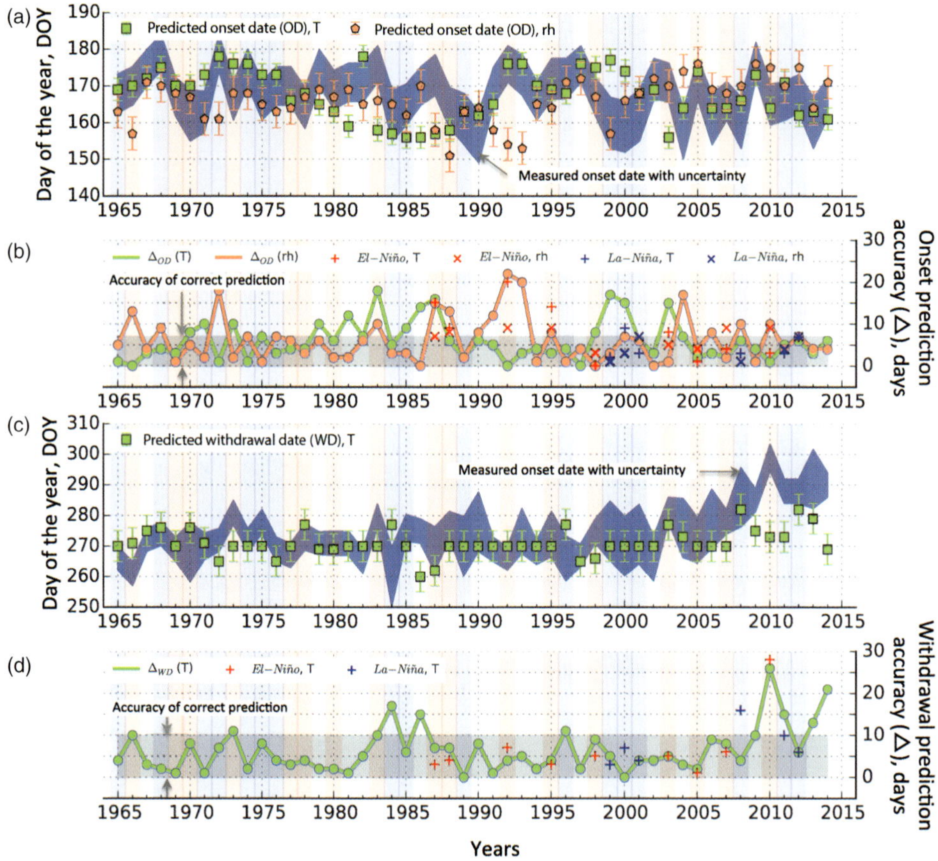

Figure 9.4 Monsoon OD and WD prediction based on T (green) and rh (orange) and measured (dark blue) OD (a) and WD (c). Red and light-blue shading indicate El Niño and La Niña years. Also shown is the difference (Δ) between the real onset or withdrawal and predicted dates in days ($\Delta_{OD(T)}$: prediction based on temperature, $\Delta_{OD(rh)}$: on relative humidity). Gray shading indicates a range of seven days, within the prediction this is considered accurate (b). The accuracy of prediction of the WD has a range of ten days (gray shadow) (d). Markers "+" (T) and "x" (rh) show improved prediction based on the training period of 14 years only from preceding El Niño (red) and La Niña years (b and d). Source: Stolbova et al. (2016).

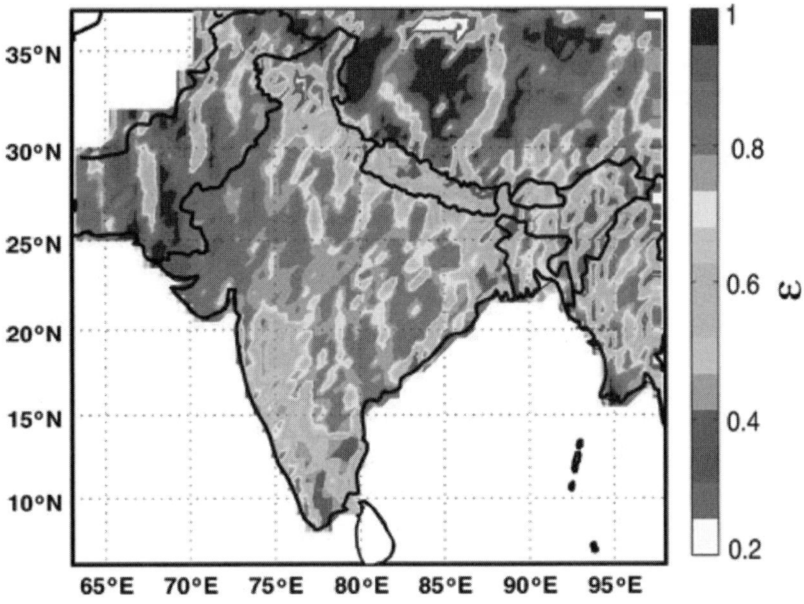

Figure 6.16 The accuracy of prediction ϵ on a scale from 0 to 1. Source: Malik et al. (2010). (A black and white version of this figure appears in some formats. For the color version, please refer to the plate section.)

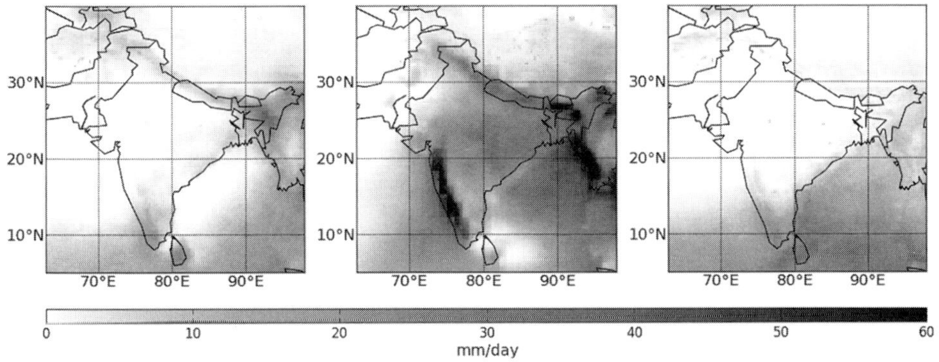

Figure 6.17 The 90th percentile of daily rainfall amounts for pre-monsoon (MAM), summer (JJAS), and post-monsoon (OND) periods for the TRMM data set. Source: Stolbova et al. (2014).

The 90th-percentile of the rainfall amount for all three time periods, which was used as a threshold for the extreme rainfall definition, is shown in Fig. 6.17.

Stolbova et al. (2014) showed that the topology of the network of extreme precipitation changes significantly with seasons, and comparison of networks of the synchronized extreme precipitation for the pre-monsoon, ISM, and post-monsoon seasons can help reveal typical spatial patterns which influence extreme rainfall synchronization, specifically during the ISM: North Pakistan (NP), Eastern Ghats (EG), and Tibetan Plateau (TP).

These spatial patterns of synchronized extreme rainfall events over the Indian subcontinent, which are spatial signatures of the ISM season, occur before the monsoon, develop during the monsoon season, and disappear after the monsoon. The large number of connections and long average link length of the NP, TP, and EG regions during the ISM season implies that these regions strongly affect extreme rainfall event synchronization all over the Indian subcontinent. Also, average and maximal link lengths of these regions are significantly increased during ISM in comparison to pre-monsoon and post-monsoon periods. The detailed analysis of the climate network based on ES for signature patterns of the ISM season reveals the role of these patterns in the organization of the extreme rainfall over the Indian subcontinent, and comparison of the links of the dominant patterns with wind fields allows us to establish a linkage to atmospheric processes (see Fig. 6.18).

Specifically, two of the determined patterns – EG and TP – were previously known as areas that influence the ISM dynamics due to the intricate topographies of these regions. The complex network approach shows that NP pattern also plays an important role in the extreme rainfall organization during the ISM, because it is strongly influenced by western disturbances and may serve as a key region to infer interaction between the ISM system and western disturbances.

The prediction of the ISM rainfall intensity and the prediction of the extreme rainfall events is vital for the population of the Indian subcontinent, as well as in areas around the globe that are indirectly affected by the ISM strength. Spatial domains that are used for the prediction of the ISM strength are either too large, or, probably, do not include all necessary information for the ISM strength forecasting. Recently, Stolbova et al. (2016) developed an approach for the prediction of onset and withdrawal of the ISM based on the patterns described above. The method allowed them to predict onset two weeks earlier and withdrawal dates more than a month earlier than existing methods. Moreover, it correctly forecast the monsoon duration for some anomalous years, often associated with ENSO.

6.7 South American Monsoon

6.7.1 Air–Sea Interaction in the South Atlantic Convergence Zone

Tirabassi et al. (2015a) used the complex network approach to study one of the principal actors of the South American Monsoon system: the South Atlantic

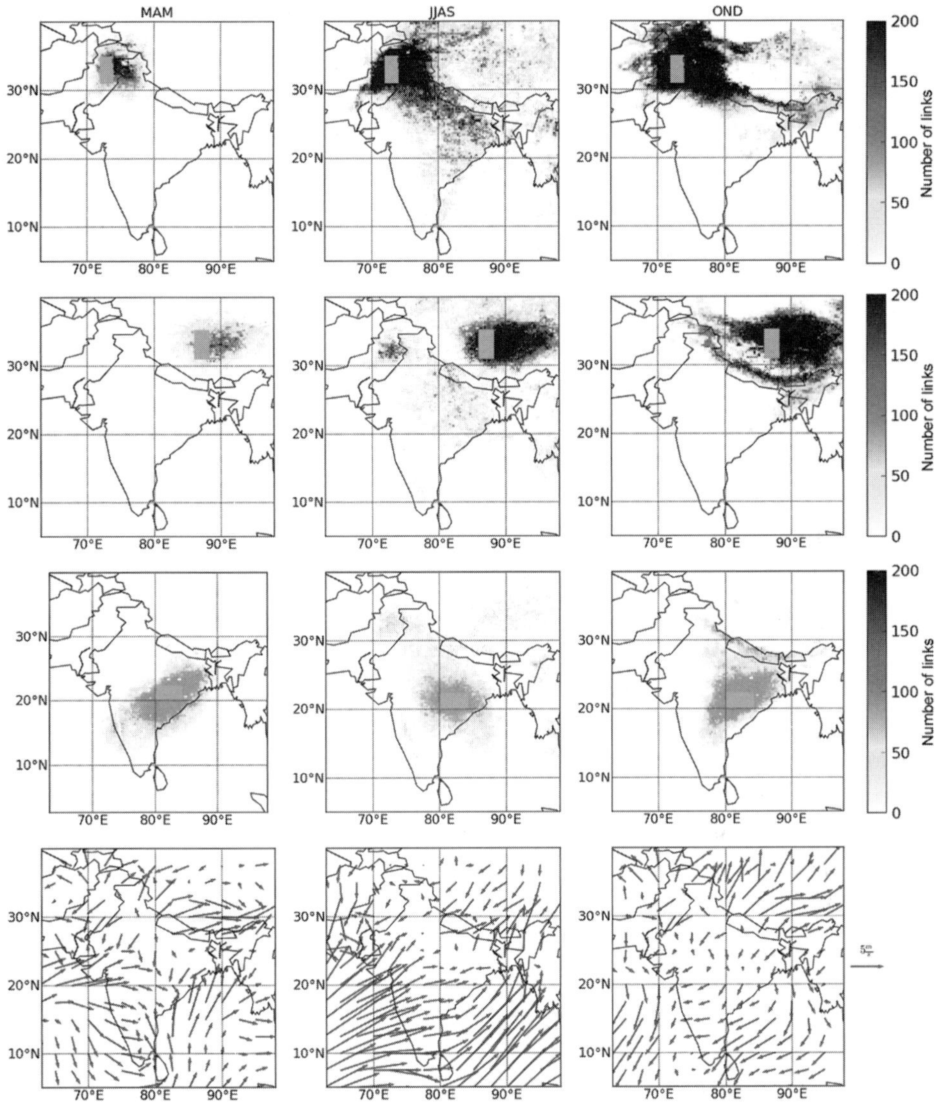

Figure 6.18 Links between a set of 153 reference grid points to other grid points for pre-monsoon (MAM), summer (JJAS), and post-monsoon (OND) seasons. From top to bottom: North Pakistan (NP), Tibetan Plateau (TP), Eastern Ghats (EG) (TRMM). The bottom panels display the mean surface winds between 1982 and 2012 for the different seasons (NCEP/NCAR). Source: Stolbova et al. (2014). (A black and white version of this figure appears in some formats. For the color version, please refer to the plate section.)

Convergence Zone (SACZ). In particular, they focused on the air–sea interaction and how the regional ocean can influence the SACZ variability on short timescales.

As explained in Section 2.5 the SACZ is a convective pattern that extends from the Amazon rainforest to the subtropical South Atlantic ocean, oriented in a northwest–southeast direction. A relevant question is to determine the local ocean influence on SACZ variability. Understanding how much the ocean affects the SACZ dynamics can allow improving precipitation prediction over very populated areas of South America, such as the Brazilian regions of Minas Gerais, Sao Paulo, and Rio de Janeiro.

To investigate the relationship between ocean and precipitation dynamics, two variables are needed to describe the state of the two systems. As a proxy for SACZ precipitation, Tirabassi et al. (2015a) used the vertical velocity ω at 500 hPa, while for the state of the ocean the natural choice was the SST. The study of Tirabassi et al. (2015a) focused on austral summer and, as usual, the seasonal cycle was removed prior to analysis.

Using daily mean data from ERA Interim reanalysis, Tirabassi et al. (2015a) calculated the Granger causality estimator (GCE, see Section 3.3.4) between ω and SST time series at each geographical location, both in the "upwards" (SST$\rightarrow$ ω) and "downwards" (ω $\rightarrow$SST) directions. Since the data, in each grid point, are not formed by a unique long time series, but rather by a collection of short series of summer records, the GCE must be computed via a multi-trial Granger causality estimation.

Results of this procedure are presented in Fig. 6.19. It can be observed that the main two regions in which the ocean forces the atmosphere are the deep tropics and the subtropical waters off Brazil. The deep tropics is expected to be a region of oceanic forcing, and the methodology consistently captures this fact. The ocean off southeast Brazil corresponds to the SACZ region and is within our region of interest. Moreover, Fig. 6.19a clearly shows that the atmosphere strongly forces the surface ocean also off Brazil. Thus, this ocean region can play both an active and a passive role, suggesting the presence of feedback loops.

This analysis captures the *local* features of the two-way interaction among the two systems. In order to get a more complete picture of the interaction, a *directed bipartite network* is constructed.

As a first step in this direction, the cross-GC is computed, that is, the GC is computed for every couple of ocean–atmosphere nodes in both directions. This allows investigating also the presence of non-local forcing. Once we have computed all the possible pairs of values we are left with two square matrices, one for the "upward" GC and one for the "downward" GC.

At this point the two matrices are pruned using an F-test, keeping only the GC values significant at 99% confidence level. Significant values are set to 1 and

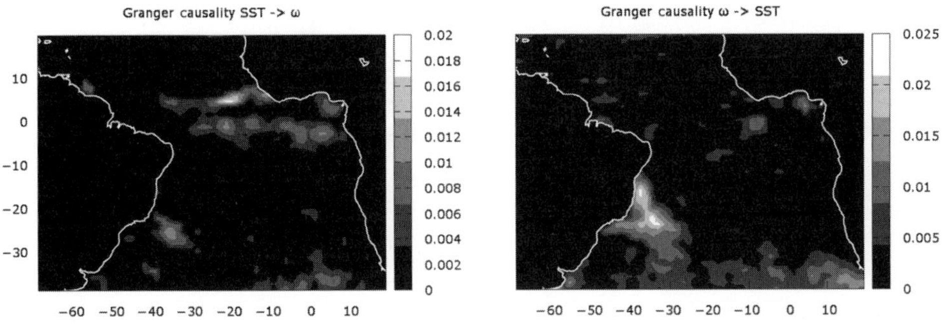

Figure 6.19 Local Granger causality among daily ω at 500 mb and SST during summer. Source: Tirabassi et al. (2015a).

all others to 0, and the results are two binary adjacency matrices, one describing "upward links" (that is, links going from the ocean to the atmosphere), and one describing "downward links" (that is, links going from the atmosphere to the ocean). This constitutes a *directed bipartite network*, that is, a network composed of two distinct sets of nodes in which the links are directed and can only take place among nodes of two different sets.

As in any network, in a bipartite network it is also possible to calculate the AWC. Since the network is directed, we can both consider the outward links and the inward links when computing the AWC. Here the analysis is restricted to outward links only, as we want to understand the mastering action of one system onto the other. Thus, the outward AWC is computed separately for the two adjacency matrices, obtaining the results depicted in Fig. 6.20.

It is worth noting that in the construction of such a network the information about the actual values of GC are lost, that is, it is not possible to know which areas have a strong coupling and which have a weak one. In the present case it could be expected that the tropical interaction would be mainly local (vertical) and strong, while the extratropical interaction should be weak and spread, and this is what we find in the AWC maps. In fact, as can be seen in Fig. 6.20a, which reports the AWC values for the SST$\rightarrow$ ω links, the strongest signal is present in the waters off the shore of southern Brazil, while a weaker one is detectable in the tropical Atlantic. Thus, while in the tropical Atlantic there are small values of AWC and high values of local GC, in the subtropical part of the ocean the situation is the opposite, with many points presenting high AWC but relatively low values of GC (compared with the tropical ones).

Regarding the AWC map of $\omega \rightarrow SST$, it is clear that the connectivity is generally large over the whole region, stressing the dominant role of atmospheric circulation in forcing SST anomalies. Moreover, there is a large, highly connected

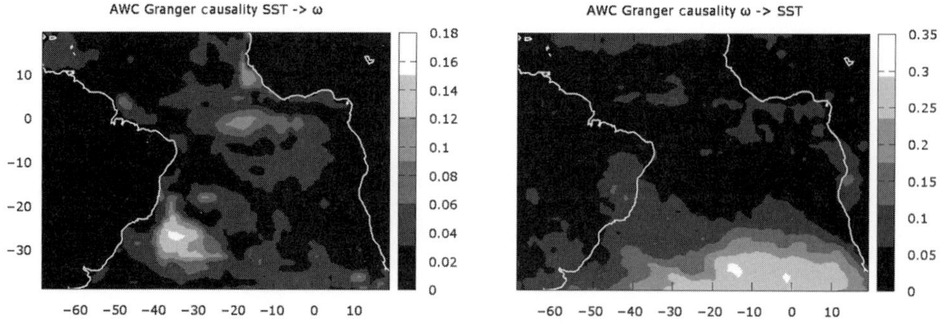

Figure 6.20 AWC of a bipartite climate network composed of ω at 500 mb and SST layer. The links are given by significant values of Granger causality. Source: Tirabassi et al. (2015a).

region south of 30°S which corresponds to a largely non-local pattern of connectivity (there is a very weak signal in the local map in Fig. 6.19). The location of the maximum suggest that it could be related to the influence of the south Atlantic anticyclone on SST through changes in latent heat fluxes. More in detail, it is possible that changes in the South Atlantic high-pressure system can affect the circulation over the subtropical South Atlantic, changing surface winds and thus temperature through evaporation. These results reveal that the SACZ region displays complex air–sea interactions, and further analysis using an alternative approach for quantifying GC in the temporal domain was done by Tirabassi et al. (2017).

6.7.2 Moisture Sources of Southeastern South America

Southeastern South America (SESA) is a region that covers Uruguay and portions of northeastern Argentina and southern Brazil. Precipitation plays a key role in the generation of hydroelectric energy and in the local agricultural economy. Moisture that could lead to rainfall over the region can come from two different sources: (1) water vapor advection from other regions, and/or (2) local recycling. The advection of water vapor in turn depends on the atmospheric circulation and can have continental or oceanic origin.

The main oceanic moisture sources over the La Plata basin (LPB) during summertime are the southwestern South Atlantic, the tropical north Atlantic, and the surrounding Atlantic Ocean located in eastern to central Brazil (Drumond et al., 2008). Martin-Gomez and Barreiro (2016, 2017) used the network methodology to study how SST anomalies in these three tropical oceans can work together to induce springtime rainfall variability over SESA in present and future climate.

They constructed a climate network considering indices of SST over the tropical oceans and of SESA precipitation as network nodes, and focused on the detection of synchronization periods during the twentieth century and how they can change under a scenario of anthropogenic forcing. Synchronization was understood as those periods in which several of the networks' component were significantly interacting with each other. Focusing on the summer season, Martin-Gomez et al. (2016) found a strong decadal variability in the impact of the oceans on SESA rainfall, which is accompanied by large changes in moisture sources.

Martin-Gomez et al. (2016) constructed the network, considering as the network's nodes the following five tropical oceanic indices: Niño3.4, Tropical North Atlantic (TNA), Tropical South Atlantic (TSA), Equatorial Atlantic (ATL3), and Indian Ocean Dipole (IOD), as well as a precipitation index over SESA (PCP SESA). The election of the indices takes into account all the tropical basins that are known to influence SESA precipitation during the austral summertime. The oceanic indices are defined considering the monthly mean SST from the Extended Reconstructed Sea Surface Temperature database (ERSSTv3b; Smith et al., 2008) with a resolution of $2^\circ \times 2^\circ$. The precipitation index is defined using the monthly mean observed data from the GPCCv5 (Global Precipitation Climatology Center; Schneider et al., 2011) with a resolution of $1^\circ \times 1^\circ$. The period of study is 1901–2005.

The methodology to construct the network (also used by Martin-Gomez and Barreiro (2016)) consists of the following steps:

1. The climate indices are calculated for individual trimesters: September to November (SON) for the case of the Nino3.4 index and December to February (DJF) for the other indices (TNA, TSA, ATL3, IOD, and PCP). The lag time of three months between Niño3.4 and the rest of the nodes was established in order to allow them to respond to the atmospheric anomalies generated by the equatorial Pacific.
2. For each year, a climate network is constructed by computing the Spearman correlation coefficient among each pair i, j of time series (see Section 3.2.1) using a sliding window of length $\Delta t = 11$ years. The mean network distance is computed as a measure of synchronization among the nodes:

$$d(t) = \frac{2}{N(N-1)} \sum_{i<j} \sqrt{2(1 - \rho_{ij}^t)} \tag{6.5}$$

   where N is the number of the network's nodes (in this case, six). The network is completely synchronized when the distance is zero and uncorrelated nodes give a value of $\sqrt{2}$ to the distance.
3. The statistical significance (see Section 3.4) of the mean network distance is calculated employing the Montecarlo method under the following criterion: A

red-noise null model is considered for those nodes with auto-correlation coefficient at lag 1 significant at the 5% level in a one-tailed t-test. In the opposite case white noise is considered. Under this assumption, only the TNA and TSA can be considered as red noise. Then, 1000 surrogate time series of each index are generated under these null hypotheses and a network distance time series is computed considering a sliding window of 11 years. In this way, 1000 surrogate time series of the mean network distance are constructed, which allows determining the 5% significance level. A statistically significant synchronization event is defined when the mean network distance is below this threshold for more than seven consecutive years.
4. A relative precipitation weight (RPW) index was defined to quantify the degree of interaction of the precipitation index (PCP) with the rest of the nodes.

The spatial distribution of the moisture sources of SESA is calculated considering a Lagrangian particle dispersion model (FLEXPART; Stohl et al. (2005), described and used in Section 6.5) driven by the six-hour forecast from Climate Forecast System Reanalysis (NCEP-CFSR; Saha et al., 2010) with a resolution of $0.5^\circ \times 0.5^\circ$ during the period 1979–2000. FLEXPART tracks the trajectories of the atmospheric moisture running forward and backward in time while dividing the atmosphere into a large number of particles. Each particle represents a mass of air with a given mass (m) which is transported by the 3D wind field which includes modeled turbulence. As in Drumond et al. (2008), Martin-Gomez et al. (2016) selected the vertical atmospheric column located over SESA, from where they release 50 000 particles per simulation, with a vertical distribution proportional to the air density. Five simulations were performed per month (DJF), releasing the particles on the following days of each month: 12nd, 16th, 20th, 24th, and 28th. All these particles are transported by FLEXPART backward in time for ten days and tracked, recording their positions and specific humidity every six hours. Then, the net budget evaporation (e) minus precipitation (p) of each particle i with mass m is computed through changes in the specific humidity (q) along its trajectory. The last step involves defining a $1^\circ \times 1^\circ$ grid and calculating for each grid cell the parameter $(E - P)^{10}$, which represents the net budget evaporation minus precipitation in the whole vertical column located over an area A (the area of the grid cell) averaged over the ten days of trajectory of the particles going toward SESA (Martin-Gomez et al., 2016). The positive (negative) values of $(E - P)^{10}$ will represent the regions where particles when passing gain (lose) moisture on average over the ten days of trajectory toward SESA, and therefore these regions will represent sources (sinks) of moisture.

Figure 6.21 shows the network distance (solid black line) and the PCP index on DJF (dashed black line) during the last century. Clearly, the network distance is characterized by interannual and interdecadal variability. Moreover, during the

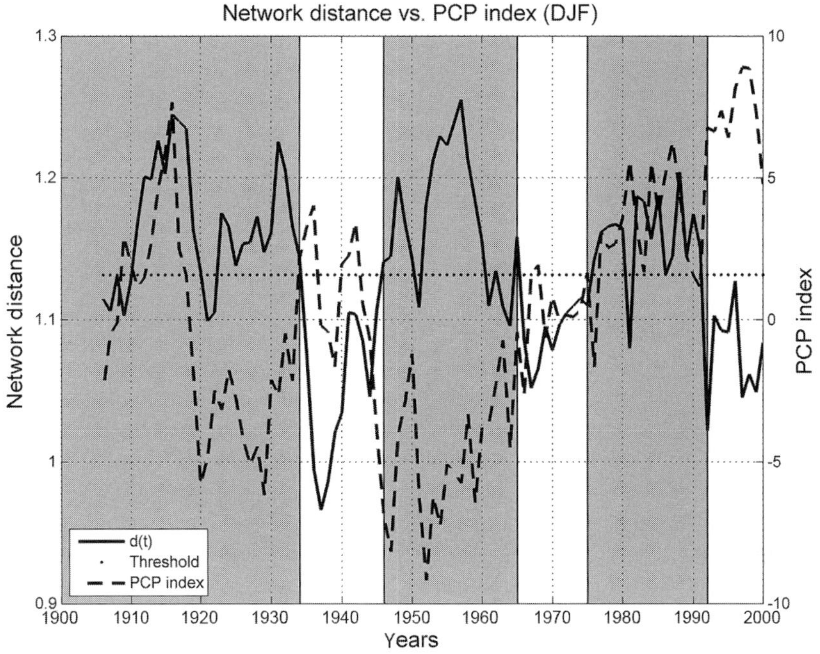

Figure 6.21 Temporal evolution of the mean network distance ((6.5), solid line) during the twentieth century. The dashed line represents the precipitation index over SESA in austral summer (DJF) and the horizontal black dot line represents the 5% significance level below which the distance identifies significant synchronization. Each point of the network distance time series represents the value of the mean network distance computed considering a sliding window of 11 years. Source: Martin-Gomez et al. (2016).

last century there were synchronization periods during 1934–1946, 1965–1975, and 1992–2000.

The mean network distance and a precipitation index constructed taking averages of 11-year sliding windows from the SESA PCP index are (Spearman) correlated at −0.25, which is statistically significant at the 5% level in a one sided *t*-test. This suggests that an increase of the network distance (a decrease in the synchronization among the network's nodes) is associated with less precipitation over SESA. The anti-correlation is evident in Fig. 6.21. Additionally, the Spearman correlation coefficient between the RPW and the mean network distance, −0.22, statistically significant at the 5% level in a one sided *t*-test, suggests that a larger connectivity of the precipitation index is associated with a smaller network distance (larger synchronization of the network). These results suggest that an increment of the precipitation in SESA is related to a larger influence of the tropical oceans on SESA, which in turn is associated with a larger degree of synchronization.

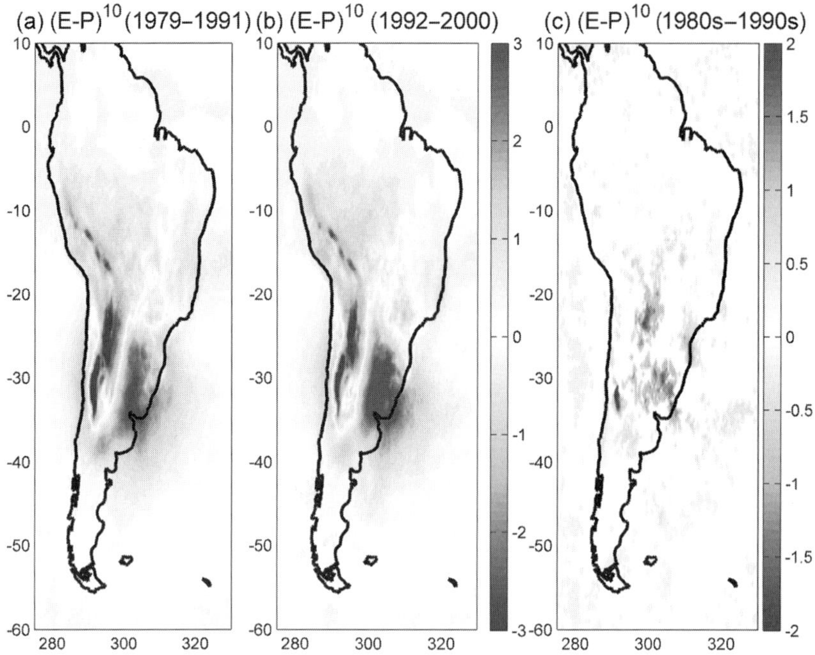

Figure 6.22 Ten-day average of the net budget evaporation minus precipitation, $(E - P)^{10}$, during (a) the 1980s (1979–1991), (b) the 1990s (1992–2000). (c) Difference between the 1980s and 1990s. Only locations with anomalies significant at the 90% confidence level are colored. Units: mm/day. Source: Martin-Gomez et al. (2016). (A black and white version of this figure appears in some formats. For the color version, please refer to the plate section.)

Given the availability of ERA Interim and NCEP-CFSR data, the characterization of moisture sources focuses on the differences between the 1980s (1979–1991) and 1990s (1992–2000). This allows comparing a period of nonsignificant synchronization (1980s) with another of statistically significant synchronization (1990s) among network components. Figures 6.22a, b represent the ten-day average of the net budget evaporation minus precipitation, $(E-P)^{10}$, over the periods 1979–1991 and 1992–2000, respectively. Regions with positive (negative) values of this variable are associated with a net gain (loss) of moisture of the particles when passing along their trajectories toward SESA, and therefore these regions will represent the main moisture sources (sinks) of SESA. The figures show that the main moisture source regions (with positive values of $(E - P)^{10}$) are the recycling over SESA, the central-eastern shore of Brazil together with its surrounding Atlantic Ocean, and the south Atlantic Ocean surrounding the SESA shore. Results are in close agreement with fig. 1(d) in Drumond et al. (2008).

Figure 6.22(c) shows the difference in $(E - P)^{10}$ during the 1980s and 1990s that are significant at the 10% level. It suggests that during the 1980s the central-eastern shore of Brazil acted as a stronger moisture source, while during the 1990s the intensity of the recycling was greater. Moreover, during the 1990s the region at 60°W between 20 and 25°S acted as a moisture source, not clearly present during the 1980s. These changes in the moisture sources were found to be consistent with EOF analysis for the net budget $(E - P)^{10}$.

Therefore, the analysis of Martin-Gomez et al. (2016) suggests that during the 1980s there were northerly anomalies along the coast of central-eastern and southern Brazil associated with a cyclonic circulation at (15°S, 50°W) that helped the transport of moisture toward SESA. During the 1990s the situation was the opposite: a low-level anticyclonic circulation developed over central-eastern Brazil that did not favor the advection of moisture from the central-east shore of Brazil toward SESA, decreasing the contribution of this region as a moisture source. Instead, it favored the transport of moisture from the Amazon basin, and could explain the extension of the region acting as a moisture source toward the north of SESA during that decade.

This chapter has presented applications of climate networks to the study of atmospheric variability at very different timescales, including teleconnections associated with climate phenomena such as ENSO or transport processes associated to blockings and monsoon systems. After recovering what is known about these phenomena, the network analysis has allowed us to further understand the mechanisms involved (such as identification of Rossby waves or location of key geographical regions) that can contribute to improving the prediction of atmospheric anomalies. It is worth noting that the studies presented in this chapter used network measures that have a straightforward physical interpretation. Other, more elaborate measures will likely help in advancing the understanding of atmospheric variability, provided they can be understood in physical terms.

7

Applications to Oceanic Variability

In this chapter, results from the application of network theory to ocean variability will be presented. The two first sections use climate networks to study propagation phenomena in the ocean: Section 7.1 analyzes wave propagation relevant to the El Nino Southern Oscillation (ENSO; see Section 2.4) and Section 7.2 analyzes the westward propagation of sea-surface temperature (SST) anomalies, which is linked to the Atlantic Multidecadal Oscillation (AMO; Section 2.6). The last two sections use flow networks (Section 4.4) to investigate transport phenomena in the Mediterranean Sea, and characterize the mixing and coherent regions (Section 7.3) and transport pathways (Section 7.4).

7.1 Oceanic El Niño Wave Dynamics

ENSO (Section 2.4) is the largest climatic cycle on annual timescales, and is one of the most important processes that affect natural climate variability. El Niño has a profound impact not only on the tropical atmospheric and oceanic conditions, but also on remote regions, including North America, the Indian Monsoon region, and Antarctica (Clarke, 2008; Sarachik and Cane, 2010; Turner, 2004).

Oceanic waves play an important role in El Niño dynamics, and the equatorial region can be regarded as a natural waveguide of several-degrees latitude-width, centered at the equator (Gill, 1982). Tropical waves, such as Kelvin and Rossby waves (Gill, 1982; see also Section 2.2) are propagating in this waveguide. The intraseasonal Kelvin waves have been extensively studied using various observational and modeling methods due to their importance to the theory of ENSO (Roundy and Kiladis, 2006). For example, the 1997/1998 strong El Niño event was initiated by a group of very pronounced Kelvin waves which were generated by energetic westerly wind bursts (McPhaden, 1999). Satellite altimetry data have become an invaluable resource to study equatorial wave dynamics (Chelton et al., 2003; Wakata, 2007). The velocity of waves in the equatorial Pacific region can be estimated using the so-called Hovmöller (longitude–time) diagrams by transferring the data to wavenumber–frequency space, or by decomposing the data into several leading modes of equatorial Kelvin and Rossby waves (Delcroix et al., 1994; Wakata, 2007). Generally speaking, typical phase speeds of the first baroclinic mode of equatorial Kelvin and Rossby waves are $\sim$2.8 m s^{-1} and $\sim$0.9 m s^{-1}, respectively (Delcroix et al., 1994; Wakata, 2007). As discussed in Section 2.2, Kelvin waves propagate eastward over the equator, while Rossby waves propagate to the west.

Tropical oceanic waves connect different, sometimes very far, regions. In spite of the importance of oceanic dynamics for the climate system, relatively few studies have applied the climate network approach to oceanic data (Feng and Dijkstra, 2014; Feng et al., 2014). Also, the studies of El Niño using climate networks have focused so far on surface atmospheric data (Gozolchiani et al., 2011; Ludescher et al., 2013; Yamasaki et al., 2008).

Wang et al. (2016a) used *ocean observational data* to address El Niño. By transforming the daily and relatively long ($\sim$20 years) satellite-based altimeter (i.e., sea surface height, or SSH) records into climate networks, they found that the time delays, velocities, and directions associated with climate network links can be attributed to equatorial Kelvin, Rossby, and tropical instability waves (TIW). The topology of the climate network is strongly affected by El Niño, which implies that the relative number of links in the eastern equatorial Pacific is significantly higher during El Niño times compared to normal times. Notably, the network properties (e.g., degree distribution) reveal the locations of the wave initiation, as well as its strong dissipation. The climate network approach provides a novel way to quantify new aspects of El Niño-related oceanic waves, including the concentration of the waves in specific regions and at specific times, and the location of the decay of the Kelvin waves ($\sim$120°W).

Wang et al. (2016a) analyzed the daily SSH anomaly data (AVISO, 2013), focusing on the tropical Pacific Ocean between 7°S and 7°N, and between 120°E to 280°E (80°W). The time extent of the data was from January 1, 1993, to December 31, 2010, and the spatial resolution was 1° and 4° in the meridional and zonal directions, respectively. In the area analyzed there are 615 grid points, which were regarded as the nodes of a climate network. For each node, daily SSH anomaly values were analyzed after subtracting the seasonal cycle. The value of the SSH anomaly at location s and time t is denoted by $H_s(t)$. Time series were normalized as

$$\Theta_s(t) \equiv [H_s(t) - \langle H_s(t)\rangle] \Big/ \Big\langle (H_s(t) - \langle H_s(t)\rangle)^2 \Big\rangle^{1/2}, \qquad (7.1)$$

where $\langle \cdots \rangle$ represents the temporal average.

Then, a climate network was constructed for each calendar year y. In order to define links between each pair of sites s_1 and s_2 on year y, the lagged cross-correlation function $X^y_{s_1,s_2}(\tau \geq 0) = \langle \Theta_{s_1}(t)\Theta_{s_2}(t+\tau) \rangle$, and $X^y_{s_1,s_2}(\tau \leq 0) = X^y_{s_2,s_1}(-\tau)$ was calculated (see Section 3.2.1), where τ is the time lag parameter that was considered up to 200 days, and the averages are over times t inside year y. To eliminate the effect of slow modulations on correlation estimates, a two-year high-pass filter was used. Then, optimal time lags τ^* and positive link weights $W^{y,+}_{s_1,s_2}$, were assigned to each pair (s_1, s_2) and year y by considering the lag at which $X^y_{s_1,s_2}(\tau)$ is maximal (see (4.10) in Section 4.2). When s_1 is to the west of s_2 and the time lag is positive, the link direction is to the east.

Climate networks were constructed for three subregions of the tropical Pacific: equatorial Pacific (2°S to 2°N), northern tropical Pacific (4°N to 7°N), and southern tropical Pacific (4°S to 7°S). To identify a threshold below which link weights W can be statistically disregarded (see Section 3.4), Wang et al. (2016a) applied a shuffling procedure in which the order of calendar years was shuffled while the order of days within each year remained unchanged.

Figure 7.1a depicts the temporal average of the positive link weights, $\langle W^+ \rangle$, versus the geographical distances of the links, D, for the real and for shuffled data in the equatorial Pacific regions (from 2°S to 2°N). High positive average link weight values which exist in the real data but not in the shuffled data are less likely to occur by chance. Thus, only links that are separated from the shuffled links (indicated by the horizontal dashed lines and vertical arrows in Fig. 7.1a were considered. In addition, short-distance links (indicated by the vertical dashed line and the horizontal arrow in Fig. 7.1a exhibit trivial correlations due to the "proximity effect" (Wang et al., 2013) and were also ignored. Significant links with $p \leq 0.02$ include 26.6% of the weighted links in the equatorial Pacific (2°S to 2°N), 6.5% for 4°N to 7°N, and 11.7% for 4°S to 7°S.

One of the main characteristics of equatorial Kelvin waves is their phase speed (which is equal to their group velocity since these waves are non-dispersive), and the links' statistics can be used to estimate this velocity. Long–Rossby waves are also non-dispersive and thus their phase speed is equal to their group velocity. Assuming that the delay time τ^* is a good estimator for a dynamical interaction time between two nodes (Cimponeriu et al., 2004), the phase velocities can be estimated by dividing the geographical distance D between the nodes by the delay time τ^*, i.e., $v \approx D/\tau^*$. The PDF of phase speeds (of links with $D \geq 500$ km and $p \leq 0.02$) is shown in Fig. 7.1c. For the equatorial Pacific, the phase speed is positive (eastward) and peaks around $3\,\mathrm{m\,s^{-1}}$, in agreement with the value and direction of Kelvin waves (Wakata, 2007; Toshiaki et al., 2008).

The topological properties of the climate network in the equatorial Pacific region show distinct behavior during El Niño and normal times. The climate network

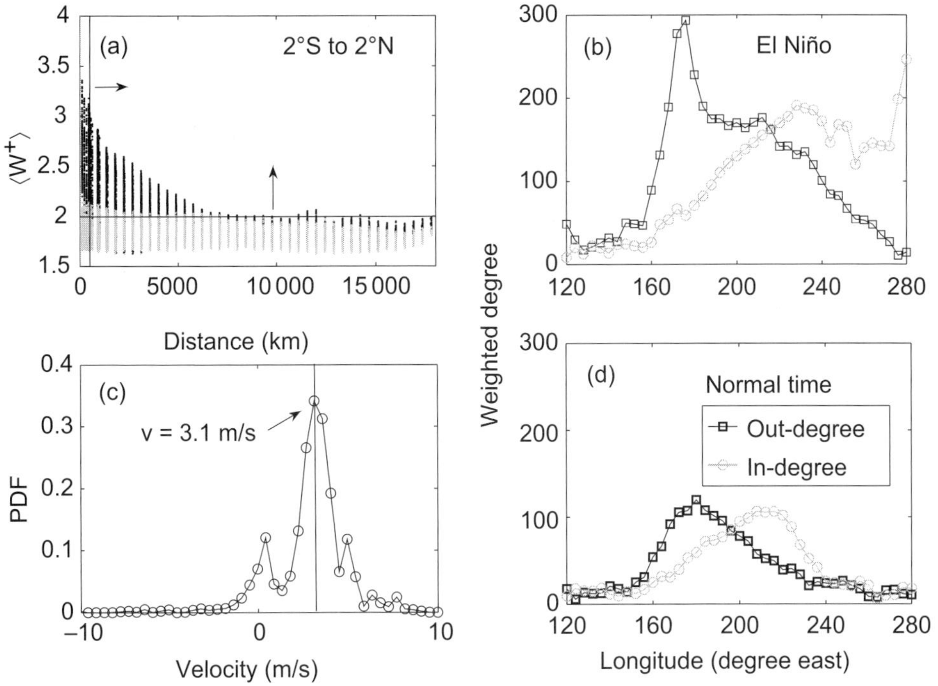

Figure 7.1 Equatorial Pacific network (2°N to 2°S) analysis. (a) The mean weight of positive links $\langle W^+ \rangle$ (average over years) versus their distance for both real (black dots) and shuffled (gray dots) data. (b) The "in" (grey circles) and "out" (black squares) weighted degree as a function of longitude for El Niño years, for positive links ($p \leq 0.02$ and $D \geq 500$ km). (c) The probability density function (PDF) of phase speeds that are based on positive links with $p \leq 0.02$ and distances larger than 500 km. (d) Same as (b) but for normal (non-Niño) years. Source: Wang et al. (2016a).

considered above is directed – positive and negative τ^* indicates eastward and westward propagation, respectively. One can distinguish between a link that is pointing toward a node or away from the node; the former/latter is related to the total link weight pointing to/from a specific node and was referred to as weighted "in"/"out"-degree. Figures 7.1b, d depict the in- and out-weighted degree versus longitude, averaging over all the latitudes from 2°S to 2°N during El Niño and normal times, respectively. In this plot we first note that the out-weighted degree peaks around 170°E at the equator, with a much more pronounced peak during El Niño, which indicates enhanced Kelvin wave activity, consistent with previous analysis (McPhaden, 1999). Indeed, stronger Kelvin wave activity during El Niño around 170°E was associated with strong westerly wind bursts (Eisenman et al., 2005), which is enhanced wind activity that underlies the generation of equatorial Kelvin waves. Second, there is a peak in the in-weighted degree curve at ~220°E during both El Niño and normal times, where for El Niño times the in-degree curve is high all the way to the coast of South America, while during normal times it decays for longitudes larger than $\sim$220°E. Wang et al. (2016a) interpreted this decay as strong dissipation of the Kelvin waves, suggesting that the dissipation of Kelvin waves is much weaker during El Niño.

The attenuation of the Kelvin waves to the east of $\sim$220°E during El Niño was implied based on wave packets (found using Hovmöller diagrams) in model data (Oceanic General Circulation Model simulations; Benestad et al., 2002; Mosquera-Vásquez et al., 2014). This attenuation may be attributed to several mechanisms, including the strong vertical current shear during La Niña, strong wind at $\sim$220°E (Benestad et al., 2002), scattering of waves, and partial reflection of Kelvin waves into Rossby waves (Mosquera-Vásquez et al., 2014).

Wang et al. (2016a) next applied the climate network methodology to the northern and southern tropical Pacific to study Rossby (and other) tropical waves, respectively. The statistical properties of positive links in the climate networks of the northern tropical Pacific (4°N to 7°N) and in the southern tropical Pacific (4°S to 7°S) are shown in Fig. 7.2. The phase speed that is calculated based on links above significance level ($p \leq 0.02$ and $D \geq 500\,\mathrm{km}$) is negative (westward) and their PDF peaks around 0.5 $\mathrm{m\,s^{-1}}$ and 0.6 $\mathrm{m\,s^{-1}}$ respectively (Figs. 7.2c, d), in general agreement with the phase velocity of TIW and Rossby waves (Chelton and Schlax, 1996; Chelton et al., 2003; Kessler, 1990; Lyman et al., 2005; Wakata, 2007; Willett et al., 2006). The structure of link weight versus distance is different for the northern and southern tropical Pacific (Figs. 7.2a, b), (see Chelton et al., 2003). Network analysis based on northern tropical Pacific is presented in Fig. 7.3. The westward propagation of waves is evident during both normal and El Niño times, with most of the wave activity concentrated between 200°E to 250°E (see also Chelton et al., 2003). Evidently, extratropical waves are less pronounced during El Niño times (Fig. 7.3). The enhanced wave activity during normal times can be associated with the TIW. The characteristic phase speed, location, and wave activity of TIW during El Niño indeed fits the network observations (Willett et al., 2006).

The TIW are not directly related to the theory of El Niño even though they are possible sources of random perturbations that affect both the mean state and the interannual climate variability. Therefore, they were filtered out with a 60-days low-pass filter on the SSH data as their characteristic period is about 30 days, which is much smaller than the period of Rossby waves (Lyman et al., 2005). After this filtering, the climate network was reconstructed and the corresponding wave characteristics recalculated. This "filtered" climate network is expected to be related to Rossby waves, which play an essential role in the El Niño dynamics. Figure 7.4 depicts the weighted degree at 4°N of this filtered network during El Niño and normal times. The Rossby waves signal, represented here by the weighted degree of

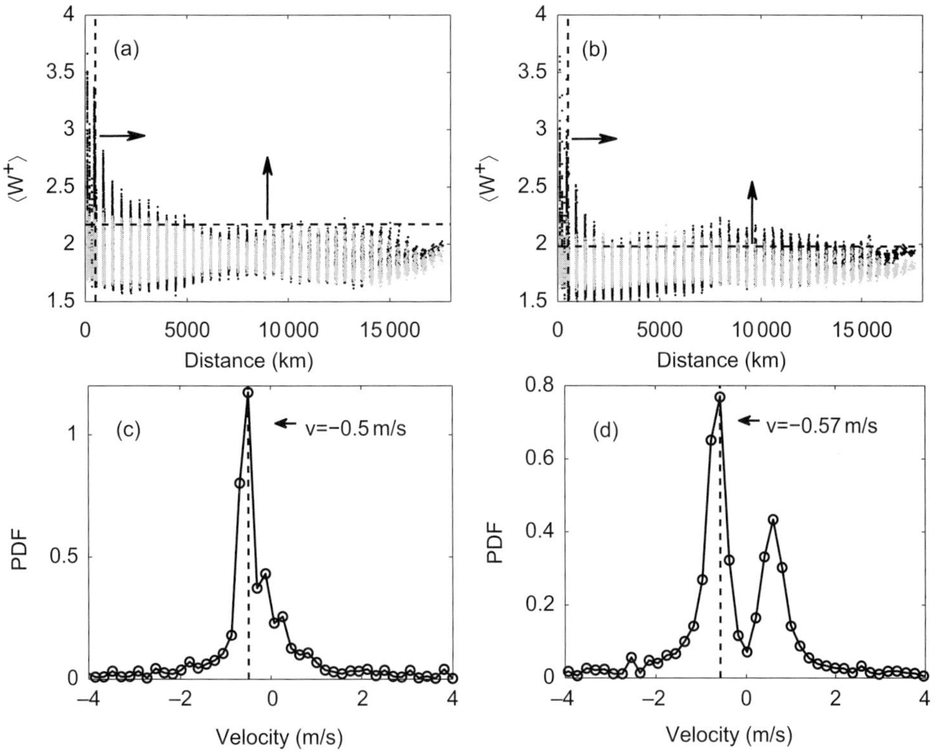

Figure 7.2 The temporal average link weight $\langle W^+ \rangle$ versus geographical distance D in (a) the northern tropical Pacific (4°N to 7°N) and (b) the southern tropical Pacific (7°S to 4°S). Black points are real data, grey points are shuffled data. (c) The PDF of the phase velocity in the northern tropical Pacific. (d) The PDF of the phase velocity in the southern tropical Pacific. Source: Wang et al. (2016a).

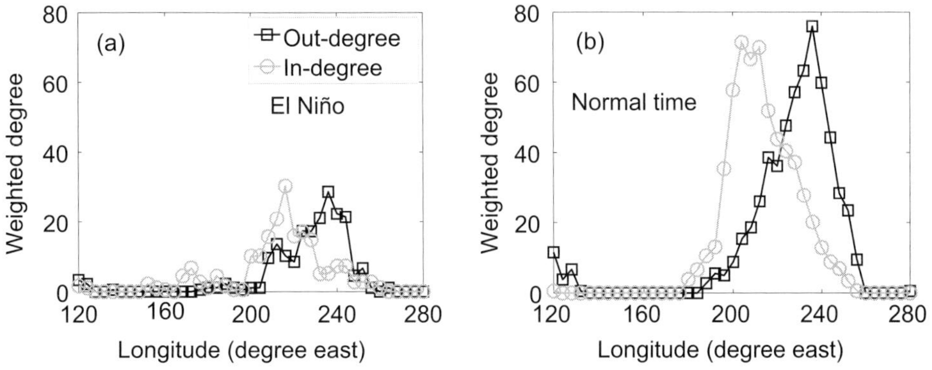

Figure 7.3 The out- (black squares) and in- (gray circles) weighted degree of nodes in the climate network from 4°N to 7°N during (a) El Niño times and (b) normal times. Source: Wang et al. (2016a).

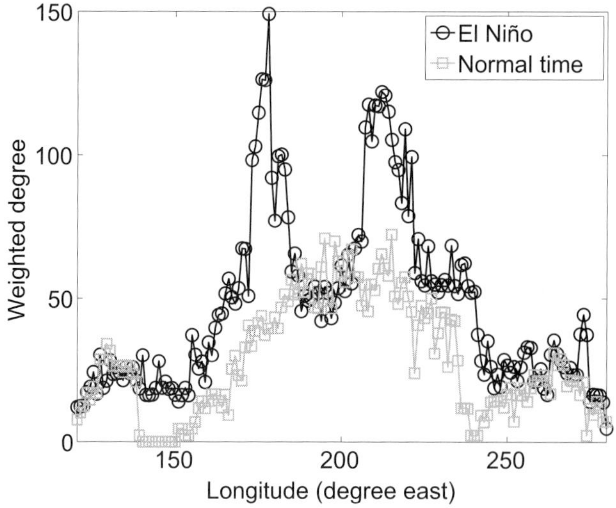

Figure 7.4 The weighted degree (60-day low-pass filtered network) at 4°N for El Niño (black circles) and normal (grey boxes) times. Source: Wang et al. (2016a).

the climate network, is only clearly observed to the west of ~240°E (120°W) (see Fig. 7.4), although some Rossby waves close to the eastern boundary (~280°E) are the reflection of Kelvin waves. Note that Rossby waves are more pronounced during El Niño periods: Around the dateline (180°E) there is a peak in the weighted degree, which can be associated to strong westerly wind bursts in this region. In addition, there is another peak in the weighted degree distribution between ~200°E to ~230°E, which can be associated with the enhanced wind-stress curl in this region. The result supports the suggestion that the enhanced wind-stress curl is related to the Rossby wave activity (Chelton et al., 2003).

To summarize, from the study by Wang et al. (2016a) of a climate network of the tropical Pacific Ocean based on daily satellite SSH data, the most dominant long-distance links in the network share the same characteristics (speed and direction) as equatorial Kelvin, Rossby, and tropical instability waves, indicating that the climate network approach allows us to characterize and to quantify oceanic wave activity. This conclusion was further strengthened by the observation that the location of the hubs of the network coincide with the location of initiation and dissipation regions of equatorial oceanic waves. Moreover, the climate network was found to have different features during El Niño and normal times, such as the pronounced equatorial Kelvin and Rossby waves, and suppressed TIW during El Niño. In addition, the dissipation of Kelvin waves was found to be much weaker during El Niño.

7.2 Multidecadal North Atlantic SST Anomalies

The AMO (see Section 2.6) is characterized by low-frequency variability in SST occurring in the North Atlantic Ocean (Delworth and Mann, 2000; Schlesinger and Ramankutty, 1994). The AMO is thought to have significant effects on summer temperatures and precipitation variations in Western Europe and North America (Enfield et al., 2001; Sutton and Hodson, 2005), and possibly hurricane development in tropical Atlantic regions (Andronova and Schlesinger, 2000). There is an ongoing debate on whether the North Atlantic SST anomalies in the twentieth century are the expression of an internal climate mode, or an imprint of external forcing (Ting et al., 2009). Thus, there is a need to improve the understanding of AMO, and the theory should explain the processes responsible for the multidecadal timescale and the spatial patterns of the SST anomalies (Dijkstra et al., 2008).

In one of the theories of the AMO (Te Raa and Dijkstra, 2002), the variability arises due to internal ocean dynamics through an oscillatory instability of the background Meridional Overturning Circulation (see Section 1.4.2). A sketch of this so-called thermal Rossby mode mechanism is displayed in Fig. 7.5. A warm anomaly in the north-central part of the basin causes a northward meridional perturbation temperature gradient, which induces – via the thermal wind balance – an anticlockwise or westward surface zonal flow (Fig. 7.5a). Because of a southward background temperature gradient, the anomalous anticyclonic circulation around the warm anomaly causes southward (northward) advection of cold (warm) water to the east (west) of the anomaly, resulting in westward phase propagation of the warm anomaly. Due to this westward propagation, the zonal perturbation temperature gradient becomes eastward, inducing an anticlockwise or southward surface meridional flow (Fig. 7.5b). The resulting upwelling (downwelling) perturbations along the northern (southern) boundary cause a southward meridional perturbation

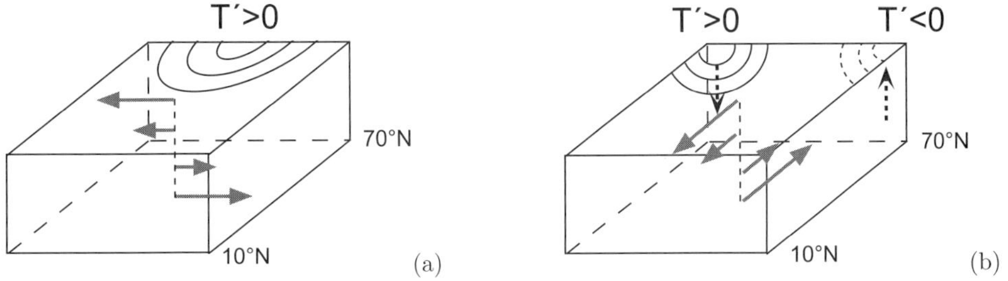

Figure 7.5 Schematic diagram of the oscillation mechanism associated with the thermal Rossby mode caused by the westward propagation of the temperature anomalies T'. The phase difference between (a) and (b) is about a quarter period.

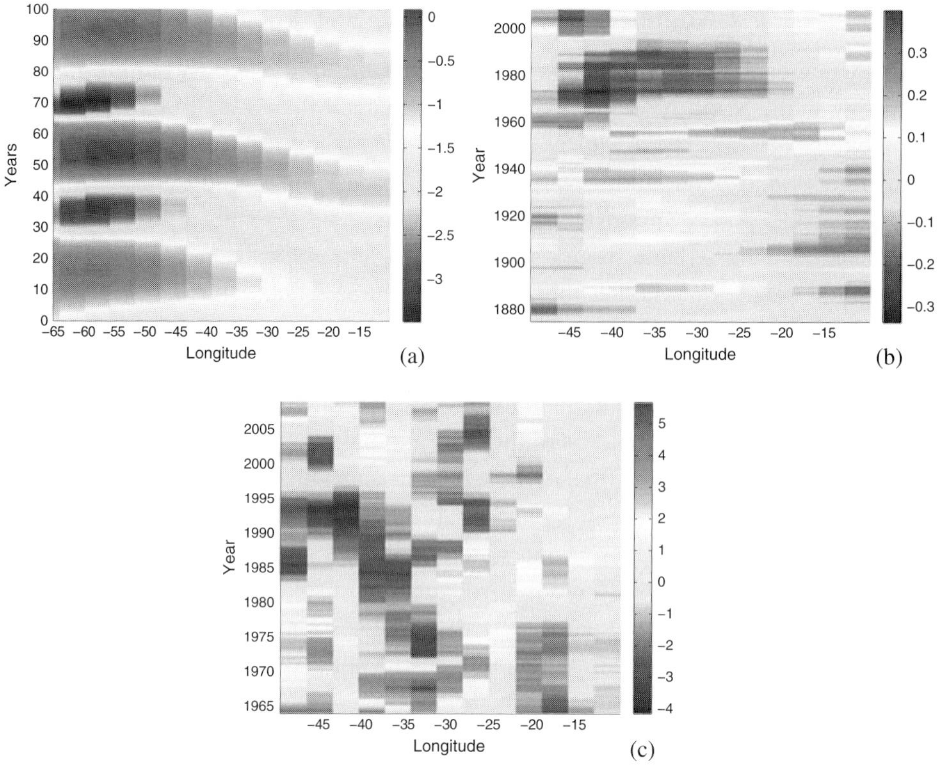

Figure 7.6 Hovmöller diagram of the 30–60°N averaged North Atlantic SST anomalies from (a) the idealized ocean model simulations; (b) eight-year running mean HadISST dataset (1870–2012). (c) the same as (b) but first the surface heat flux contribution (from the ORA-S3 data set, see http://apdrc.soest.hawaii.edu) to SST is subtracted from the HadISST data set (only over the period 1959–2009). Source: Feng and Dijkstra (2014). (A black and white version of this figure appears in some formats. For the color version, please refer to the plate section.)

temperature gradient, inducing an eastward zonal surface flow, and the second half of the oscillation starts.

A Hovmöller diagram of meridionally averaged (30–60°N) SST anomalies (Fig. 7.6a) from idealized model data (Feng and Dijkstra, 2014) clearly shows the westward propagation associated with the thermal Rossby mode (Dijkstra et al., 2008). The oscillation period in the model is about 40 years, which is between the 20–30 and 50–70 dominant periods in the observations (Frankcombe et al., 2010). Note that the warm phase of the oscillation is slightly longer than the cold phase, reflecting the nonlinear effects in the variability.

Indications for westward propagation were found in observed subsurface temperature fields (Frankcombe et al., 2008) and sea-level records (Frankcombe and

Dijkstra, 2009; Vianna and Menezes, 2013). However, this property was difficult to distinguish from SST observations because many other processes affect SST in the North Atlantic (Booth et al., 2012; Sutton and Hodson, 2005). When a running eight-year mean is applied to the Hadley Centre Sea Ice and Sea Surface Temperature data set (HadISST), a similar Hovmöller diagram (Fig. 7.6b) does not clearly indicate westward propagation (Delworth and Mann, 2000; Tourre et al., 1999). As part of the SST response is determined by the surface heat flux, one can attempt to remove this contribution. Using the surface heat flux field from the ORA-S3 reanalysis data set over the period 1959–2009 to remove this contribution from the HadISST data, the result (Fig. 7.6c) still does not show a clear westward propagation. The correction cannot be performed over the whole HadISST data set and, in addition, the correction may also remove components of the thermal Rossby wave as the surface heat flux also depends on SST.

To improve the propagation detection, Feng and Dijkstra (2014) used network techniques to investigate the existence of the westward propagation of observed North Atlantic SST anomalies. Two types of measures of spatial correlations between SST variations were used to construct climate networks (see Section 4.2): linear Pearson correlation at lag τ, resulting in Pearson correlation climate networks (PCCNs), and nonlinear mutual information at lag τ, resulting in mutual information climate networks (MICNs). The analysis of the topological properties of both networks showed that MICNs better captured the main features of propagating patterns from the noisy SST data than did PCCNs, and that westward propagation of multidecadal SST anomalies indeed seems to occur in the North Atlantic.

For the MICN, the Hovmöller diagram of the relative degree (the degree divided by the maximum degree of the MICN), averaged over latitudes 30–60°N, for the idealized model data is plotted in Fig. 7.7a. The largest lag used in the reconstruction of the MICN is 30 years. For smaller lags (0–10 years), the nodes with relatively high degree emerge from the eastern part, and gradually high-degree nodes are found more westward. After approximately 15 years, the nodes with maximal degree appear at the western boundary of the domain. For the next 15 years (lags 15–30 years) there are only low-degree nodes in the eastern part, while the degrees of the nodes in the western part remain high. This is consistent with Fig. 7.6a, where high spatial correlations are found for about 15 years in the west during the oscillation. Similar propagating features can also be found in the Hovmöller diagram of the relative degree field of the MICN constructed from the HadISST data, as seen in Fig. 7.7b. There is again a westward propagation of high degree over the lags 0–10 years, and the maximum occurs around year 15, near 40°W. For the following 15 years, relatively high-degree nodes only appear in the western part in contrast to the lower-degree nodes in the eastern part.

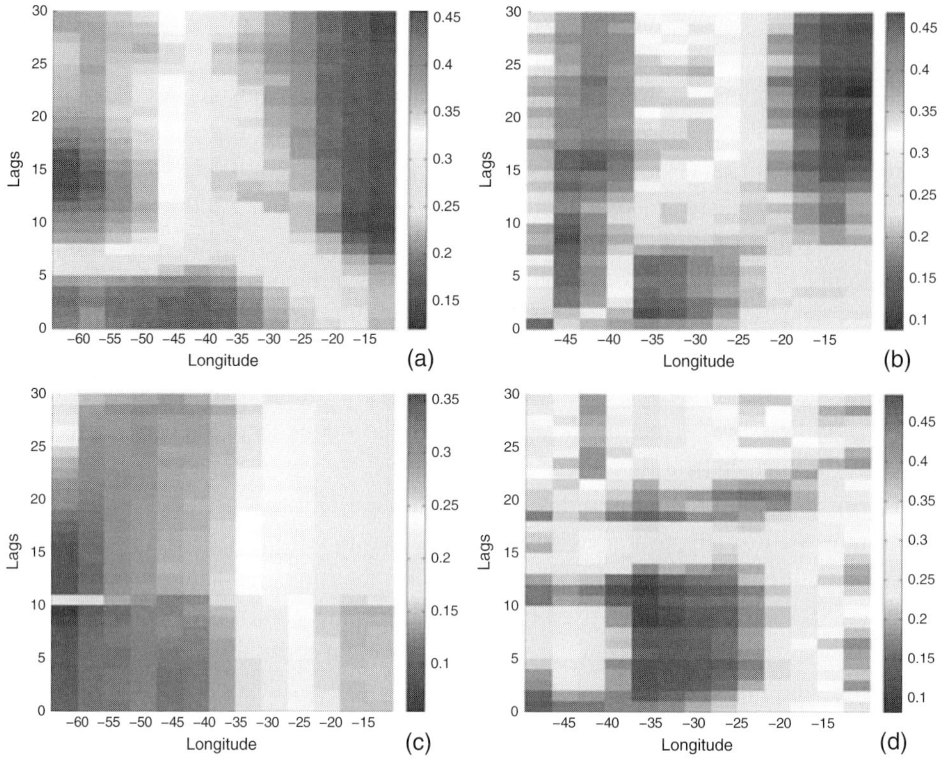

Figure 7.7 Hovmöller diagram of the 30–60°N averaged relative degree fields of (a) the idealized model data, (b) the observations, (c) the idealized model data, and (d) the observations. The networks of (a) and (b) are constructed by using mutual information under a link density of 0.2. The networks of (c) and (d) are constructed by using Pearson correlation also under a link density of 0.2. The time lag (in years) used to reconstruct the networks is in the vertical axes. Source: Feng and Dijkstra (2014). (A black and white version of this figure appears in some formats. For the color version, please refer to the plate section.)

Therefore, the use of a nonlinear measure of correlation such as mutual information seems to be crucial to identify the westward propagation behavior in the observations. For the PCCN of the idealized model data, the Hovmöller diagram of the degree field is plotted in Fig. 7.7c. The relative degree field (Fig. 7.7c) shows the same features as in the MICN (compare with Fig. 7.7a), but the position of the high-degree nodes shifts much more abruptly to the west at year 10. The relative degree field of the PCCN constructed from the observations (Fig. 7.7d) does not show any westward propagation tendencies. In fact, the patterns of degree do not correspond to the one from the PCCN of the idealized model data (Fig. 7.7c), or to the one from the MICN of the observations (Fig. 7.7b). Apparently, the PCCN

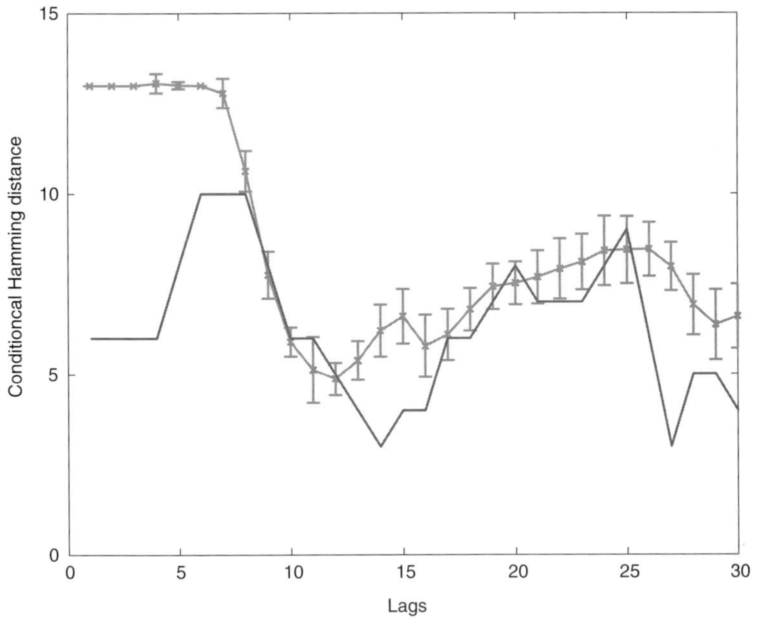

Figure 7.8 Conditional Hamming distance H_α between the relative degree field of the idealized model data and that of the observations (dark line), for $\alpha = 0.15$, and between the relative degree field of the idealized model data and that of 200 surrogate data sets derived from the observations through shuffling (line with error bars, which represent the standard deviation in H_α. Source: Feng and Dijkstra (2014).

analysis fails to distinguish the main features of the westward propagation pattern in the noisy SST signal from the observations.

To quantify the similarity between the patterns of the degree fields of interaction networks (for example, Figs. 7.7a, b), a *conditional Hamming distance* H_α was also applied by Feng and Dijkstra (2014). The Hamming distance measures the number of substitutions required to change from one string into another (Hamming, 1950). The conditional Hamming distance used by Feng and Dijkstra (2014) computed the distance between two degree fields, obtained from two networks, by summing the number of nodes where the degree was different, understanding that two degree values are significantly different only if their relative absolute difference exceeds a fraction α. The dashed curve in Fig. 7.8 is the conditional Hamming distance between the relative degree fields of the idealized model data and the observations, allowing for a relative difference $\alpha = 0.2$. At small lag, H_α is small but increases up to about a lag of 6 years; a second local minimum occurs around a lag of 15 years.

To evaluate the significance of this similarity measure, surrogate data obtained by 200 shuffled time series (see Section 3.4) of the observed SST anomalies was also used by only randomizing the sequence of the years without shuffling the data within a particular year. Hence the probability distribution function of the observed data is preserved under this shuffling scheme. However, the scheme removes the lagged correlations in the spatial pattern and hence any westward propagation. The solid curve in Fig. 7.8 shows the same conditional Hamming distance between the relative degree of the MICN constructed from the surrogate observations, where the error indicates the standard deviation. When the lower values of the error bars of the solid curve are above the dashed curve, the value of the conditional Hamming distances reveals a significant similarity between the degree fields of the observations and the idealized model data, and hence indicates the existence of westward propagation. Significance is most prominent near the minima of the dashed curve, which are in the 10–15-year and 25–30-year timescale, and this result is robust over quite a large range of α values.

Feng and Dijkstra (2014) demonstrated that tools from interaction networks are useful detecting spatial propagation of patterns of variability in the climate system. These tools can be directly applied to results from General Circulation Models and can also be extended to look at (lagged) spatial correlations between different fields.

7.3 Mediterranean Sea Surface Flow Network

Understanding how ocean flows connect different parts of the Earth system is important to characterize heat, salt, or nutrient flows, with implications, depending on the time and space scales, from climate to local ecosystem behavior. Here we summarize an application of the methodology of flow networks presented in Section 4.4 to the surface of the Mediterranean Sea: Ser-Giacomi et al. (2015b) considered a relatively short transport timescale (of the order of months), although some climatological characterization of the resulting patterns was done.

7.3.1 Network Construction

The input velocity field originates from the Mediterranean Forecasting System Model (physics reanalysis component). It is a hydrodynamic model supplied by the Nucleus for European Modelling of the Ocean (NEMO; www.nemo-ocean.eu), which solves primitive equations in spherical coordinates. NEMO has been implemented in the Mediterranean at a horizontal resolution of $1/16$ degrees, and 72 unevenly spaced vertical levels (Oddo et al., 2009). It also slightly extends into the Atlantic in order to resolve the Strait of Gibraltar. The model uses vertical partial cells to fit the bottom depth shape. It is forced by momentum, water, and heat fluxes interactively computed by bulk formulae using the six-hours, 1/4-degree horizontal-resolution operational analysis and forecast fields from ECMWF while precipitation and river runoffs are provided by monthly mean data sets. The Dardanelles inflow is parameterized as a river and the climatological net inflow rates are considered. Assimilated data include sea-level anomaly, remotely sensed SST and in-situ temperature and salinity profiles. Daily horizontal velocity fields generated by the model in the whole Mediterranean basin during ten years of simulation (2002–2011) were used, selecting only one layer at a nominal depth of 7.9 m. This layer extends in fact between 4.58 and 11.55 m depth, so that it has a vertical extension of 6.97 m. For the integration timescales considered (values of τ always below three months) one can reasonably neglect motion to other layers and consider only horizontal dynamics (d'Ovidio et al., 2004; Rossi et al., 2014).

To switch from continuous space to discrete nodes the above-described horizontal near-surface Mediterranean layer was partitioned into 3270 two-dimensional square boxes. To impose an equal-area constraint, as in Section 6.5, cells were defined in a sinusoidal projection given by coordinates $x = \varphi \cos\theta$ and $y = \theta$, where φ and θ are the standard longitude and latitude, respectively. In these x, y coordinates, boxes are squares of side length 0.25°, or $\Delta = 27.78$ km. The area Δ^2 of each box is 771.9 km^2. The "amount of water in a box B_i" is then related to its area Δ^2 through a simple multiplication by the layer thickness (6.97 m), giving the value of 5.38×10^9 m^3 per box.

The resolution of the model-generated velocity field is much finer than the discretization used for network construction. In this sense the dynamics represented in the flow network is a coarse-graining of the simulated Mediterranean flow, keeping the effect of the small scales only in a statistical sense. The most energetic features of the Mediterranean flow are mesoscale structures (Millot and Taupier-Letage, 2005) ranging from 10 km to a few hundred kilometers. With the value of Δ used by Ser-Giacomi et al. (2015b), the network description displays most of the mesoscale range and neglects submesoscales, which anyway are only marginally resolved by the NEMO implementation.

To characterize the transport phenomena, $N_i = 500$ ideal fluid particles were released in each box B_i, providing enough statistics to estimate the transport matrix $\mathbf{P}(t_0, \tau)$ (see Section 4.4). The motion of these $3270 \times 500 = 1.635 \times 10^6$ particles was simulated by integrating the trajectories in the velocity field using a fourth-order Runge–Kutta algorithm. The velocity at any arbitrary point in the sea was computed with a bilinear interpolation from the input data. A time step of one day (the same resolution as the data) was used. The two key parameters of the simulations are the *starting time*, t_0, and the *tracking time*, τ.

The simulation provides the initial and final positions for each particle, allowing us to compute the transport matrix $\mathbf{P}(t_0, \tau)$ from (4.15). A directed link is established from node i to node j if and only if $\mathbf{P}(t_0, \tau)_{ij}$ is non-vanishing. In that case its value gives the weight of the link.

Note that the Lagrangian integration is done under the full resolution of the velocity field (1/16 degrees). This means that particle trajectories contain the small-scale features produced by the model during the time interval τ. While such details are not explicitly present in the network description $\mathbf{P}(t_0, \tau)$ after coarse-graining the initial and final positions to the box size Δ, their effects have been incorporated in a statistical way.

As explained in Section 4.4 the transport matrices $\mathbf{P}(t_0, \tau)$, for several values of t_0 and τ, are interpreted as the adjacency matrices of directed and weighted flow networks. All the standard quantities characterizing the topology of a network, such as the degree, the clustering, the betweenness, etc. can be calculated (see Section 4.1.1). Ser-Giacomi et al. (2015b) focused on the measures with potential to yield insight in (horizontal) dispersion and mixing processes, and in the identification of coherent regions.

7.3.2 Dispersion and Mixing

Section 4.4.2 introduced definitions of quantities capturing different aspects of dispersion and mixing, and described the network quantities defined by Ser-Giacomi et al. (2015b) to quantify them. Figure 7.9 displays the in- and out-degree at the geographical locations defined by the nodes of the Mediterranean network for particular values of t_0 and τ. High values of the degrees appear to be associated to the strong and unstable currents in the southern part of the basin (Millot and Taupier-Letage, 2005). Low-degree values are observed in regions where the circulation is rather slow, such as the Tunisian continental shelf and the semi-enclosed seas (e.g. Adriatic and Aegean). Generally, the values of the in- and out-degree tend to increase with τ. With respect to the dependence on t_0, degree values tend to be slightly higher in winter than in summer.

The validity of the heuristic arguments of Section 4.4.2 relating network measures with mixing quantifiers can be checked by comparing the values of $K_O(i)$ obtained from the flow network and the right-hand side of (4.19), which involves the finite-time Lyapunov exponent (see Section 4.4.2). Figure 7.10a shows an example of FTLE field obtained with t_0 = July 1, 2011 and τ = 15 days. Figure 7.10b shows the clear correlation between the two sides of (4.19). Three values of τ are plotted to appreciate the general validity of the relationship. The deviations with respect to the exact identity are attributed to the fact that the filament-type

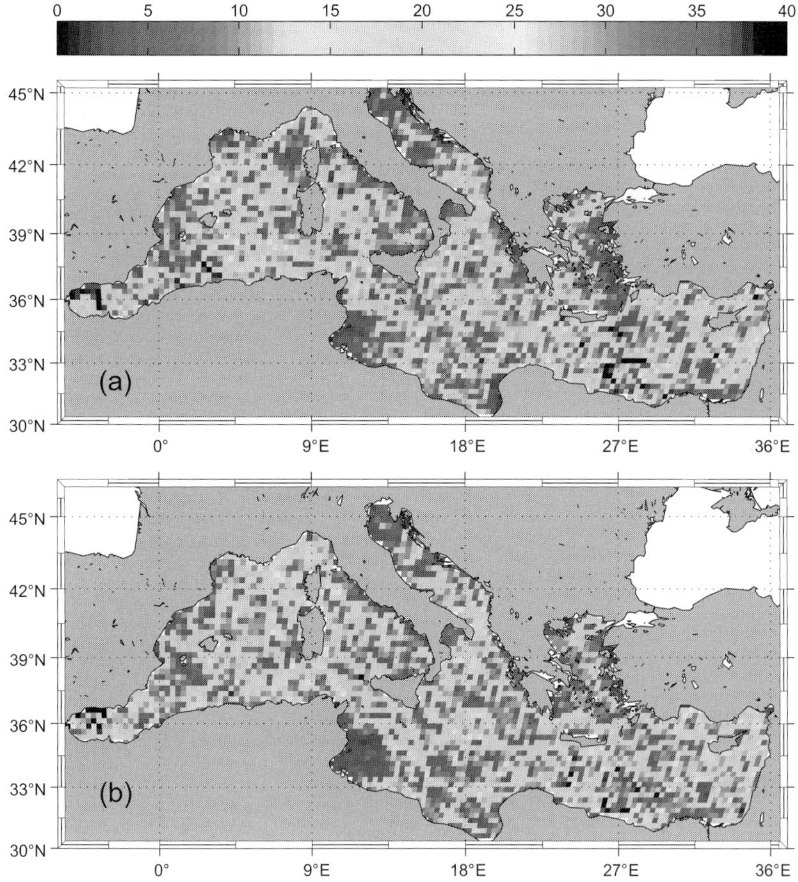

Figure 7.9 Degree of the nodes in the flow network defined by $\mathbf{P}(t_0, \tau)$, for t_0 = July 1, 2011 and $\tau = 15$ days. (a) The in-degree $K_I(i)$. (b) The out-degree $K_O(i)$. Source: Ser-Giacomi et al. (2015b). (A black and white version of this figure appears in some formats. For the color version, please refer to the plate section.)

arguments of Section 4.4.2 are only valid for sufficiently long τ and in flow regions dominated by strain. Also, the argument neglected the presence of filament foldings that sometimes would occupy the same box, and of associated saturation effects. In addition, quantization effects arising from the discrete nature of the degree K_O are visible at small degree values.

Figure 7.11a shows the spatial distribution of the finite-time entropy (4.23) in the Mediterranean Sea for particular values of t_0 and τ. The standard Pesin-like results relating the metric or Kolmogorov–Sinai entropy to the sum of positive Lyapunov exponents (Boffetta et al., 2002; Castiglione et al., 2010; Cencini et al., 2010) suggest that, at least for large τ, the entropy H_i^1 gives a good approximation to the

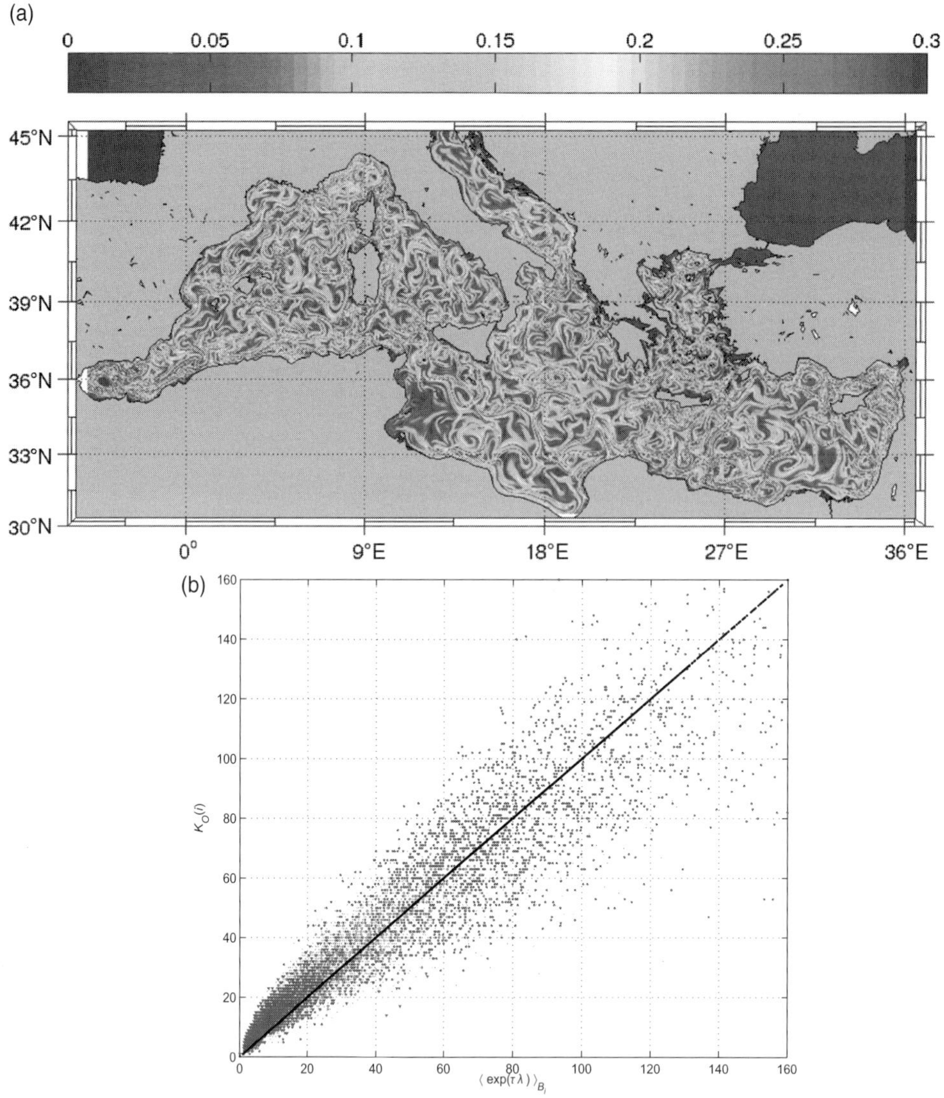

Figure 7.10 (a) An example of forward FTLE field $\lambda(\mathbf{x}_0, t_0, \tau)$ at t_0 = July 1, 2011, and $\tau = 15$ days. Color bar in day^{-1}. (b) Values of the out-degree $K_O(i)$ of each node i versus the average value of the stretching factor $e^{\tau\lambda}$ in that node. t_0 = July 1, 2011. Blue symbols are from $\tau = 15$ days, green from $\tau = 30$ days, and red from $\tau = 60$ days. Black line is the main diagonal. Source: Ser-Giacomi et al. (2015b). (A black and white version of this figure appears in some formats. For the color version, please refer to the plate section.)

values of the FTLE field averaged over each box B_i: $\lambda_i(t_0, \tau) \equiv \langle \lambda(\mathbf{x}_0, t_0, \tau) \rangle_{B_i} \approx H_i^1(t_0, \tau)$. Section 4.4.2 presents calculations supporting this claim in a heuristic way. Figure 7.11b shows the geographical distribution of $\lambda_i(t_0, \tau)$ and the lower

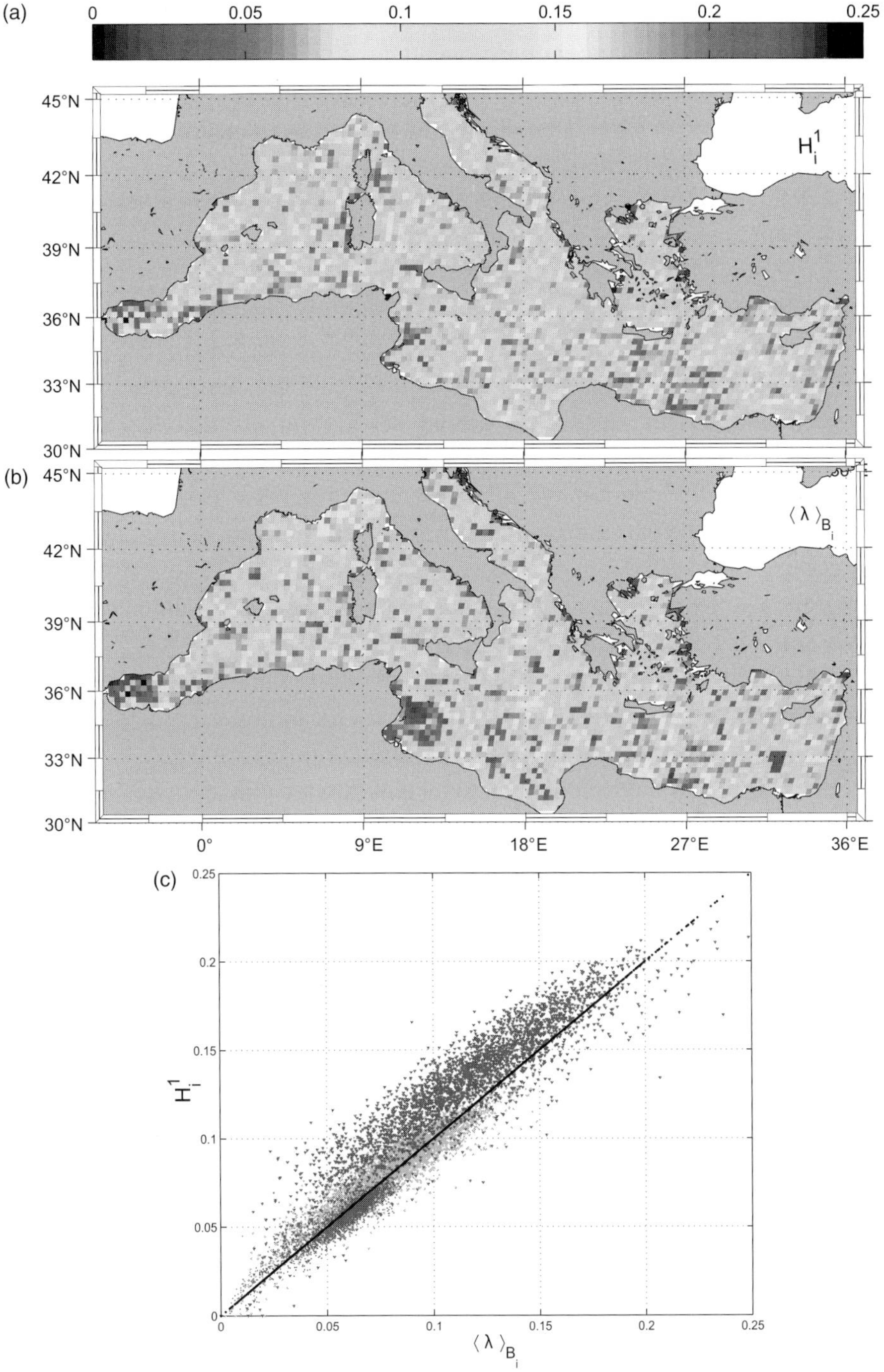

Figure 7.11 (a) The network entropy $H_i^1(t_0, \tau)$. (b) Coarse-graining of the Lyapunov field in Fig. 7.10a into the discretization boxes: $\lambda_i(t_0, \tau) \equiv \langle \lambda(\mathbf{x}_0, t_0, \tau) \rangle_{B_i}$.

panel compares both quantities for several values of τ. The entropies tend to be slightly larger than the Lyapunov exponents for $\tau = 15$ days, but both quantities approach each other and become well correlated for larger τ.

For definiteness we have been discussing quantities related to the forward time evolution: out-degree, forward Lyapunov exponents, etc. The network entropies can also be defined for the backward time evolution, and they turn out to be related to backwards Lyapunov fields, which give a measure of mixing of fluid coming from different origins (Ser-Giacomi et al., 2015b).

Summarizing this section, a family of entropy-like quantities was defined (Ser-Giacomi et al., 2015b) completely in terms of the transport matrix characterization of the flow network. At least two of them, H_i^0 and H_i^1, are related to standard dispersion and mixing quantifiers in the description of fluid flows. Although we have seen that the relationships are not exact, they are valid to a good approximation. The important point is that, once the network matrix $\mathbf{P}(t_0, \tau)$ has been constructed, the entropies in (4.22) provide a computationally very cheap way to assess quantities of geophysical interest such as local dispersion, stretching, and mixing. In fact, the simplest network quantifiers such as the in- and out-degrees are already suitable for that, being related to H_i^0.

We have focused in this section on network properties at fixed initial times t_0, and relatively short transport timescales τ (as relevant for biological processes; see e.g., Dubois et al., 2016; Rossi et al., 2014). However, one can calculate averages over different initial times, which could yield insight into the climatologies of stretching and mixing, as the ones computed, for example, by d'Ovidio et al. (2004), Waugh and Abraham (2008), Rossi et al. (2009) or Hernández-Carrasco et al. (2012) directly from Lyapunov exponents.

7.3.3 Communities in the Mediterranean Surface Flow

One of the advantages of formulating processes of geophysical transport in terms of flow networks is the possibility of using the powerful tools of community detection to locate regions that are well mixed by the flow, but remain poorly connected with their surroundings (see Sections 4.3 and 4.4.3). This is of particular relevance when

---

Figure 7.11 *(Cont.)* In both cases, data from the network constructed for $t_0 =$ July 1, 2011, and $\tau = 15$. Color bar in day^{-1}. (c) Values of the network entropy $H_i^1(t_0, \tau)$ of each node i versus the average value of the Lyapunov exponent in that node, $\lambda_i(t_0, \tau)$. $t_0 =$ July 1, 2011. Blue symbols are from $\tau = 15$ days, green from $\tau = 30$ days, and red from $\tau = 60$ days. Black line is the main diagonal. Source: Ser-Giacomi et al. (2015b). (A black and white version of this figure appears in some formats. For the color version, please refer to the plate section.)

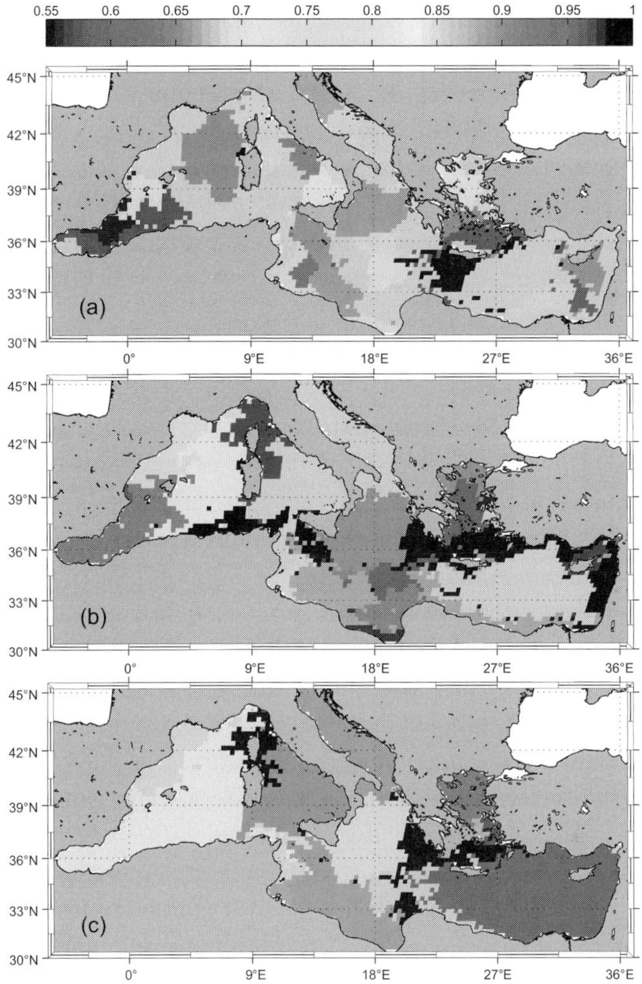

Figure 7.12 *Infomap* partition of flow networks in the Mediterranean Sea, defined by $\mathbf{P}(t_0, \tau)$, into communities or provinces for increasing values of τ. Each province is colored by its coherence ratio value from (4.27), as given in the color bar. In all panels t_0 = July 1, 2011. (a) $\tau = 30$ days; the number of communities is $p = 56$, the global coherence (see Section 4.4.3) $\rho_{t_0}^{\tau}(\mathcal{P}) = 0.76$, and the global mixing $M_{t_0}^{\tau}(\mathcal{P}) = 0.47$. (b) $\tau = 60$ days; $p = 33$, $\rho_{t_0}^{\tau}(\mathcal{P}) = 0.73$, $M_{t_0}^{\tau}(\mathcal{P}) = 0.54$. (c) $\tau = 90$ days; $p = 22$, $\rho_{t_0}^{\tau}(\mathcal{P}) = 0.80$, $M_{t_0}^{\tau}(\mathcal{P}) = 0.59$. Source: Ser-Giacomi et al. (2015b). (A black and white version of this figure appears in some formats. For the color version, please refer to the plate section.)

considering environmental conservation strategies and marine reserves (Nilsson-Jacobi et al., 2012; Rossi et al., 2014; Thomas et al., 2014), where it is important to investigate the connectivity among marine zones, or provinces (Rossi et al., 2014).

The idea of bounded regions that remain weakly mixed with other zones can be formalized using the approach of almost-invariant sets (Froyland, 2005; Froyland and Dellnitz, 2003; Speetjens et al., 2013), which employ spectral techniques for graph partition (Newman, 2010). They use the eigenvectors or singular vectors of the transport matrix (or other matrices derived from it). The methodologies used by Froyland and Dellnitz (2003) and Froyland (2005) found almost-invariant sets, in the sense that loss and gain of fluid was minimized. However, the condition of strong internal mixing, which is important in geophysical applications, was not imposed.

Here we present examples of community detection in the surface of the Mediterranean Sea, from Rossi et al. (2014) and Ser-Giacomi et al. (2015b), where the *Infomap* algorithm (Rosvall and Bergstrom, 2008) was used. The method (see also Section 4.3) is based on the probability flow of random walks in the network moving with transition probabilities given by the adjacency matrix $\mathbf{P}(t_0, \tau)$, and on exploiting the properties of information compression in the description of that probability flow. *Infomap* finds the network partition minimizing the average size of the codeword needed to describe inter- and intra-community transitions. Details of the method and code implementations can be found at www.mapequation.org. This methodology is especially suited to partition flow networks for the following reasons: First, it takes into account the "direction" and "weight" of each link, which are crucial for defining the flow network. In contrast, standard spectral methods and most modularity-optimization algorithms take as input an undirected version of the network. Second, *Infomap* does not require to fix *a priori* the number of communities forming the domain partition. Third, *Infomap* does not impose similar sizes to the communities so that it does not suffer from the "resolution limit" (Fortunato and Barthélemy, 2007) restricting the minimum community size detectable by most algorithms, including spectral methods. This is important in geophysical flow networks since ocean structures of different sizes coexist in the sea, some of them arising from geographical accidents, bathymetry, etc.

However, *Infomap* has some limitations. One of them is the "field of view limit" (Schaub et al., 2012) due to the use of a single-step transition matrix $\mathbf{P}(t_0, \tau)$. In general this imposes that the detected communities are only those with intense intra-community connections (clique-like). For the application of the algorithm to the flow network, this feature was convenient since *Infomap* identified as communities only regions well mixed internally by the flow.

Since *Infomap* considers random walkers exploring the network with the transition probabilities in the matrix $\mathbf{P}(t_0, \tau)$, one could be tempted to confuse these walkers with the Lagrangian particles advected by the flow. However, this is not correct. $\mathbf{P}(t_0, \tau)$ contains relationships between initial and final positions of particles after a time τ, but does not describe the trajectories at intermediate times.

In addition, it cannot be used beyond that time since in time-dependent velocity fields flow connectivity will change with the initial time t_0, defining the dynamic network. *Infomap* unveils the graph structures present in the single matrix $\mathbf{P}(t_0, \tau)$ by releasing random walkers that evolve in a virtual time not directly related to the physical time.

The output of the *Infomap* algorithm applied (Ser-Giacomi et al., 2015b) to the flow network defined by the Mediterranean $\mathbf{P}(t_0, \tau)$, for increasing values of τ, is displayed in Fig. 7.12 (see also Fig. 4.11). Each community A_k is colored with the value of its coherence ratio $\rho^{\tau}_{t_0}(A_k)$ (values close to 0 mean a very leaky community, whereas values close to 1 indicate a very robust and unconnected region, see Section 4.4.3). We see that most coherence values are rather high. The global mixing parameter (see Section 4.4.3) has only moderate values (see caption to Fig. 7.12), but it increases with τ. The main coast-constrained regions appear clearly outlined (the Tyrrhenian, the Adriatic, and the Aegean), but also other areas defined only by persistent circulation patterns (the three-gyre system in the Adriatic and the Balearic front). We refer the reader to Rossi et al. (2014) for a thorough interpretation of the hydrodynamic provinces in relation to surface circulation patterns and known eco-regionalization of the Mediterranean basin. Note that there is no obvious relationship between the size of a community and its coherence. Both large and small provinces may have indeed moderate (<0.6) or high (>0.8) coherence ratios. The detection of small communities confirms that *Infomap* is not affected by the "resolution limit" (Fortunato and Barthélemy, 2007). With increasing τ, communities merge and on average become larger, thus their number decreases.

Because of the turbulent nature of oceanic motions, the community decomposition changes with t_0. Some communities (even of small size) are repeatedly observed while some others appear and disappear when changing t_0. In order to identify persistent communities, i.e., those whose limits are relatively stable in space and time, Ser-Giacomi et al. (2015b) explored two averaging procedures leading to a mean "climatological" community partition. In a first approach a number of matrices $\mathbf{P}(t_0, \tau)$ were averaged, corresponding to the same starting date (e.g., January 1) for the ten different years of the data set (e.g., January 1, 2002, January 1, 2003, etc. until January 1, 2011). Figure 7.13 shows the *Infomap* partition of the network defined by the average matrix $\overline{\mathbf{P}(t_0, \tau)}$ made with the ten matrices $\mathbf{P}(t_0, \tau)$, using the same starting date for each of the ten years (2002–2011). An example of t_0 in winter and another one in summer are displayed. The figure shows the most persistent communities for a particular month, averaging out the variability that occurs over ten years. We note that some communities have a rather small size (most of them reflecting shallow oceanic regions such as continental shelves), and that there is some interseasonal variability.

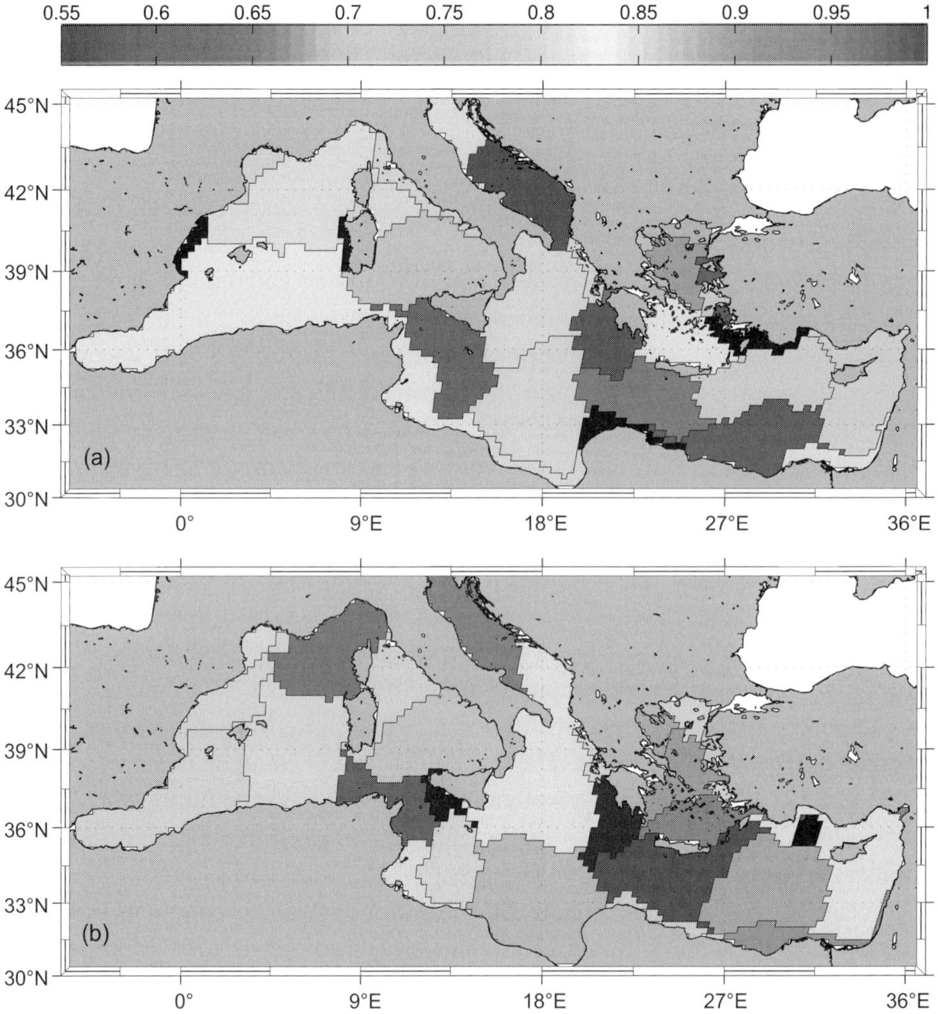

Figure 7.13 *Infomap* communities obtained from the average networks given by $\overline{\mathbf{P}(t_0, \tau)}$, with $\tau = 30$ days. Each community is colored by its coherence ratio. (a) The average is over the ten matrices corresponding to $t_0 =$ January 1 in ten years (2002–2011) of simulation; the number of communities is $p = 34$, the global coherence $\rho^{\tau}_{t_0}(\mathcal{P}) = 0.78$, and the global mixing $M^{\tau}_{t_0}(\mathcal{P}) = 0.68$. (b) The average is over the ten matrices corresponding to $t_0 =$ July 1 in the ten years 2002–2011; $p = 30$, $\rho^{\tau}_{t_0}(\mathcal{P}) = 0.77$, $M^{\tau}_{t_0}(\mathcal{P}) = 0.69$. Source: Ser-Giacomi et al. (2015b). (A black and white version of this figure appears in some formats. For the color version, please refer to the plate section.)

A second approach to obtain average or climatological description of the community partition is illustrated in Fig. 7.14. Instead of applying *Infomap* once on an averaged transport matrix, it is applied ten times separately on the ten

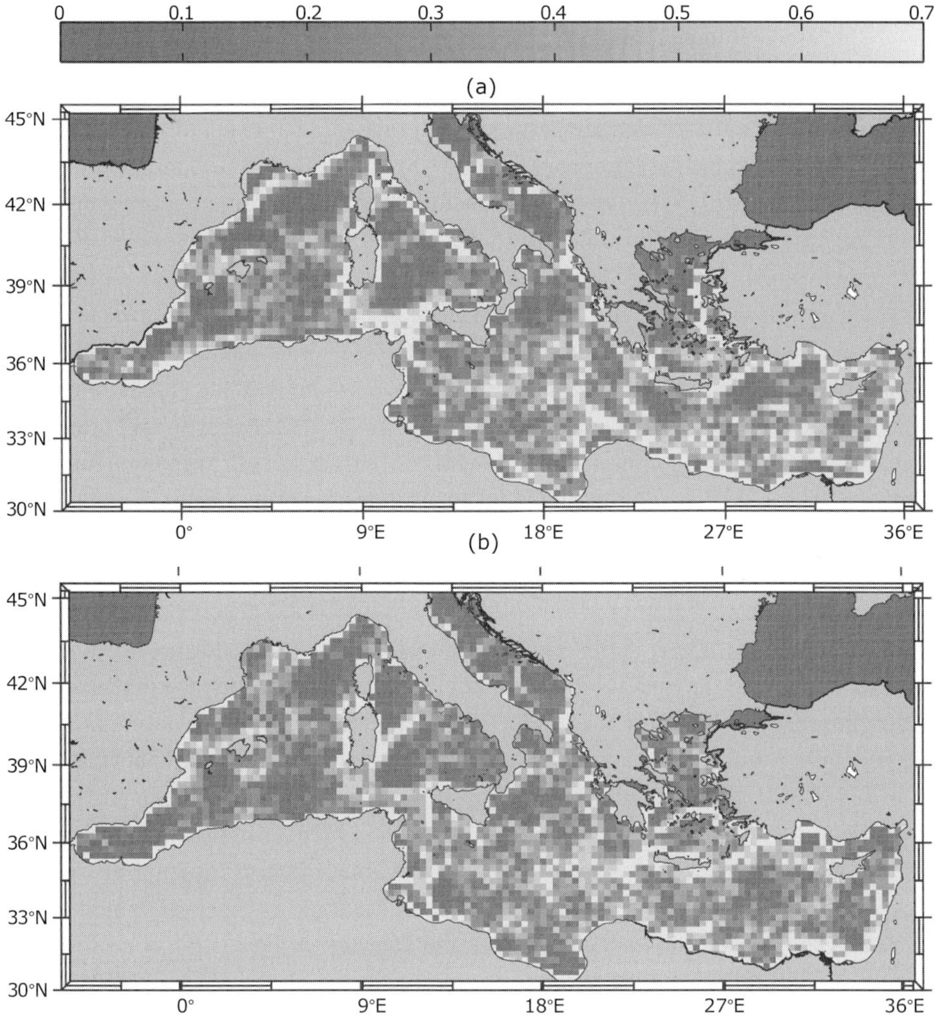

Figure 7.14 Persistence of community borders over time. Gray-level code indicates the proportion of times one of the borders between communities has appeared at a given location, with lighter shades indicating larger proportions. $\tau = 30$ days. (a) $t_0 =$ January 1 of (2002–2011). (b) $t_0 =$ July 1 of (2002–2011). Source: Ser-Giacomi et al. (2015b).

transport matrices corresponding to the same starting date for each of the ten years (2002–2011). The color at a particular location of Fig. 7.14 indicates the frequency of occurrence (in these ten partitions) at that location of a border between communities. Then, lighter gray indicates a more persistent community border. The strongest lines would represent true "barriers to transport" which remain fixed in space. Fuzzier lines may indicate intermittent border appearance, but also a larger wandering amplitude. Figures 7.14a, b display the situation in the same winter and summer days as in Fig. 7.13.

Different methods based on the spectral properties of transport matrices have been previously used to identify and locate almost-invariant sets in flows (Dellnitz et al., 2009; Froyland, 2005; Froyland and Dellnitz, 2003; Froyland et al., 2007; Speetjens et al., 2013). They exploit the fact that for a set to remain almost invariant after the effect of the flow, it has to be related with eigenvectors of $\mathbf{P}(t_0, \tau)$, with eigenvalues close to 1. Ser-Giacomi et al. (2015b) compared the partitioning obtained by *Infomap* with the one from those spectral methods, specifically with the method described by Froyland and Dellnitz (2003). The technique in this last paper obtains a partition $\mathcal{P}$ minimizing in an approximate way the global coherence $\rho_{t_0}^{\tau}(\mathcal{P})$ (4.28). To this end it computes eigenvectors associated to nearly vanishing eigenvalues of the Laplacian matrix (Newman, 2010) obtained from the symmetric part of $\mathbf{P}(t_0, \tau)$, and combines them using a fuzzy c-means clustering algorithm (Froyland and Dellnitz, 2003). Note that this approach eliminates any directionality information present in the transport network. Also, the c-means clustering can define as a single community pieces of the ocean which are geographically disjoint or in fact quite far apart, if this enhances the coherence defined in (4.28). In this approach, one has to specify the number of eigenvectors being combined (ten were chosen by Ser-Giacomi et al. (2015b)) and the number of communities in the partition. Figure 7.15 shows the results using the same average matrix $\overline{\mathbf{P}(t_0, \tau)}$ as in Fig. 7.13b, and imposing a partition in 10 and 14 communities. The change in the number of communities leads to rearrangements in the Tyrrhenian, the central Mediterranean, the Aegean, and the Levantine basin. In panel a, some of the communities are made of disjoint pieces. A larger number of communities decreases the global coherence ratio (see the caption to Fig. 7.15). If one tries to increase the number of communities to approach the one given by *Infomap*, the clustering algorithm becomes unstable. Instabilities also occur when the number of links in the transport network becomes too high (as occurring for example when increasing τ beyond one month).

When compared with the *Infomap* decomposition we see that several of the boundaries coincide. But there are important differences, such as the wider range of community sizes and the sharper details revealed by *Infomap*. This is because a constraint of similar sizes for the communities associated to the same eigenvector needs to be imposed in the spectral method. When clustering several eigenvectors together, this limitation is partially bypassed but still not removed. The values of the coherence ratio are of the same order or somehow larger for the spectral method, but note that the number of spectral communities has been kept much smaller to avoid the instabilities in the clustering algorithm. Since merging two communities into a single one increases the global coherence, joining some of the

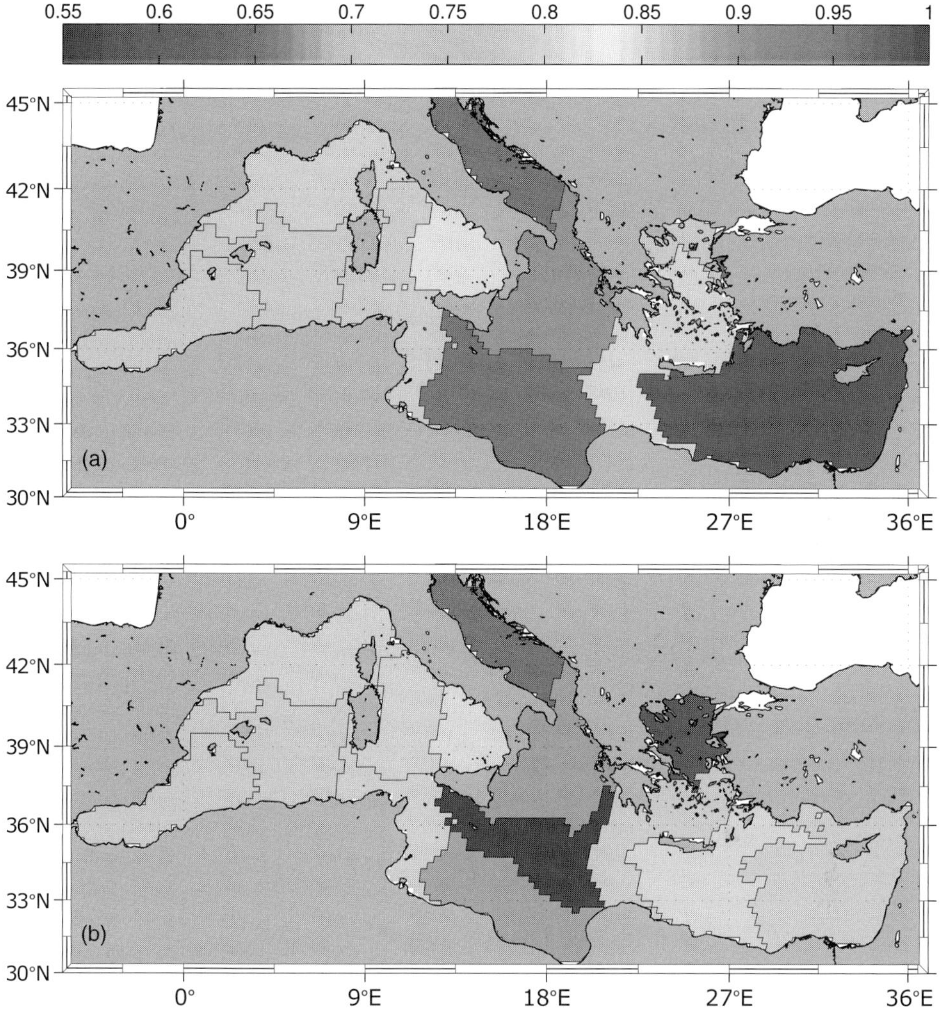

Figure 7.15 Community decomposition by the spectral method with fuzzy c-means clustering described by Froyland and Dellnitz (2003). The matrix used is the same average $\overline{\mathbf{P}(t_0, \tau)}$ as in Fig. 7.13b, i.e., with $t_0 =$ July 1, averaged over the ten years 2002–2011, and $\tau = 30$ days. Ten eigenvalues are used. (a) The number of communities is fixed to be $p = 10$; the global coherence is $\rho^{\tau}_{t_0}(\mathcal{P}) = 0.85$; and the global mixing is $M^{\tau}_{t_0}(\mathcal{P}) = 0.62$. In the Aegean, the southern yellow community is the only independent one: The portions of the Aegean further north are clustered by the c-means algorithm as being part of the same province as areas in the central Mediterranean with the same color. (b) $p = 14$; $\rho^{\tau}_{t_0}(\mathcal{P}) = 0.78$, $M^{\tau}_{t_0}(\mathcal{P}) = 0.64$. Source: Ser-Giacomi et al. (2015b). (A black and white version of this figure appears in some formats. For the color version, please refer to the plate section.)

Infomap communities in Fig. 7.13 until arriving to 10 or 14 communities as in Fig. 7.15 would give rather large values of $\rho_{t_0}^{\tau}(\mathcal{P})$. As expected, the global mixing parameter is larger for the *Infomap* partition, but only by a small amount, reflecting that, even if internal mixing is not imposed in the spectral method, it is achieved to a reasonable extent.

From the methodological point of view, *Infomap* presents the advantage of determining itself the number of communities in the partition, whereas this needs to be fixed *a priori* (as well as the number of eigenvectors to be clustered) in the spectral approach. On the other hand, the spectral method is formulated as an algorithm to minimize the global coherence ratio, a quantity with a clear physical meaning. The quantity optimized by *Infomap* is a codeword length (Rosvall and Bergstrom, 2008), an abstract information-theoretic object without a clear physical meaning. The heuristic interpretation of the optimization process leads to the "large internal–small external connectivity" property for the communities.

The results of this section indicate that the *Infomap* methodology proposed by Rossi et al. (2014) and Ser-Giacomi et al. (2015b) to identify coherent fluid regions seems more appropriate than spectral methods when a wide range of community sizes is expected, when internal mixing is a key parameter, or to minimize user input (such as entering the number of communities). Spectral methods seem appropriate when one is looking precisely for the sets defined mathematically as almost-invariant, the coherence ratio describes well the desired properties of the partition, and one expects a limited range of sizes.

7.4 Optimal Mediterranean Flow Paths

The concept of optimal paths in flow networks (Section 4.4.4; see also Section 6.5) has been also applied to networks describing surface water transport in the Mediterranean Sea. Ser-Giacomi et al. (2015c) computed (from the same NEMO velocity data as in the previous sections) most probable paths (MPPs) of M steps, with $6 \leq M \leq 9$. The time step was $\tau = 10$ days so that during the total path duration (i.e., in between two and three months), the horizontal-flow assumption remains a good approximation. Ten temporal networks (i.e., sequences of M flow networks of duration τ) were built, each one having t_0 as January 1 of each of the years available in the simulation database (2002–2011).

Figure 7.16 shows the set of all the MPPs of $M = 9$ time steps starting from a given node in one of the temporal networks (the one corresponding to 2011). Notice how many different connections are possible from a single starting node. The P_{IJ}^{M} values span several orders of magnitude and this behavior is typical for the distribution of probability across MPPs. We stress here that MPPs do not coincide

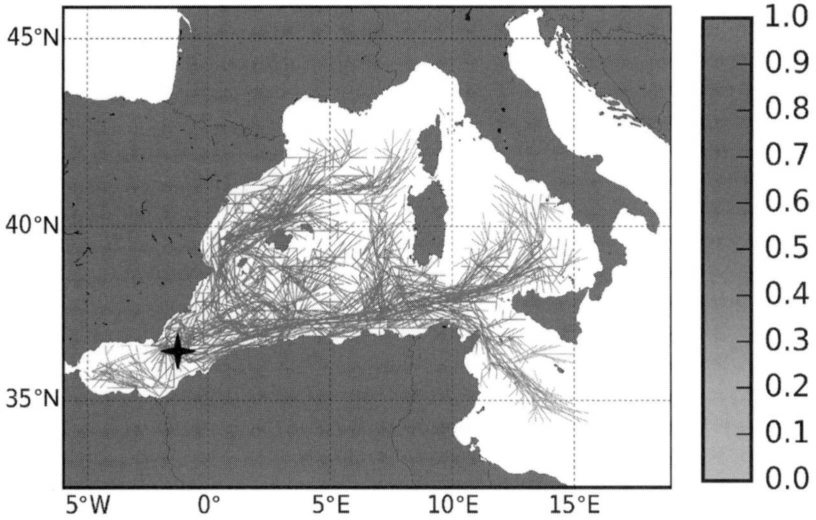

Figure 7.16 Paths of $M = 9$ steps (three months) in the Mediterranean flow network with starting date January 1, 2011, represented as straight segments joining the path nodes. Most probable paths (MPPs) originating from a single node (black star) and ending in all accessible nodes. Color gives the P_{IJ}^M value of the paths in a normalized log-scale between the minimum value (10^{-15}, light turquoise) and the maximum (10^{-5}, dark pink). Source: Ser-Giacomi et al. (2015b). (A black and white version of this figure appears in some formats. For the color version, please refer to the plate section.)

in general with fastest paths: The fastest connection among two nodes is not always the most probable one (Ser-Giacomi et al., 2015c), stressing the importance of a weighted description of the network.

To assess how representative of the whole dynamics are MPPs such as the ones shown in Fig. 7.16, we show in Fig. 7.17 the distribution of λ_{IJ}^M and $\lambda_{IJ}^M(\epsilon)$ for two values of M. These quantities (see Section 4.4.4) are the fraction of the total flow which is carried by the MPP and by a highly probable path (HPP) set, respectively. The distributions are collected from the λ-values of the whole set of accessible pairs (I, J) in the ten temporal networks. For small M most of the MPP have significant λ-values, but as M increases the peak in the distribution of λ_{IJ}^M shifts toward zero as a consequence of the huge increase with M of the number of available paths between two nodes. Then, it becomes important to consider larger sets of paths such as $\mathcal{K}_{IJ}^M(\epsilon)$. For the cases plotted, i.e., $M = 5, 6$ and $\epsilon = 0.1$, the mean values of $\lambda_{IJ}^M(\epsilon)$ are around 0.5. This means that, despite that $\mathcal{K}_{IJ}^M(\epsilon)$ may not contain the full set of paths which individually carry a probability larger than ϵP_{IJ}^M, it is large enough so that, for most of the (I, J) pairs, it contains globally over 50% of the connection probability. This result further gains meaning if we consider that the

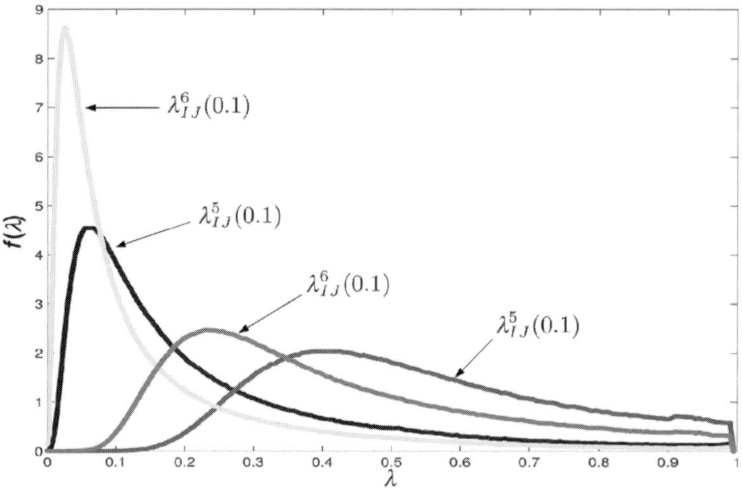

Figure 7.17 Normalized histogram $f(\lambda)$ of coefficients λ^M_{IJ} for $M = 5$ and $M = 6$, and $\lambda^M_{IJ}(\epsilon = 0.1)$ for $M = 5$ and $M = 6$. The statistics are compiled from all connected pairs of nodes (I, J) and the ten temporal flow networks corresponding to the first months of the ten years of velocity data. The mean values are: $\langle \lambda^5_{IJ} \rangle = 0.24$; $\langle \lambda^6_{IJ} \rangle = 0.16$; $\langle \lambda^5_{IJ}(0.1) \rangle = 0.52$; $\langle \lambda^6_{IJ}(0.1) \rangle = 0.42$. Source: Ser-Giacomi et al. (2015c).

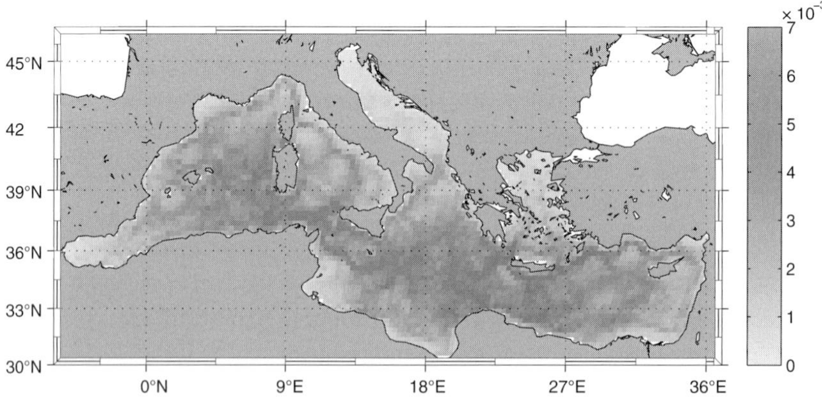

Figure 7.18 Multi-step MPP-betweenness $\mathcal{B}_K^{[6,9]}$ at each geographical node K, computed for each of the ten (2002–2011) temporal networks and then averaged over them. Source: Ser-Giacomi et al. (2015c).

number of paths in $\mathcal{K}^M_{IJ}(\epsilon)$ is on average well below 1% of the total number of paths of $M = 5$ and 6 steps. Hence, despite the strong particle dispersion characterizing our flow networks it is true that only a small subset of paths contribute significantly

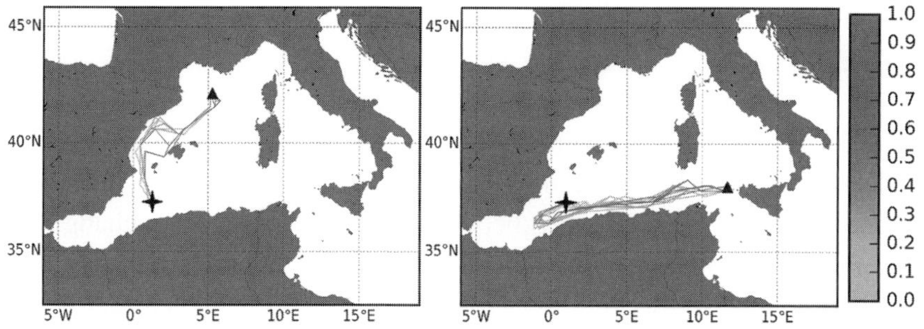

Figure 7.19 All the paths of $M = 9$ steps (three months) in the $\mathcal{K}_{IJ}^{M}(\epsilon)$ set with $\epsilon = 0.1$, represented as straight segments joining the path nodes, initial point marked by a cross and final point marked by a triangle. Starting date January 1, 2011. (a) The 18 HPPs, out of a total of 54/,276 paths between the two sites. The MPP, with $P_{IJ}^{M} = 3 \times 10^{-9}$, is displayed in dark red, whereas the other paths are colored with a normalized logarithmic scale according to their $(p_{IJ}^{M})_{\mu}$ values in $[\epsilon P_{IJ}^{M}, P_{IJ}^{M}]$. (b) shows the 39 HPPs, out of a total of 61×10^{6}, in a similar logarithmic scale normalized in $[\epsilon P_{IJ}^{M}, P_{IJ}^{M}]$ with $P_{IJ}^{M} = 1.4 \times 10^{-6}$. Source: Ser-Giacomi et al. (2015c). (A black and white version of this figure appears in some formats. For the color version, please refer to the plate section.)

to the main transport features. This conclusion will also show up when studying global network properties, such as the betweenness centrality measure.

In Section 4.4.5 we introduced the concept of betweenness of a flow network as a quantity measuring the number of optimal paths passing through a particular node, and thus highlighting the regions acting as bottlenecks for the flow. In Fig. 7.18 we show the multistep MPP-betweenness $\mathcal{B}_{K}^{[6,9]}$, averaged over the ten networks considered, i.e., representative of the (2002–2011) period. Ser-Giacomi et al. (2015c) noticed that the distribution of the betweenness decreases exponentially at large betweenness values so that there are not strong hubs in the network. Spatial patterns determined by the transport dynamics of the flow are clearly evident in the figure, where high betweenness areas are organized in one dimensional-like structures corresponding to the main corridors of transport, i.e., narrow pathways that connect different regions of the ocean. Main paths of the Mediterranean Sea are observed, like Cyprus and Rhodes Gyres, surrounding the Ionian basin, the Algerian current, and those along the Sicily strait, etc. Note that because of the ten-year average, individual short-lived mesoscale features (eddies and fronts) are averaged out. We note, however, that, despite this inter-year average, MPP-betweenness maps remain dependent on the starting date t_0. These observations highlight the seasonal and inter-annual variability of the flow and justifies further the need for a time-dependent network approach as opposed to a fully averaged static network description.

Ser-Giacomi et al. (2015c) addressed the robustness of the methodology by checking the stability of the results on MPP-betweenness under different conditions. First, dealing with temporal networks it is important to understand how much the results are affected by the choice of the time step duration τ (Ribeiro et al., 2013). This issue was checked considering different M and τ values but the same total duration. Results for the MPP-betweenness remain nearly unchanged when keeping a constant total duration $M\tau$, confirming robustness under variations of the temporal resolution. Second, Ser-Giacomi et al. (2015c) noted that the MPP-betweenness does not significantly change when computed with just the 50% of MPPs with larger values of λ (i.e., when using a threshold to retain only the most significant MPPs). Finally, to support the interpretation of most probable paths as main carriers of connectivity, betweenness was also calculated from HPP subsets, i.e., $g^M_{IJ;K}$ is now the number of times node K appears in the set $\mathcal{K}^M_{IJ}(\epsilon)$ of HPPs between I and J, with $\epsilon = 0.1$. No important differences were found between this calculation and the one involving only MPPs. Indeed the Pearson correlation coefficient between the two betweennesses was larger than 0.9. Hence, despite the MPPs representing a small portion of the paths in the $\mathcal{K}^M_{IJ}(\epsilon)$ subsets (between 3 and 10% for $\epsilon = 0.1$, depending on the value of M), which is itself a very small fraction of the full set of paths in the network, they seem to be representative of the main spatiotemporal structures describing the global dynamics. To further substantiate this, Fig. 7.19 displays sets of HPP between two nodes and shows that most of the relevant paths remain spatially close to the MPP. This observation is confirmed by calculations of the spatial dispersion between paths in $\mathcal{K}^M_{IJ}(\epsilon)$, whose average turns out to be of the order of the size of the boxes defining the nodes.

In this chapter we have seen how network techniques give information about oceanic processes which are hard or impossible to obtain by other techniques. Propagation of waves or anomalies leaves a trace into the correlations between climatic variables at different locations, which are exploited by the network approach. Also, the methodology of flow networks allows us to quantify transport, mixing, communities, and pathways in marine systems. All of this is achieved with rather standard network measures for which a clear physical interpretation is often available in the ocean context.

8

Climate Tipping Behavior

The occurrence of sharp transitions to different states or regimes during the evolution of complex systems is a phenomenon of great importance both from the fundamental point of view and for practical implementations of control and management. Examples of such abrupt changes can be found in ecology (Scheffer et al., 2001), economy (Yan et al., 2010), electrical engineering (Dobson et al., 2007), physiology (van de Leemput et al., 2014), or climate (Dakos et al., 2008; Lenton et al., 2012; Thompson and Sieber, 2011). Variables or components in a complex system which are prone to undergo such abrupt changes are called *tipping elements*. In this chapter, we will show applications of network methods to several climate tipping elements. In Section 8.1, an introduction to climate tipping elements is given. This is followed by general considerations on critical slowdown as an indication that a tipping point is being reached (in Section 8.2), followed by sections each discussing network techniques to look at the tipping phenomena of the Atlantic Meridional Overturning Circulation (MOC) (Section 8.3), desertification (Section 8.4) and the El Niño-Southern Oscillation (ENSO) (Section 8.5).

8.1 Climate Tipping Elements

It came as a surprise (Clement and Peterson, 2008) from proxy data that temperatures in the past (e.g., in Greenland) have fluctuated drastically. During so-called Dansgaard–Oeschger events, mean temperature changes of about 5°C occurred in Greenland over a period of about 1500 years, with a relatively long cooling phase followed by relatively rapid warming.

Since evidence of abrupt climate change was found in ice cores (Dansgaard, 1993; Johnsen et al., 1992), there have been a number of review articles on the mechanisms behind these changes (Broecker, 2006; Clement and Peterson, 2008; Marotzke, 2000; Overpeck and Cole, 2006) and the ramifications for abrupt change in the future (Alley et al., 2003; McNeall et al., 2011). A number of components in the climate system with the potential to undergo abrupt change have been identified, mostly in idealized models, and are now often referred to as *climate tipping elements* (Lenton et al., 2008; Levermann, 2011). Each tipping element is conjectured to undergo abrupt change when an external parameter reaches a certain critical level, the "tipping point" (Gladwell, 2000).

Interestingly, abrupt climate change was often first found in models, before reconstructions had revealed that such changes occurred in reality. Among these models were idealized representations of the Atlantic MOC (Stommel, 1961), global ice coverage (Budyko, 1969; Sellers, 1969), vegetation–atmosphere interaction in North Africa (Brovkin and Claussen, 1998), and the Asian Monsoon system (Levermann et al., 2009). Each of these models captures a positive feedback that results from two processes enhancing each other, for example ice loss and warming, or vegetation loss and drying. Feedbacks are often defined in terms of interactions between different variables of a system (DeAngelis et al., 1986). However, in open systems like the climate, one is interested in the response of the system's long-term state to changes in external conditions.

Positive feedbacks increase the susceptibility of a system to such external changes. For example, the increase in atmospheric water vapor and the loss of ice surface on Earth amplify global warming. However, they do not necessarily imply a nonlinear response or abrupt change, which requires the balance of feedbacks to be state-dependent. In the models mentioned above, such state-dependence is often due to bounded variables like surface cover fractions, or to nonlinear processes. If a positive feedback is very strong, alternative stable states can exist under the same external conditions. The parameter values at which the system dynamics suddenly change (e.g., at which one state disappears or becomes unstable so that the system jumps to another state) can be conceptualized as bifurcation points. Fig. 8.1 shows a hypothetical system with bifurcation points at L_1 and L_2. If the control parameter is driven to pass through one of these values, an abrupt and irreversible shift toward a different equilibrium will occur.

It is no coincidence that we find this iconic behavior in many idealized models: If a nonlinear system is simplified by neglecting terms of high order in the distance to the bifurcation point, only a small number of generic and structurally robust bifurcations are possible (Guckenheimer and Holmes, 1990; Kuehn, 2013), and only some of them are catastrophic, i.e., allow for abrupt change (Thompson and Sieber, 2011). Despite this generic mathematical behavior, it is obvious that reality is much more complicated, for example because of spatial heterogeneity and non-stationarity. Even deciding if an abrupt jump has really occurred in a given noisy time series sometimes requires nontrivial techniques (Andersen et al., 2009).

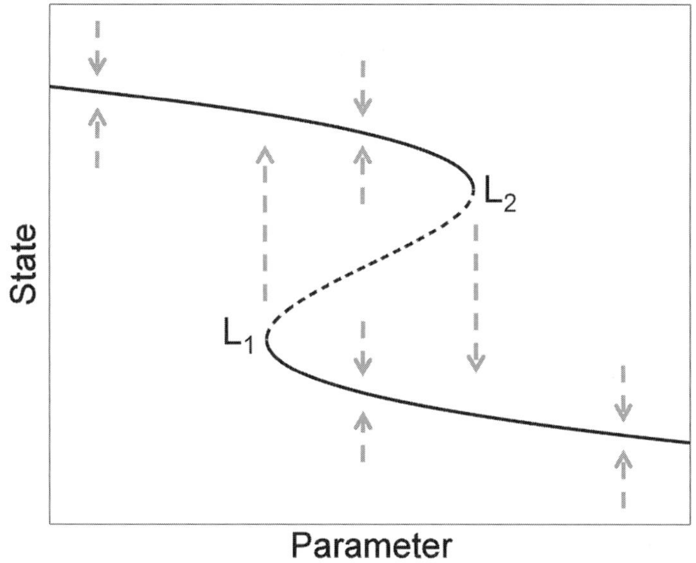

Figure 8.1 Sketch of the occurrence of multiple states and bifurcation points. The equilibria of the system's state are shown as black lines (continuous: stable; dashed: unstable), depending on some control parameter. The dynamics toward a stable state are shown as dashed gray arrows. L_1 and L_2 indicate the saddle-node (or fold) bifurcation points, also referred to as limit or turning points. Source: Bathiany et al. (2016).

In principle, two approaches can be distinguished. The first approach addresses the climatic response to a gradual change in conditions that are considered as external to the system, like orbital forcing or CO_2 concentration. A response is considered abrupt if the climatic change suddenly accelerates at a certain point due to mechanisms internal to the climate system, and under only a gradual change in forcing over time. While this definition is useful in the context of anthropogenic climate change, it neglects phenomena that are unforced, or where the forcing is not well known. To this end, a purely phenomenological alternative definition is helpful, which classifies abrupt change as an event that separates two rather stationary-in-time episodes with a distinctly different climate (Collins et al., 2013). In agreement with this second definition, shifts that are due to natural low-frequency climate variability are sometimes also presented as abrupt changes (Overpeck and Cole, 2006). Although the term *abrupt* is not always compelling in the case of natural variability, this second definition reminds us that the contributions of forcing and variability cannot be clearly separated (for example, because external forcing also affects climate variability). An overview of the different tipping elements, as a result of a search in the CMIP5 (cmip.llnl.gov/cmip5/) model database for transitions, is shown in Fig. 8.2.

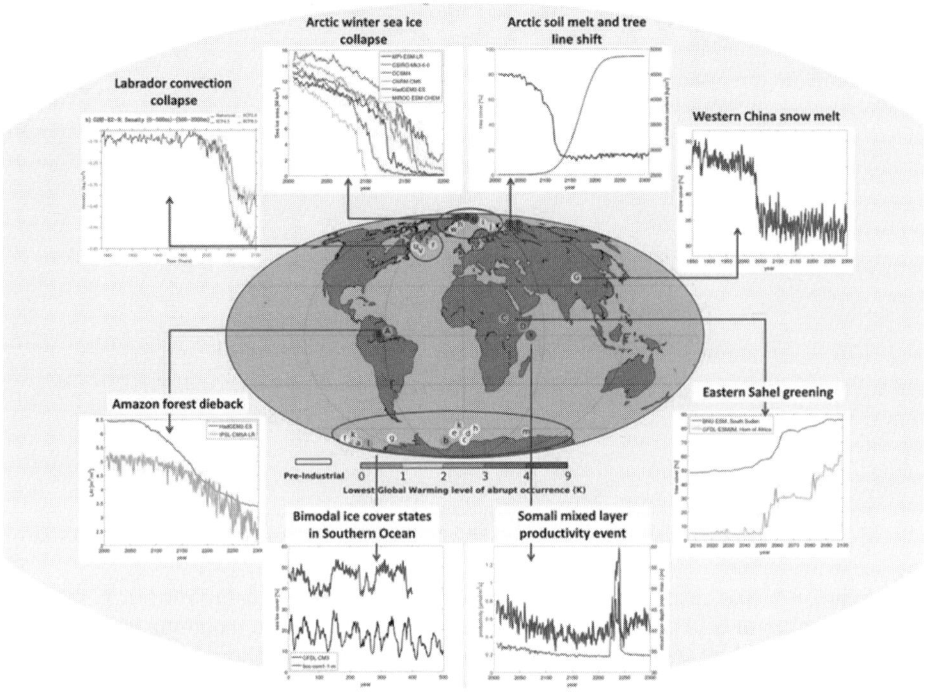

Figure 8.2 Abrupt shifts in complex climate model simulations. Source: Bathiany et al. (2016). (A black and white version of this figure appears in some formats. For the color version, please refer to the plate section.)

8.2 Critical Slowing Down

Over the last few years there has been a growing interest in developing early warning indicators for proximity of tipping points in systems for which the external conditions change very slowly with time (Barnosky et al., 2012; Dakos et al., 2015; Kuehn, 2011; Scheffer et al., 2009, 2012; Wang et al., 2012), including experimental verification in living and environmental systems (Quail et al., 2015; Veraart et al., 2012; Wang et al., 2012). Early warning indicators for regime shifts in ecosystems have been based on either temporal or spatial correlations (Donangelo et al., 2010). However, early warning indicators of the approach to climatic tipping points, as for example the Atlantic MOC, have so far been mostly based on temporal characteristics of the single time series (Held and Kleinen, 2004) or an ensemble of such time series (Livina and Lenton, 2007). The techniques currently used (for an overview, see Lenton, 2011) are based on the concepts of critical slowing down (degenerate fingerprinting; Held and Kleinen, 2004), detrended fluctuation analysis (Livina and Lenton, 2007), and the existence of multiple equilibria (potential analysis; Livina et al., 2010, 2011).

Critical slowing down has been a key phrase in the detection of tipping points in ecosystems (Dakos et al., 2011) and the climate system (Lenton, 2011). Indeed, in many natural and man-made systems there is an increase in response times to perturbations as a tipping point is approached with a parallel increase in related quantities such as auto correlation and variance, which can be connected to the distance to the tipping point. Hence these quantities are frequently proposed as early warning indicators. Critical slowing down and related concepts rely on the loss of resilience occurring generically when dynamical systems approach most (although not all) types of bifurcation points. Recovery rates from perturbation become small, leading to *critical slowing down* of the dynamics, increased memory, long temporal auto-correlations, and to growth of the temporal variance (Dakos et al., 2015; Scheffer et al., 2009; Thompson and Sieber, 2011).

From a dynamical viewpoint, these critical slowing down phenomena appear when the eigenvalue of the Jacobian matrix describing the rate of relaxation toward the attractor approaches zero close to bifurcation points. In many cases the increase in temporal correlation is accompanied by the growth of spatial correlations (Dakos et al., 2010, 2011; Guttal and Jayaprakash, 2009). This, with the associated increase in spatial variance and response functions, is actually a standard method to characterize phase transitions in thermodynamic physical systems (Chaikin and Lubensly, 1995). To understand why the temporal slowing down occurring close to bifurcations impacts spatial correlations, we have to resort to the *center manifold theorem* (Guckenheimer and Holmes, 1990), or to the *slaving principle* (Haken, 1977): Only the linear modes of the Jacobian matrix associated to the eigenvalue which vanishes at the bifurcation evolve slowly; the other ones relax faster and in fact adapt to the slowly evolving one. Thus, if we denote by $X_{t,i}$ the time-dependent deviation of the system variables at location i with respect to the pre-bifurcation state, we should have

$$X_{t,i} \approx f(t) A_{1i} + \cdots , \tag{8.1}$$

where $f(t)$ is a slowly evolving function of time and A_{1i} is the component at location i of the linear bifurcating eigenmode (assuming there is only one). The corrections are unimportant close enough to the bifurcation.

Two observations are now in order: First, if an empirical orthogonal decomposition (Section 3.2.4) is attempted close to the bifurcation, comparison with (3.8) implies that the decomposition will be dominated by a single empirical orthogonal function (EOF), which will be related to the linear mode of the bifurcating eigenvalue $\vec{A}_1$, being $f(t)$, the associated principal component. The existence of one particular EOF capturing most of the variance indicates the possibility of being close to a bifurcation. The second observation is that if the zero-lag Pearson correlation coefficient C_{ij} is calculated between pairs of variables X_i^t and X_j^t given by (8.1), then it is easy to see that $C_{ij} = \pm 1$ (it is enough to notice that the variables are linearly related, $X_{t,i} = (A_{1i}/A_{1j})X_{t,j}$): The spatial configurations described by (8.1) are fully coherent and then correlations extend to the whole system. If a functional network is constructed by linking the locations with strong correlations, a fully connected network results since all nodes have maximal correlation. Deviations from full connectivity only arise from the corrections (the terms in "...") to the complete coherence of (8.1).

One thus expects a growing connectivity in climate networks when the underlying system approaches a bifurcation. In the rest of this chapter we develop these ideas and illustrate them with applications which have been successful to address climatic regime changes and tipping phenomena.

8.3 Atlantic MOC Collapse

The Atlantic MOC is a crucial part of the climate system because of its associated meridional heat transport (Ganachaud and Wunsch, 2000; Johns et al., 2011). Since Stommel's seminal paper (Stommel, 1961), many model studies have shown that the Atlantic MOC may be sensitive to changes in the freshwater balance of the northern North Atlantic (Bryan, 1986). The Atlantic MOC has been identified as one of the important tipping elements in the climate system (Lenton et al., 2008), as it may collapse to a state with a strongly reduced northward heat transport (Bryan, 1986; Rahmstorf, 2000). When an anomalous freshwater flux is gradually applied over a broad swath in the subpolar North Atlantic, the Atlantic MOC collapses in many ocean–climate models (Rahmstorf et al., 2005). Freshening of the surface waters in the Nordic and Labrador Seas inhibits the production of North Atlantic Deep Water (NADW), which feeds the deep southward branch of the Atlantic MOC. The Atlantic MOC collapse occurs due to the existence of a saddle-node bifurcation associated with the salt-advection feedback (Dijkstra and Ghil, 2005; Lenton, 2011).

Although there are studies that argue that such tipping point events are unlikely to occur within the twenty-first century (Tietsche et al., 2011), we should be modest in our confidence in this statement as state-of-the-art General Circulation Models (GCMs) have not been extensively tested for this behavior (Collins et al., 2013; Huisman et al., 2010). In addition, the MOC intensity has been weakening in recent times (Caesar et al., 2018; Thornalley et al., 2018).

The consideration of spatial correlations leading to climate networks (CNs) has led to a novel perspective for finding transition precursors (see Section 8.2).

Several network-based warning-signals have been proposed. Specifically the values of the mean degree, its kurtosis, assortativity, and clustering (see definitions in Section 4.1.1) increase when approaching a tipping point (Feng et al., 2014; Tirabassi et al., 2014; van der Mheen et al., 2013).

In the following we provide an overview of these CN-based early warning indicators for the collapse of the Atlantic MOC, which at a fundamental level are based on spatial correlations. In Section 8.3.1, we give a general overview of the two models of the Atlantic MOC used by van der Mheen et al. (2013) and Feng et al. (2014) to construct the CNs, whereas in the next two sections these networks are built and analyzed with different diagnostics.

8.3.1 Models and Data

To be able to develop an early warning indicator for detecting a tipping point of the MOC, it is beneficial to have a model where such tipping points can be explicitly computed. The study of van der Mheen et al. (2013) used the two-dimensional (meridional-depth) model of the Atlantic MOC, which is presented by Den Toom et al. (2011). It is the B-grid version of the ThermoHaline Circulation Model (De Niet et al., 2007), which is a fully implicit ocean-only model based on the hydrostatic primitive equations in spherical coordinates. It has been shown that the bifurcation diagrams of two-dimensional and three-dimensional models (when only steady states are considered) are qualitatively the same (Weijer and Dijkstra, 2001). There are two active tracers: temperature T and salinity S, which are related to the density ρ by a linear equation of state

$$\rho = \rho_0(1 - \alpha_T(T - T_0) + \alpha_S(S - S_0)), \tag{8.2}$$

where α_T and α_S are the thermal expansion and haline contraction coefficients, respectively, and ρ_0, T_0, and S_0 are reference quantities.

In order to eliminate the longitudinal dependence from the problem, a purely buoyancy-driven two-dimensional flow on a non-rotating Earth is considered, and free-slip conditions at the lateral boundaries (which avoids the necessity of boundary layer formation) are prescribed. Inertia is also neglected in the meridional momentum equation. The mixing of momentum and tracers due to eddies is parameterized by simple anisotropic diffusion. The equations for the meridional velocity v, vertical velocity w, pressure p, and the tracers T and S are given by (e.g., Den Toom et al., 2011)

$$\frac{1}{\rho_0 r_0}\frac{\partial p}{\partial \phi} - A_V\frac{\partial^2 v}{\partial z^2} = \frac{A_H}{r_0^2}\left(\frac{1}{\cos\phi}\frac{\partial}{\partial \phi}\left(\cos\phi\frac{\partial v}{\partial \phi}\right) + (1 - \tan^2\phi)v\right), \tag{8.3a}$$

$$\frac{1}{\rho_0}\frac{\partial p}{\partial z} = g(\alpha_T T - \alpha_S S), \tag{8.3b}$$

$$0 = \frac{1}{r_0 \cos\phi}\frac{\partial v \cos\phi}{\partial \phi} + \frac{\partial w}{\partial z}, \tag{8.3c}$$

$$\frac{\partial T}{\partial t} + \frac{v}{r_0}\frac{\partial T}{\partial \phi} + w\frac{\partial T}{\partial z} = \frac{K_H}{r_0^2 \cos\phi}\frac{\partial}{\partial \phi}\left(\cos\phi\frac{\partial T}{\partial \phi}\right) + K_V\frac{\partial^2 T}{\partial z^2} + \mathrm{CA}(T), \tag{8.3d}$$

$$\frac{\partial S}{\partial t} + \frac{v}{r_0}\frac{\partial S}{\partial \phi} + w\frac{\partial S}{\partial z} = \frac{K_H}{r_0^2 \cos\phi}\frac{\partial}{\partial \phi}\left(\cos\phi\frac{\partial S}{\partial \phi}\right) + K_V\frac{\partial^2 S}{\partial z^2} + \mathrm{CA}(S). \tag{8.3e}$$

Here, t indicates time, ϕ latitude, z the vertical coordinate, r_0 the radius of Earth, g the acceleration due to gravity, A_H (A_V) the horizontal (vertical) eddy viscosity, and K_H (K_V) the horizontal (vertical) eddy diffusivity; the term CA represents the convective adjustment. The introduction of CA is the only change to the model analyzed in section 3a of the paper by Weijer and Dijkstra (2001). It acts solely on the tracers and is implemented as enhanced diffusion, which for a tracer C takes the form

$$\mathrm{CA}(C) = K_V F_0 \frac{\partial}{\partial z}\left(\mathcal{F}\left(\frac{D}{\rho_0}\frac{\partial \rho}{\partial z}\right)\frac{\partial C}{\partial z}\right). \tag{8.4}$$

Here, F_0 is the efficiency of convection and $\mathcal{F}(x) = \max(0, \tanh\left[(\epsilon x)^3\right])$ is a continuous approximation of the Heaviside step function. The steepness of the transition from 0 to 1 is determined by the selectivity ϵ. In the limit $\epsilon \to \infty$ the function $\mathcal{F}$ tends to the step function.

The equations are solved on an equatorially symmetric, flat-bottomed domain. The basin is bounded by southern and northern boundary latitudes ϕ_S and ϕ_N, and has depth D. In order to calculate transports in Sverdrup (1 Sv = 10^6 m^{3}s^{-1}), a longitudinal width of 64° is assumed. The numerical resolution is 3.75° in the horizontal and 250 m in the vertical (a meridional grid of $N_\phi = 32$ points and a vertical grid of $N_z = 16$ points). Free-slip conditions apply at the lateral walls and at the bottom, and rigid lid conditions are assumed at the surface. The wind-stress is zero everywhere, and "mixed" boundary conditions apply for temperature and salinity:

$$K_V\frac{\partial T}{\partial z} = \frac{H_m}{\tau_T}(T_S(\phi) - T) \tag{8.5a}$$

$$K_V\frac{\partial S}{\partial z} = S_0 F_S(\phi), \tag{8.5b}$$

with $T_S(\phi) = 10.0\cos(\pi\phi/\phi_N)$. This formulation implies that temperatures in the upper model layer (of depth H_m) are relaxed to a prescribed profile T_S at a rate τ_T^{-1}, while salinity is forced by a net freshwater flux F_S, which is converted to an

Table 8.1 *Fixed model parameters of the idealized ocean model used in van der Mheen et al. (2013).*

$D = 4.0 \times 10^3$	[m]	$H_m = 2.5 \times 10^2$	[m]
$\phi_N = -\phi_S = 60^\circ$		$\rho_0 = 1.0 \times 10^3$	[kg m^{-3}]
$r_0 = 6.371 \times 10^6$	[m]	$T_0 = 15.0$	[°C]
$g = 9.8$	[m s^{-2}]	$S_0 = 35.0$	[psu]
$A_H = 2.2 \times 10^{12}$	[m^{2}s^{-1}]	$\alpha_T = 1.0 \times 10^{-4}$	[K^{-1}]
$A_V = 1.0 \times 10^{-3}$	[m^{2}s^{-1}]	$\alpha_S = 7.6 \times 10^{-4}$	[psu^{-1}]
$K_H = 1.0 \times 10^3$	[m^{2}s^{-1}]	$\epsilon = 2.0 \times 10^3$	
$K_V = 1.0 \times 10^{-4}$	[m^{2}s^{-1}]	$\tau_T = 75.0$	[days]

equivalent virtual salinity flux by multiplication with S_0. The surface freshwater forcing F_S is chosen as

$$F_S(\phi) = \frac{(\gamma + \eta)}{\cos\phi} \cos \pi \frac{\phi}{\phi_N} + \beta F_p(\phi) - Q, \tag{8.6}$$

where γ (in m/year) is the strength of the background freshwater forcing and η must be a white-noise term, with $\langle \eta(t)\eta(t')\rangle = \eta_0 \delta(t - t')$. Furthermore, β is the strength of an anomalous freshwater flux which is only applied over the area [40°N, 60°N] (where $F_p = 1$; it is zero elsewhere). Given the area over which the freshwater is applied, the total freshwater flux is easily converted to m^3/year or to Sv. The constant Q appears due to the requirement that the surface integrated salt flux is zero for all parameter values. Although in a crude way, these equations express that ocean–atmosphere interaction affects temperature and salinity in a fundamentally different way. The numerical values of the model parameters used (apart from those in the freshwater flux) are summarized in Table 8.1.

The meridional overturning stream function Ψ is defined by

$$w = \frac{1}{r_0 \cos\phi} \frac{\partial \Psi}{\partial \phi}, \tag{8.7a}$$

$$v = -\frac{1}{\cos\phi} \frac{\partial \Psi}{\partial z}, \tag{8.7b}$$

and provides a measure of the volume transport. In the following, when we refer to the *"strength of the MOC"* we refer to the values of Ψ (in Sv).

To compute the bifurcation diagrams (Dijkstra and Ghil, 2005) as a function of the strength of the anomalous freshwater flux β, a pseudo-arclength continuation technique (Keller, 1977) is used. For $\eta_0 = 0$ (deterministic forcing) and $\gamma = 0.3$ m/year, this bifurcation diagram is shown in Fig. 8.3a, which displays a measure of the steady solutions as a function of the forcing parameter β. Two saddle-node bifurcations (indicated by L_1 and L_2), connected by a branch of unstable steady states, are found. Note the similarity with the schematics in Fig. 8.1. The MOC

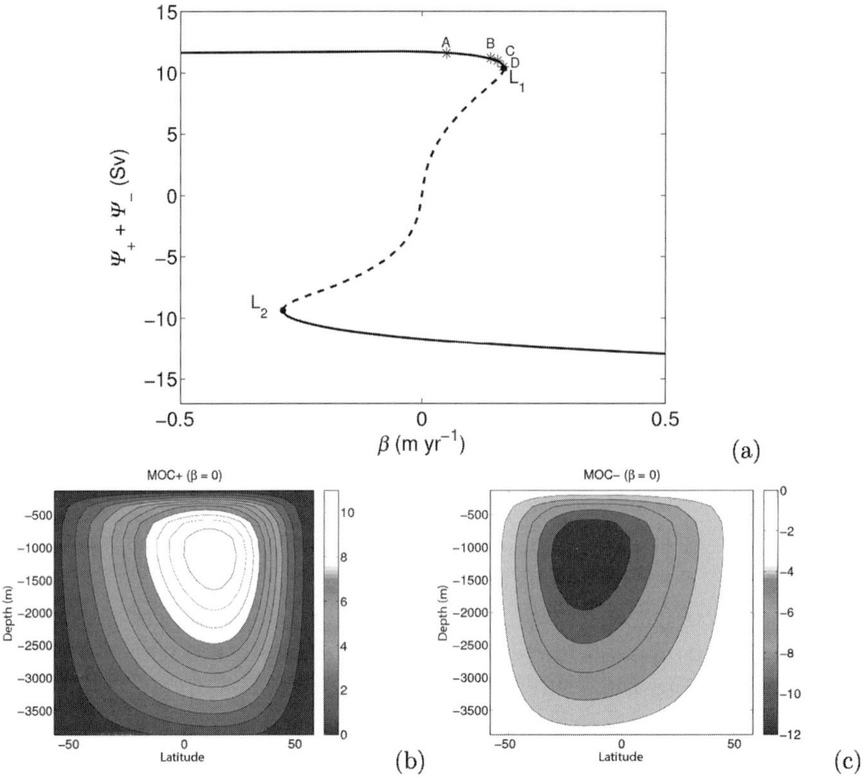

Figure 8.3 (a) Bifurcation diagram in which the sum of the maximum (Ψ_+) meridional overturning stream function and minimum (Ψ_-) meridional overturning stream function values are plotted versus the anomalous freshwater flux amplitude β. Solid (dashed) branches indicate stable (unstable) steady states. (b) Pattern of the meridional overturning stream function (Ψ, in units of $Sv = 10^{-6}$ m^{3}s^{-1}) at the upper stable branch in (a) for $\beta = 0$. (c) Pattern of the meridional overturning stream function at the lower stable branch in (a) also for $\beta = 0$. Source: van der Mheen et al. (2013).

patterns of the coexisting stable steady states for $\beta = 0$ are shown in Figs. 8.3b, c. In van der Mheen et al. (2013), four values of β were chosen close to L_1, as indicated by the labels (A, B, C, and D) in Fig. 8.3a, to construct CNs. Also, starting from the steady state solution, transient computations using a fully implicit Crank–Nicholson integration scheme with time step $\Delta t = 0.1$ year and a noise amplitude $\eta_0 = 0.1$ were run over 500 years. In these simulations the MOC streamfunction fluctuates with a typical peak-to-peak amplitude of about 5 Sv. Longer simulations were run with a time-dependent value of β (Section 8.3.2).

The Fast Met Office/UK Universities Simulator (FAMOUS) model is a version of the HadCM3 (abbreviation for Hadley Centre Coupled Model, version 3, a coupled atmosphere–ocean GCM) with a lower resolution ocean and atmosphere component. The model has a horizontal resolution of 2.5° × 3.75° and a vertical resolution of 20 levels. The details of the model are described by Hawkins et al. (2011). Feng et al. (2014) analyzed with network tools the annual mean of MOC strength data from a control simulation and from a freshwater-perturbed (from now on referred to as "hosing") simulation of the FAMOUS model. In this hosing simulation, the freshwater flux over the extratropical North Atlantic was increased linearly from zero to 1.0 Sv (1 Sv = 10^6 m^3s^{-1}) over 2000 years (Hawkins et al., 2011). Figure 8.4a displays the annual-mean time series of the Atlantic MOC streamfunction value at latitude 26°N and at 1000 m depth (the location of the maximum MOC before the collapse) for the control simulation (green curve) and the hosing simulation (blue curve). Whereas the control MOC values are statistically stationary over the 2000-year integration period, the MOC values for the hosing simulation show a rapid decrease between the years 800–1050.

The MOC streamfunction data from the FAMOUS simulations were used in which the freshwater flux was fixed after a certain integration time and the model was then integrated to the corresponding equilibrium (Hawkins et al., 2011). Time series of the MOC of the last 100 years of these simulations were analyzed and the mean MOC streamfunction values (again at 26°N and 1000 m depth) are plotted as the red dots labeled from 1 to 6 in Fig. 8.4a. Figures 8.4b–d show mean MOC streamfunction patterns of three of these equilibrium simulations (labeled as 1, 4, and 6 in Fig. 8.4a). Although the equilibrium value of the MOC decreases with larger freshwater inflow, the changes in the MOC pattern are relatively minor. For a slightly higher value of the freshwater flux than the one at point 6, the equilibrium solution is in the collapsed state (Hawkins et al., 2011). In the hosing simulation the freshwater is, however, added relatively rapidly compared to the equilibration timescale of the MOC . Hence the MOC maintains its pattern for much higher freshwater inflow and collapse occurs at a delayed time (near 900 years, see Fig. 8.4a).

To determine a collapse time τ_c more precisely, Feng et al. (2014) first determined, with a Student's *t*-test, the time τ_s at which the trend of the MOC time series of the hosing simulation deviates significantly (p = 0.05) from that of the control simulation. Subsequently, the standard deviation σ of the MOC values of the control simulation over the first τ_s years is computed, and the first time at which the MOC anomaly value (with respect to the mean of the first τ_s years) in the hosing simulation drops below 3σ, where σ is the standard deviation, is considered to be the collapsing time τ_c. In the FAMOUS model, when considering the MOC streamfunction values at 26°N and 1000 m depth, values of $\tau_s = 350$ years and $\tau_c = 874$ years (which is shown as the red dashed line in Fig. 8.4a) are found.

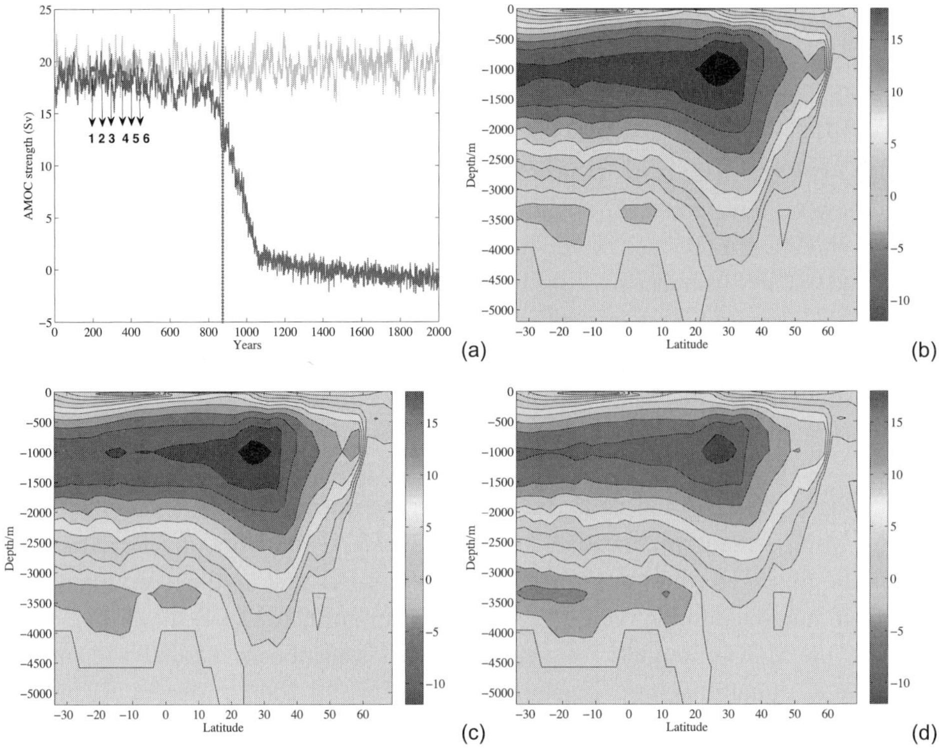

Figure 8.4 (a) Time series of the MOC (in Sv, 1 Sv = 10^6 m^{3}s^{-1}) at 26°N and 1000 m depth in the Atlantic for the control simulation (green curve) and the freshwater perturbed simulation (blue curve) of the FAMOUS model. The red dots labeled from 1 to 6 show the average values of MOC from the corresponding equilibrium simulations and the red broken line indicates the collapse time $\tau_c = 874$ years. (b) Annual mean MOC streamfunction pattern of equilibrium simulation 1. (c) Same as (b) but of equilibrium simulation 4. (d) Same as (b) but of equilibrium simulation 6. Source: Feng et al. (2014). (A black and white version of this figure appears in some formats. For the color version, please refer to the plate section.)

8.3.2 Results for the Two-Dimensional Model

In van der Mheen et al. (2013) zero-lag Pearson correlation climate networks (PCCNs, see Section 4.2) were constructed using the temperature field time series for the different values of β (locations A, B, C, and D in Fig. 8.3a). A threshold value of 0.7 for the correlations was used. The degree fields of these four PCCNs are shown in Fig. 8.5 (left column). Note that the maximum degree of a node (connections to other nodes) is $k_{max} = N_\phi \times N_z - 1 = 32 \times 16 - 1 = 511$. Relatively far from L_1, all temperature grid points have a relatively small degree. When β is increased, high degree first appears at nodes in the deep ocean. Close to

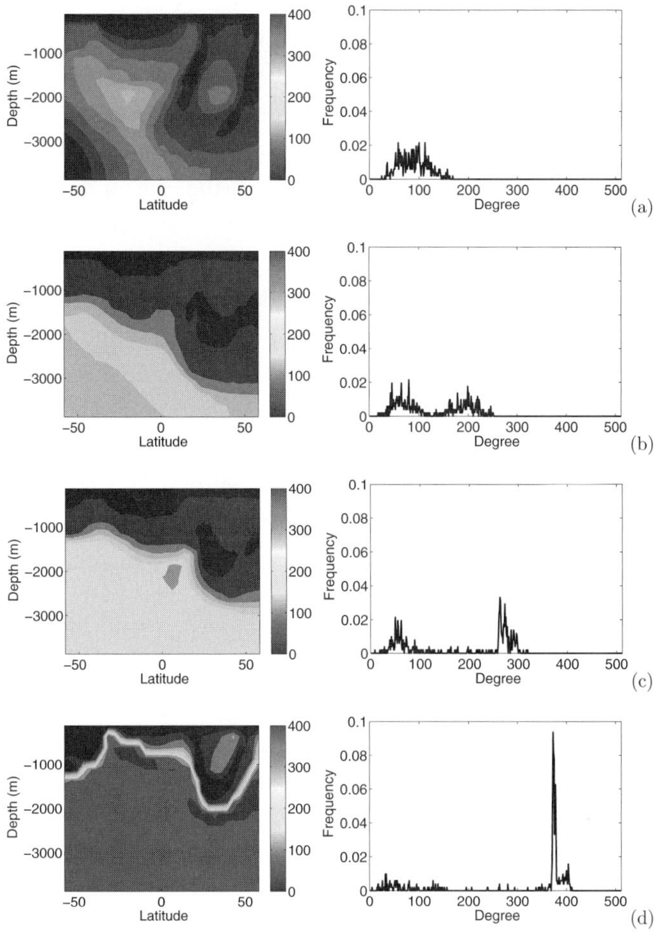

Figure 8.5 Degree fields (left column) and degree distributions (right column) of the PCCNs (with a threshold $\tau = 0.7$) for the temperature field time series obtained at different locations A, B, C, and D, as indicated in Fig. 8.3a under transient noise ($\eta_0 = 0.1$) forcing and $\gamma = 0.3$. (a) Point A: $\beta = 0.05$ m/year; (b) point B: $\beta = 0.139$ m/year; (c) point C: $\beta = 0.154$ m/year; (d) point D: $\beta = 0.166$ m/year. Source: van der Mheen et al. (2013). (A black and white version of this figure appears in some formats. For the color version, please refer to the plate section.)

L_1, there is a large area (below ~1000 m) where the degree values are very large and a relatively low degree only occurs for nodes located in the upper ocean, north of 30°N.

The histograms of the degree fields (the degree distributions) for different β are also plotted in Fig. 8.5 (right column). For relatively small β, the degree distribution is unimodal and the mean is located near $k = 100$. A second peak for $k = 200$ appears in Fig. 8.5b, where the degree distribution becomes bimodal and the maximum degree of the first peak shifts to slightly smaller values of d. This tendency strengthens in Fig. 8.5c, where the maximum of the second peak shifts to even higher degree. For location D, near L_1 in Fig. 8.3a, about 10% of the grid points have a high degree of $k = 370$ (Fig. 8.5d). Also, the clustering coefficient fields and the clustering coefficient distribution show the same pattern changes and peak shift with β (van der Mheen et al., 2013).

The results in Fig. 8.5 can be explained (van der Mheen et al., 2013) by looking at the changes with β of the first few EOFs (see Section 3.2.4 and the considerations in Section 8.2) of the temperature field. The first three EOFs account for about 90% of the variability in each case. From point A up to point C the first two EOFs of the temperature decrease in importance, while the third one increases. In point D, however, variance of EOF1 increases from roughly 45% to 60%, that of EOF2 decreases from 29% to 21%, and that of EOF3 from 17% to 13%. Hence, only very near L_1, the amount of variance explained by EOF1 increases rapidly. When the variance would be completely dominated by only one EOF, the arguments in Section 8.2 imply that the network is fully connected and hence the degree of each node is maximal. The deviation from the maximal degree $k_{max} = 511$ at each node is hence related to the relative variance explained by each of the subdominant EOFs at this node. When L_1 is approached, the amount of variance explained becomes dominated by EOF1 (in particular in the deeper ocean), and hence many nodes in the network will get a high degree (and also a high clustering coefficient).

The results in Fig. 8.5 strongly motivate the use of properties of the degree distribution as early warning indicators of the approach to the tipping point. Although there are values of β where the degree distribution is clearly bimodal, near the tipping point this distribution is changing substantially and again becomes more unimodal and peaked at a high degree value. Hence, the expectation value E_d of the distribution of the normalized degree k/k_{max} is expected to show a strong shift toward larger values as L_1 is approached. A similar measure E_c is computed from the distribution of the clustering coefficients and both are identified in van der Mheen et al. (2013) as efficient indicators of early warning. In Fig. 8.6, the indicator E_d shows a strong increase with β when the tipping point is approached (solid black curve). The same holds for the indicator E_c plotted as the dashed black curve in Fig. 8.6. As soon as $\beta > 0.14$ m/yr, a new regime in spatial correlations appears where the degree and clustering distributions become dominated by high values in the deeper ocean. Both indicators were also computed when only a part of the grid points are used to compute the degree distribution, which may be more comparable with actual observation networks deployed in the real ocean. The gray curves in Fig. 8.6 arise when only the northern and southern or the top and bottom grid points

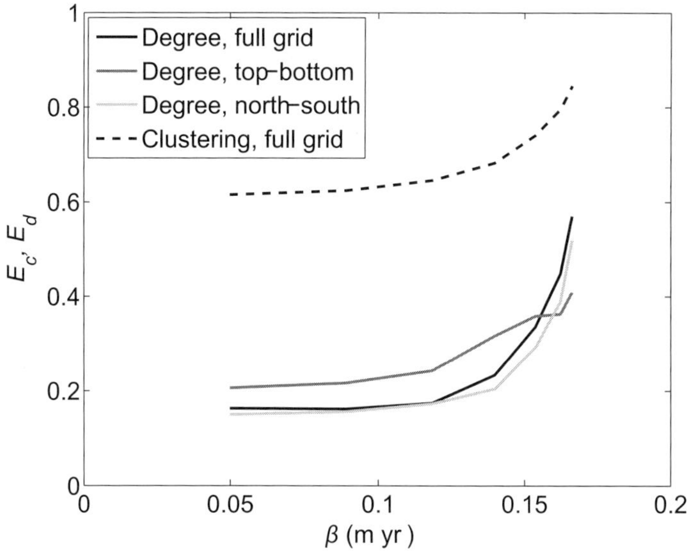

Figure 8.6 Topological properties of the PCCN constructed from the equilibrium states of the two-dimensional MOC model at different values of the freshwater forcing β. The expectation value E_d (solid) of the distribution of the normalized degree (k/k_{max}) and the expectation value E_c (dashed) of the distribution of the clustering coefficient are shown as a function of β. The black curves take into account all the grid points. The gray curves for the degree show the indicators for alternative PCCNs where only the bottom and top or the northern and southern boundary grid points of the domain are considered. Source: van der Mheen et al. (2013).

are taken into account to reconstruct the PCCN. The indicator E_d reconstructed from the north–south grid points smoothly increases with β, as when using all the model grid points, as it captures the high spatial correlations in the deep ocean. The indicator has a worse behavior when considering only the top–bottom grid points.

Besides equilibrated configurations under different β values, van der Mheen et al. (2013) also considered transient simulations under time-varying β. Over a period of 35,000 years, β was linearly increased from 0.05 up to 0.17 m/yr (Fig. 8.7a) and again white noise ($\eta_0 = 0.1$) was applied to the freshwater flux. When the zonal dimension of the basin is assumed to be 64°, this amounts to about 10^{-3} Sv/1000,yr, which is comparable to other model studies (Rahmstorf et al., 2005). A time series of the maximum value of the MOC (Fig. 8.7a) indicates that the MOC starts collapsing after about 27,000 years (indicated by the vertical dashed line in Fig. 8.7). At this time, the actual value of $\beta = 0.142$ yr^{-1} is slightly larger than the value of point B in Fig. 8.3a. Note that because point B is in the

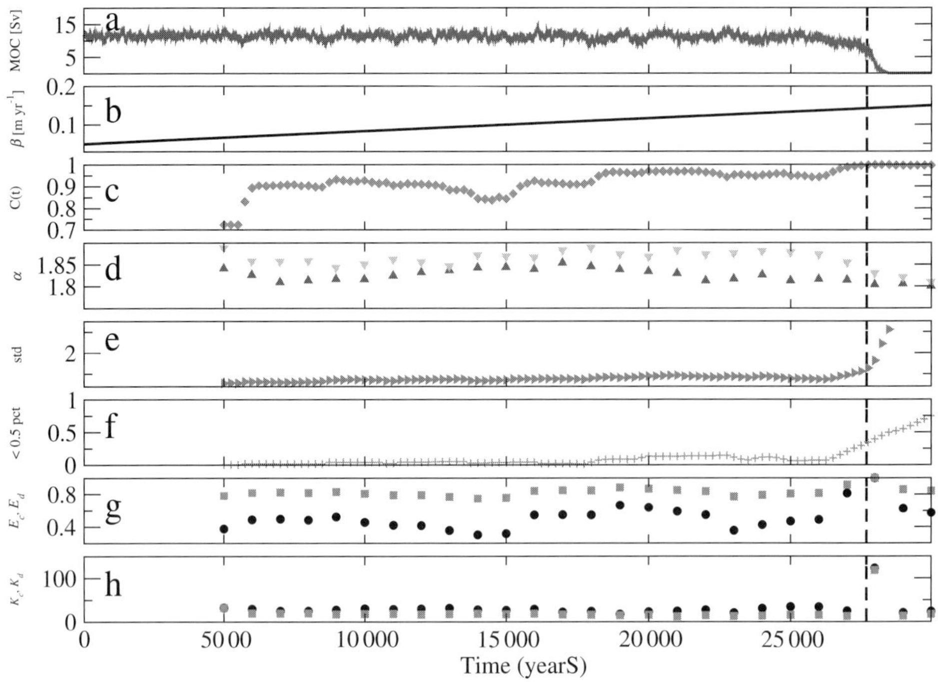

Figure 8.7 Results from transient flow simulations under time-varying forcing β. (a) Time series of the maximum value of the MOC streamfunction. The vertical dashed line indicates the tipping point. (b) Imposed transient change of β versus time, from 0.05 to 0.17 m/yr in 35 000 years. (c) The (rescaled) lag-1 autocorrelation of the projection of the time series onto the first EOF. (d) Power of the fluctuation function, as determined by the DFA procedure, using linear (up marker) and quadratic (down marker) detrending. (e) Standard deviation of the MOC time series shown in (a). (f) Fraction of the time spent below the 0.5 percentile of the MOC amplitude. (g) Results from the PCCN-based indicators E_d (circles) and E_c (squares) for the full network as in Fig. 8.6. (h) Same as (g) but now the kurtosis (K_d [circles] and K_c [squares]) of the two distributions are plotted. Source: van der Mheen et al. (2013).

multiple-equilibria regime, in the presence of noise the MOC can already jump to the low-circulation regime even before reaching L_1.

Previously suggested early warning indicators of the transition in MOC time series are are also plotted in Figs. 8.7c–f. References containing the precise definitions of all of these are listed, for example in Lenton (2011) or Lenton et al. (2012). Below we present only the results for a sliding window of 5000 years, but the robustness of the results with respect to the sliding window (5 000–15 000 years) was previously tested by using the Mann–Kendall trend test (Hamed and Rao, 1998; Lenton et al., 2012).

In Fig. 8.7c, the (rescaled) lag-1 auto-correlation of the projection of the time series onto the first EOF (this is the methodology of degenerate fingerprinting; Held and Kleinen, 2004) is plotted. The aggregation window was 50 years, the first EOF was generated from results of the first 5000 years of the simulation, and the lagged auto-correlation was calculated in a time window of 5000 years. The coefficient reaches its maximum indeed near the transition indicating a critical slowdown of the MOC . However, the indicator is not monotonic and also increases when the MOC is still far from the transition.

Fig. 8.7d shows the "power of the fluctuation function," as determined by the detrended fluctuation analysis (DFA; Livina and Lenton, 2007) methodology, using linear and quadratic detrending. The largest window taken in fitting the fluctuation function to a power law is 100 years. In most of the cases the fitting quality coefficient is close to 1 (perfect fit), and the typical error in the coefficient is 0.5%. The quality of the sampling is assessed with respect to fluctuations of the measure during time. Different orders of polynomial detrending in the DFA procedure, between 1 (linear) and 4 (quartic) were checked and the qualitative behavior is similar (van der Mheen et al., 2013). The average power coefficient is 1.82, 1.85, 1.71, and 1.62, with respect to the four detrending orders 1–4, and there is no warning of a transition.

Fig. 8.7e gives the standard deviation of the MOC record in panel a. Each point is calculated using a sliding window of 5000 years with a shift of 200 years; it indeed shows a steady increase toward the transition but it is difficult to set a threshold for an alarm. In Fig. 8.7f, the fraction of the time spent below the 0.5 percentile (estimated based on the first 5000 years) of the MOC strength is plotted using the same sliding window as in Fig. 8.7e. This indicator displays a much sharper increase but it is a rather ad-hoc measure which also depends on a calibration in the far past.

Finally, the CN-based indicators E_c and E_d are plotted in Fig. 8.7g and the kurtosis of the degree and clustering coefficient distributions in Fig. 8.7h. Both are determined by PCCNs reconstructed from the full spatial temperature field using a sliding window of 5000 years with a shift of 1000 years. Both indicators E_c and E_d exhibit similar behavior with an increase before the collapse. The kurtosis distributions (Fig. 8.7h) are even more clearly showing a strong increase before collapse. By considering reconstructed PCCNs for different sliding window lengths, it was found that the time at which the peak occurs in the indicators is not sensitive to a sliding window length in the range 5000–15 000 years. Similar peaks are found in the standard deviation and skewness of both distributions. Of all the indicators considered, the PCCN-based kurtosis indicators are least susceptible to causing false alarms of the MOC collapse.

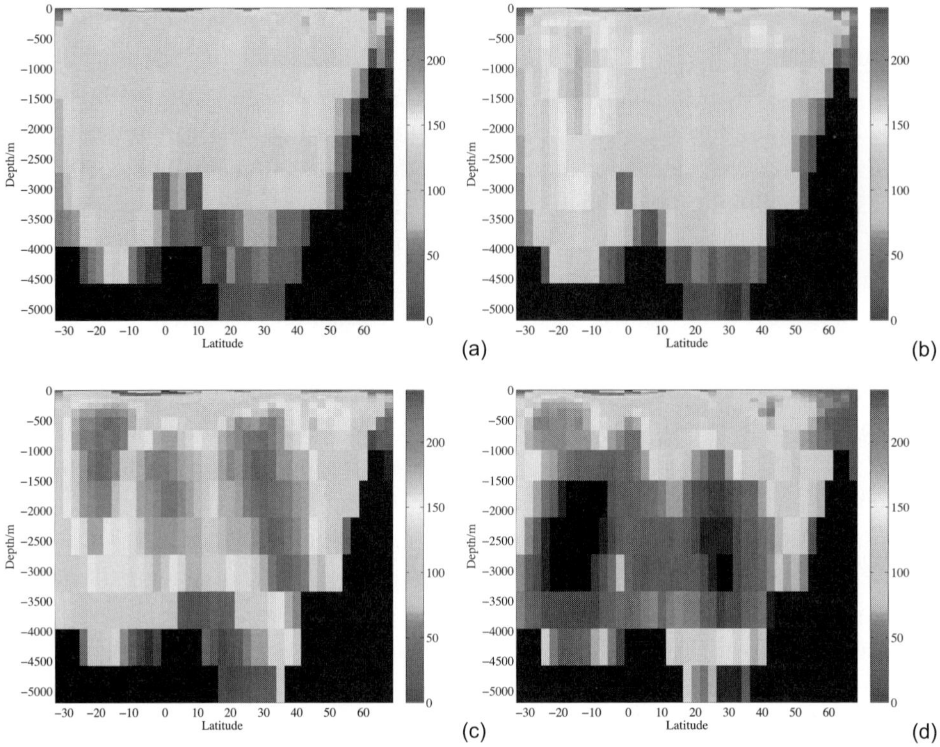

Figure 8.8 (a) Degree field of the PCCN reconstructed from the MOC streamfunction data of equilibrium simulation 1 in Fig. 8.4 using a threshold of 0.5. (b) Same as (a) but of equilibrium simulation 2. (c) Same as (a) but of equilibrium simulation 4. (d) Same as (a) of equilibrium simulation 6. Source: Feng et al. (2014). (A black and white version of this figure appears in some formats. For the color version, please refer to the plate section.)

8.3.3 Results for the FAMOUS Model

Feng et al. (2014) constructed zero-lag PCCNs with data from the more realistic FAMOUS model, using the complete Atlantic MOC streamfunction field for each of the six 100-year equilibrium simulations (red dots in Fig. 8.4a). Fig. 8.8 shows the changes in the degree field of that PCCN at the equilibrium solutions 1, 2, 4, and 6 in Fig. 8.4a. When the freshwater forcing is increased, high degree in the network – indicating high spatial MOC correlations – first appears at nodes in the South Atlantic at about 1000 m depth (Figs. 8.8a, b). It subsequently extends to the whole Atlantic with highest amplitudes in the deep ocean at midlatitudes (Figs. 8.8c, d).

The EOFs of the MOC streamfunction field are shown in Fig. 8.9. The first EOF becomes more dominant (i.e., the value of explained variance increases) when the

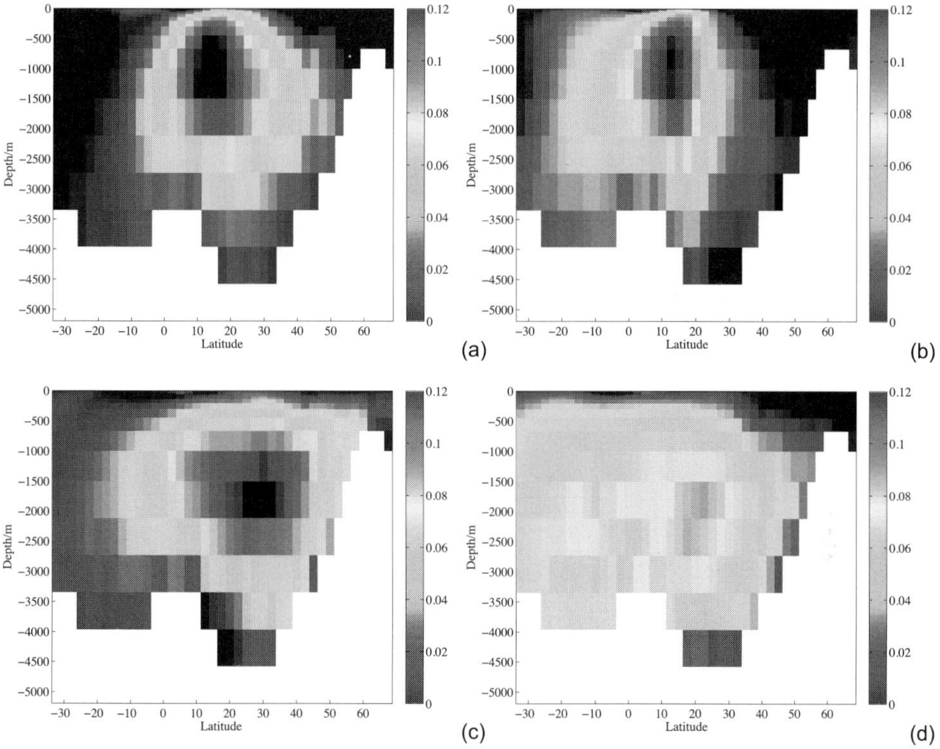

Figure 8.9 (a) The first EOF of the MOC data of equilibrium simulation 1, explaining 26.3% of the variance. (b) Same as (a) but of equilibrium simulation 2, explaining 24.1% of the variance. (c) Same as (a) but of equilibrium simulation 4, explaining 37.1% of the variance. (d) Same as (a) but of equilibrium simulation 6, explaining 37.6% of the variance. Source: Feng et al. (2014). (A black and white version of this figure appears in some formats. For the color version, please refer to the plate section.)

freshwater forcing is increased (from 1 to 6). From Fig. 8.9 one observes that for small freshwater forcing (panels a,b) most of the variability takes place near 15°N and at about 1000 m depth. However, the maximum shifts to higher latitudes and deeper locations when the freshwater is increased (panel c) and closest to the transition (panel d) the maximum occurs at about 2000 m depth at both 20°S and 20°N. In a word, once the transition is approached, one of the EOFs becomes most dominant in the variability. The network can be seen as a coarse-graining of the variability and by focusing only on the largest correlations, it is ideally suited to monitoring the changes in spatial correlations of the system once the transition is approached.

Next, the transient behavior of the hosing simulation was studied by reconstructing PCCNs at consecutive times. In Section 8.3.2, the kurtosis K_d of the degree

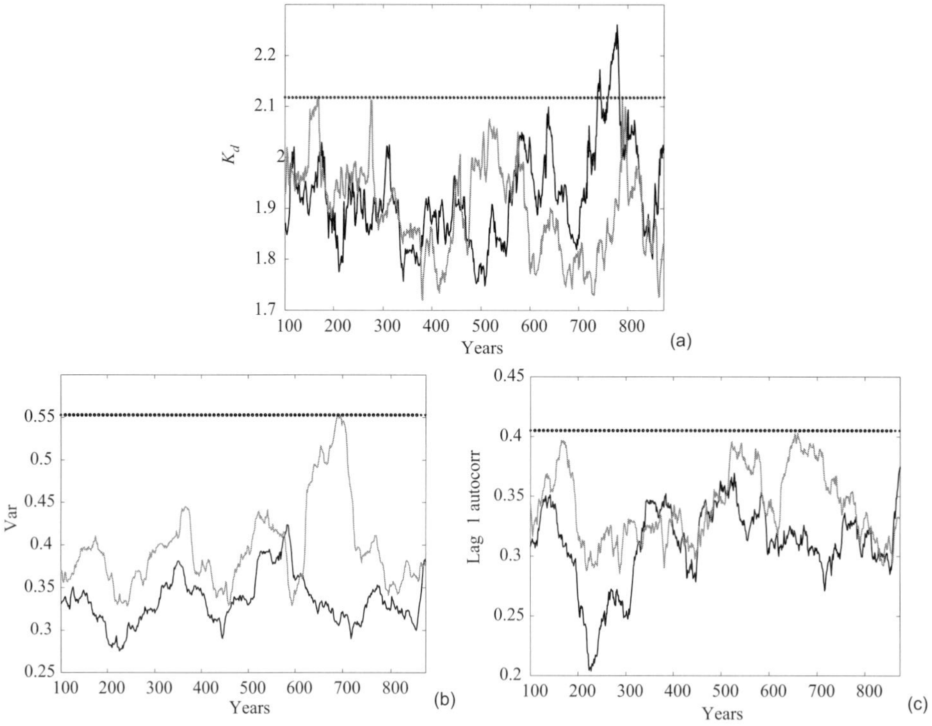

Figure 8.10 Warning indicators for the complete Atlantic MOC field. Grey curves are for the control simulation and black curves are for the hosing simulation. The dashed horizontal lines indicate the corresponding maximum values of the control simulation taken over the total time interval $[0, \tau_c]$. (a) the kurtosis indicator K_d gives the early warning signal at 738 years and lasts for 44 years. (b) The traditional variance indicator Var gives no early warning signal before the collapse time $\tau_c = 874$ years. (c) The traditional lag-1 auto-correlation indicator gives no early warning signal before the collapse time τ_c. Source: Feng et al. (2014). (A black and white version of this figure appears in some formats. For the color version, please refer to the plate section.)

distribution was introduced as an effective indicator to capture the changes in the topology of the degree field. For the complete Atlantic MOC field, the values of K_d for the FAMOUS hosing simulation (black curve) and for the control simulation (gray curve) are plotted in Fig. 8.10a. For the hosing simulation, there is indeed a strong increase of K_d to values far extending those for the control simulation significantly before the collapse time τ_c. For comparison, the critical slowdown indicators variance (Fig. 8.10b) and lag-1 auto-correlation (Fig. 8.10c) based on the complete Atlantic MOC streamfunction data (and using the same sliding window) do not show any early warning signal of the MOC transition before the collapse time τ_c.

In Feng et al. (2014), a measure of the performance of the kurtosis indicator K_d for data from different sections and combinations of sections was studied. To determine the optimal observation locations of the MOC using K_d, sections from the complete Atlantic MOC field were deleted. When the midlatitude North and South Atlantic MOC data are available, the kurtosis indicator K_d provides a strong anomalous signal at least 100 years before the transition.

The physical reason for these regions is the following. In ocean–climate models, the MOC collapse is due to a robust feedback involving the transport of salinity by the ocean circulation, the salt-advection feedback (Stommel, 1961; Walin, 1985). The subtropics are strong evaporation regions and hence the largest salinity gradients, central in the salt-advection feedback, are located in these regions. Consequently, it is expected that here the strongest response due to the salt-advection feedback appears and hence the strongest spatial correlations in the MOC field. Support for this comes from the structure of the dominant EOFs of the equilibrium solutions where the largest amplitude is also located in these midlatitude regions.

Due to the insufficient length of the available observational record, the indicator K_d is not yet applicable to observational data and hence cannot be used to determine whether the recent downward trend in the observed MOC strength at 26°N (Smeed et al., 2013) is the start of a collapse (see also Caesar et al., 2018; Thornalley et al., 2018). Such MOC behavior may just be related to natural variability such as that associated with the Atlantic Multidecadal Oscillation (AMO; Kerr, 2000; Knight et al., 2005). The strong element of an indicator such as K_d, in contrast to indicators based on a single-point time series, is that it is based on spatial correlations.

8.4 Desertification

Our previous applications of the network methodology to climate processes have focused on the atmosphere and the ocean. Here we deal with some important consequences of climate variability and dynamics on the land-based biosphere. We consider vegetation inhomogeneities in semi-arid regions. These precede the eventual abrupt transition which may occur to bare soil with diminishing rainfall. Following Tirabassi et al. (2014), we describe network techniques aimed to develop early warning indicators for these desertification transitions. The indicators are applied to results from a local positive-feedback vegetation model and are compared to classical indicators, such as the auto-correlation and variance of the biomass time series. A quantitative measure is also introduced to evaluate the quality of the early warning indicators, which show that the network-based indicators outperform the classical ones, being more sensitive to the presence of the transition point.

Table 8.2 *Parameters of the local positive feedback model (LPF) given by* (8.8a) *and* (8.8b). *Same values as in Dakos et al. (2011).*

Parameter	Meaning	Value
D	Exchange rate	0.5 $m^2 day^{-1}$
Λ	Water consumption rate by vegetation	0.12 $m^2 g^{-1}$ day^{-1}
ρ	Maximum vegetation growth rate	1 day^{-1}
B_c	Vegetation carrying capacity	10 g m^{-2}
μ	Maximum grazing rate	2 g day^{-1} m^{-2}
B_O	Half-saturation constant of vegetation consumption	1 g m^{-2}
σ_w	Standard deviation of white noise in water moisture	0.1
σ_B	Standard deviation of white noise in vegetation biomass	0.25
w_0	Water moisture scale value	1 mm
B_0	Biomass density scale value	1 g m^{-2}
τ_w	Water moisture scale time	1 day

8.4.1 Vegetation–Water Model

The local positive feedback model (LPF; Dakos et al., 2011; Guttal and Jayaprakash, 2007; Shnerb et al., 2003) is described by the following set of stochastic differential equations:

$$\frac{\partial w}{\partial t} = R - \frac{w}{\tau_w} - \Lambda w B + D\nabla^2 w + \sigma_w w_0 \xi^w(t), \tag{8.8a}$$

$$\frac{\partial B}{\partial t} = \rho B\left(\frac{w}{w_0} - \frac{B}{B_c}\right) - \mu\frac{B}{B + B_O} + D\nabla^2 B + \sigma_B B_0 \xi^B(t), \tag{8.8b}$$

where w (in mm) is the soil water amount and B (in g/m^2) is the vegetation biomass. The quantity D is a diffusivity and τ_w, μ, ρ, Λ, w_0, B_O, and B_c are additional constants explained in Table 8.2. Finally, R is the amount of rainfall, which is used as the bifurcation parameter of the system. Additive Gaussian white noise, ξ, with $\langle \xi(t)\xi(t')\rangle = \delta(t - t')$, is prescribed with amplitudes σ_w and σ_B for soil water and biomass, respectively. A characteristic spatial pattern of biomass as well as time series for different values of R are shown in Fig. 8.11.

The important mechanism in this model is a positive feedback that causes each patch to have alternative stable states. This is demonstrated in the bifurcation diagram of the LPF model, which is sketched in Fig. 8.12. The deterministic homogeneous solutions of the LPF model and their linear stability can be determined analytically. For all values of R, the trivial solution ($B = 0$, $w = \tau_w R$) exists. For the standard parameter values shown in Table 8.2, the trivial solution is linearly stable for $R < 2$ mm/day and unstable for $R > 2$ mm/day (see Fig. 8.12). At $R = 2$ mm/day, a transcritical bifurcation occurs and two additional branches of

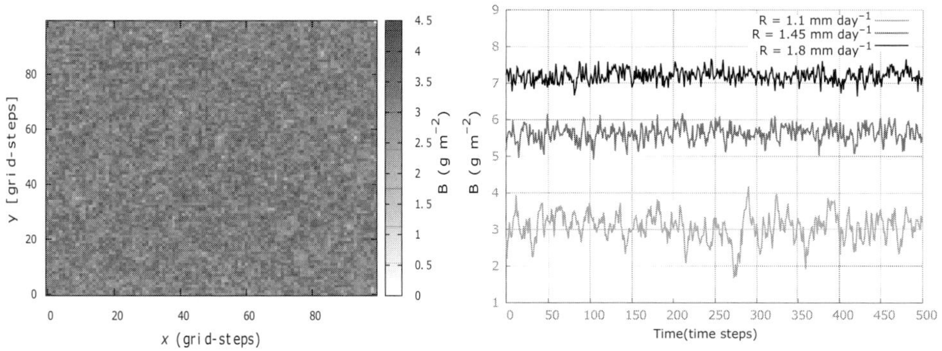

Figure 8.11 (a) A snapshot at $t = 500$ time steps of the biomass field B at $R = 1.1$ mm/day. (b) Example of biomass time series for a node (single grid cell) i and different values of R. Source: Tirabassi et al. (2014).

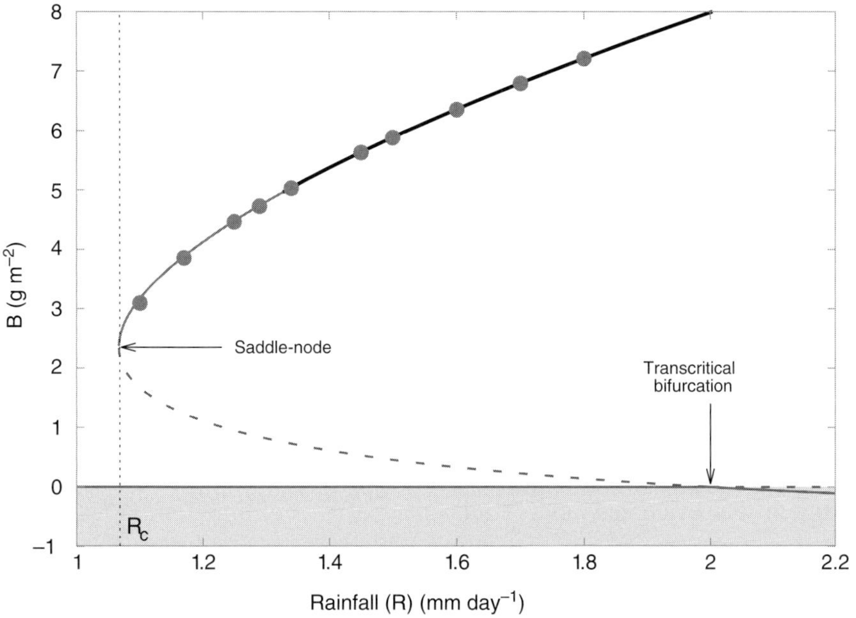

Figure 8.12 Bifurcation diagram of the LPF model given by (8.8a) and (8.8b). Curves depict steady homogeneous states, that is, determined under vanishing diffusion and noise. Linearly stable branches are denoted by solid lines, whereas linearly unstable branches are indicated by dashed lines. The black portion of the continuous line marks states for which the Jacobian matrix has eigenvalues with a non-zero imaginary part, whereas it is zero for the gray continuous line. The shaded region indicates the nonphysical negative B values. The average value of B for each simulation of biomass evolution via the full equations (8.8a) and (8.8b) is depicted by dots. Obviously, these average values are similar to those of the homogeneous solutions, hence diffusion and noise do not impact on the average state of the system. Source: Tirabassi et al. (2014).

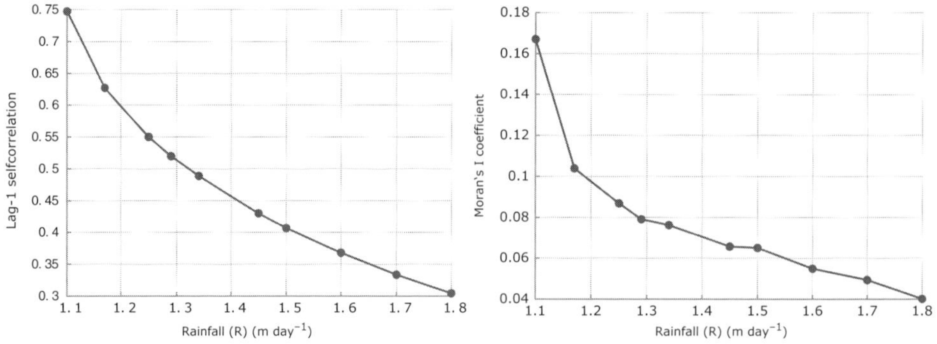

Figure 8.13 (a) Spatially averaged lag-1 auto-correlation. (b) Moran's coefficient I. Both computed using the spatial biomass distribution at the last time step of the simulation for different values of the rainfall parameter R. Source: Tirabassi et al. (2014).

steady solutions emerge. Solutions on the lower branch are not considered here because they have $B < 0$, i.e., they are physically non-realistic. Solutions on the upper branch are unstable for values of R down to $R_c = 1.067$ mm/day. At this R value a saddle-node bifurcation occurs that provides a linearly stable upper branch of solutions for $R > 1.067$ mm/day. Thus, a desertification transition occurs in this model if, starting with high values of rainfall and then of biomass, rainfall is decreased: An abrupt change to bare soil occurs when rainfall drops below R_c. Finally, a fourth homogeneous solution exists but it also has values of $B < 0$ for every R value and hence it will not be further mentioned.

In order to determine inhomogeneous vegetation patterns in the stochastic case, the model equations (8.8a) and (8.8a) are numerically solved on a periodic square grid composed of $100 \times 100 = 10^4$ grid-cells on a regular lattice with dimension $L = 100$ m. The evaluated model data consist of a set of time series (500 time steps with $\Delta t = 0.01$ days) of statistically equilibrated biomass fields B for different fixed rainfall parameters R. Time series related to ten different values of R with $1.1 \leq R \leq 1.8$ mm/day are analyzed. For $R < R_c$ only the desert-like solution, with $B = 0$ over the whole domain, is found.

The temporal and spatial mean values of the biomass distribution of each of these spatially inhomogeneous solutions are plotted as the red dots in Fig. 8.12. Obviously, the average values of the fluctuating solutions are similar to those of the steady homogeneous solutions, hence, diffusion and noise do not impact on the average state of the system.

Before going to the network description, we present some other classical indicators of the desertification transition, based on the concept of CSD. One of the most used indicators is the lag-1 auto-correlation. The spatially averaged auto-correlation of the LPF model data is shown in Fig. 8.13a for each value of R. As R decreases and approaches R_c, the system indeed experiences an increase in auto-correlation, a distinct fingerprint of CSD.

It has also been suggested that spatial statistics may be used to detect CSD (Dakos et al., 2011). In particular, an increase in spatial correlation of the system is expected when the system experiences CSD close to the transition (Section 8.2). A typical measure of spatial correlation is Moran's coefficient, I, defined here through the biomass mass field B as

$$I \equiv \frac{N}{\sum_{ij} g_{ij}} \frac{\sum_{ij} g_{ij}(B_i - \bar{B})(B_j - \bar{B})}{\sum_i (B_i - \bar{B})^2}, \tag{8.9}$$

where $g_{ij} = 1$ if i and j are two adjacent grid cells and $g_{ij} = 0$ otherwise. The spatially averaged biomass is indicated by $\bar{B}$. Moran's coefficient I is shown in Fig. 8.13b and demonstrates that the spatial correlation increases in a similar way – although strongly – to the temporal correlation as R decreases and approaches R_c.

However, despite the fact that the classical indicators are able to reflect CSD, they change in a very smooth, gradual and monotonic way. From a strictly local point of view, that is, based on a few closely spaced R values, it is not possible to estimate the proximity of the system to R_c. In other words, the correlation coefficients suffer from a lack of distinct features necessary to provide a pronounced early-warning signal of desertification.

8.4.2 Network Approach and Analysis

Tirabassi et al. (2014) associate a network to the simulated field of biomass. The nodes are defined as the $N = 100 \times 100 = 10^4$ grid cells of the discretized LPF model. In order to define the links between the nodes, the zero-lag temporal correlations between the biomass time series of all pairs of nodes are considered and a link is established between two nodes if their correlation exceeds a threshold value Θ which ensures statistical significance. To determine this value, a Student's t-test is used for the quantity

$$t = \frac{\Theta}{\sqrt{1-\Theta^2}} \sqrt{N_{\text{time-steps}} \frac{1-r}{1+r}}, \tag{8.10}$$

where $r = r(R)$ and $N_{\text{time-steps}}$ are the auto-correlation and the length of the time series, respectively. This test variable takes the effective number of degrees of freedom of the time series into account. It is found that a value of $\Theta = 0.2$ guarantees that, for each value of R, the zero-lag correlation between linked nodes is statistically significant with a p-value smaller than 0.05, and this threshold value is taken in all results below.

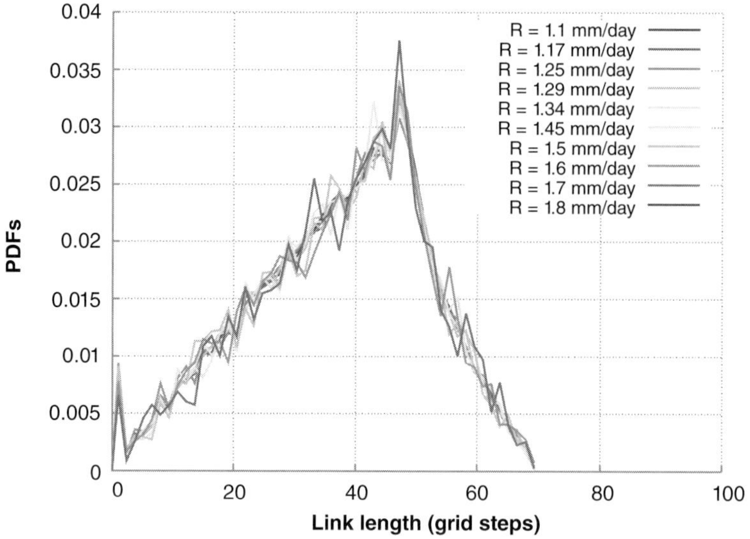

Figure 8.14 Link length PDFs for various values of rainfall R. Link length is computed as the Euclidean distance between two linked grid cells, taking into account periodic boundaries. Source: Tirabassi et al. (2014).

After the construction of the interaction network of the biomass data, changes in its topology for varying R can be studied. Tirabassi et al. (2014) considered the degree, assortativity, and the clustering coefficient. See the corresponding definitions in Section 4.1.1. Also a consideration of the link-length distribution was made.

The link length measures how close two linked nodes are in physical space, that is, the link length is the Euclidean distance between two linked nodes. The distribution of link lengths for different R is shown in Fig. 8.14. Apparently, the link length distribution does not depend on R. The smaller spike at link length of one grid spacing represents the underlying grid structure and is related to local correlations due to diffusion. The linear behavior at larger distances can be attributed to a random structure of the network: If the links are randomly distributed in physical space, the probability of two nodes i and j to be connected is independent of their distance. In fact, the number of links of length d of any node i is solely proportional to the number of nodes at distance d from node i. This number scales with the circular perimeter $2\pi d$. Hence, the probability of a certain link length is directly proportional to the link length itself, until the appearance of the boundary reduces the possible number of links again (here at a distance of 50 m, that is $L/2$). Consequently, the interaction network of the simulated biomass field appears to display a random architecture superimposed on the grid structure.

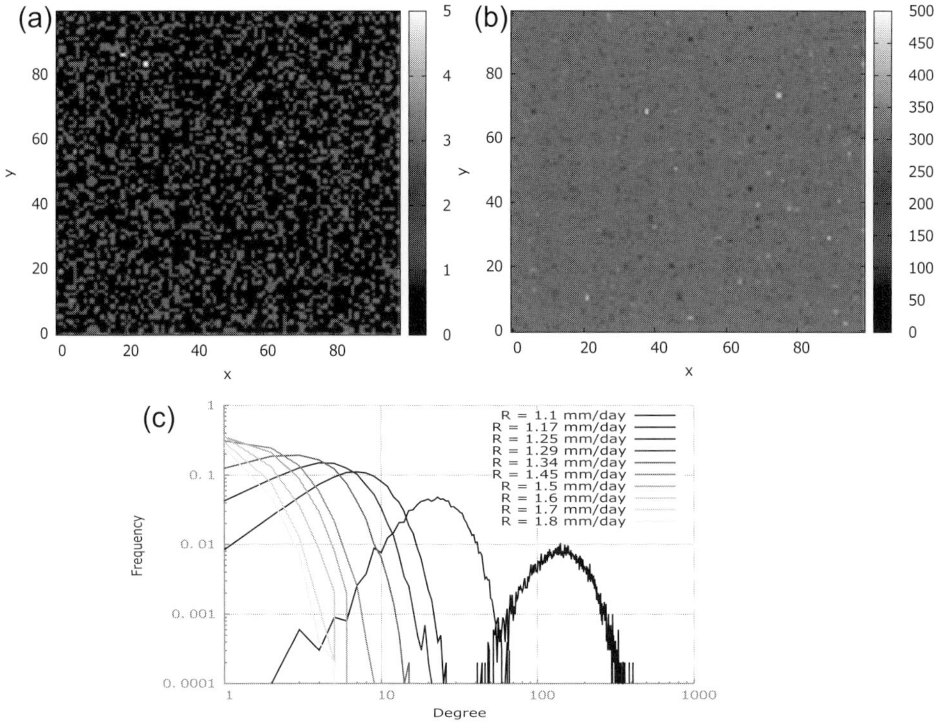

Figure 8.15 Spatial fields of degree. (a) Far from the transition, $R = 1.8$ mm/day. (b) Close to the transition, $R = 1.1$ mm/day. (c) PDF of node degree for different values of R. Source: Tirabassi et al. (2014).

Figure 8.15 shows the spatial field of the degree distribution for two values of R (a: at the transition; b: far away from the transition), together with the degree density distribution (i.e., the PDF of k_i) for different R. With decreasing rainfall R the mean number of links per node in the network increases. Consequently, the degree PDF shifts to the higher degrees when the system approaches the critical transition. The increase of network connectivity is related to higher values of the cross-correlation among the nodal time series, which occurs near the saddle-node bifurcation. This feature can also be seen in the spatial patterns of the degree field. For $R = 1.8$ mm/day (Fig. 8.15a) the network is almost disconnected and the majority of the nodes have zero or one link. With decreasing rainfall the number of connections increases, and the disconnected nodes eventually join the network. At the transition, the spatial pattern of the degree distribution shows a granular structure, with most of the nodes being highly connected (Fig. 8.15b).

The shift of the degree PDF to high degrees (Fig. 8.15c) can be used, as in the MOC analysis of previous sections, to develop an indicator of the upcoming transition. Figure 8.16a shows the mean of the degree distribution as a function of

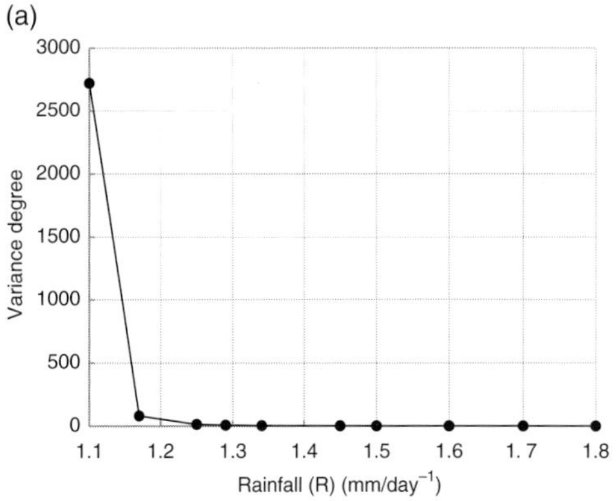

Figure 8.16 (a) Mean and (b) variance of network degree distribution as functions of the bifurcation parameter R. Source: Tirabassi et al. (2014).

R. This network measure is highly sensitive to R near criticality, showing a steep increase close to the transition. Additionally, Fig. 8.16b shows the variance of the degree distribution as a function of R. The increase in variance when approaching the transition point is even more abrupt. As with the examples in previous sections, in the proximity of the saddle-node bifurcation, variability synchronizes over the domain, a manifestation of CSD, increasing the number of connections when approaching the tipping point (Section 8.2).

Similar results were given by Tirabassi et al. (2014) for the clustering and assortativity fields. These moments of distributions can be named *scalar-based* early warning indicators, in contrast with *distribution-based* indicators that consider the changes in the shape of the distribution as a whole. As examples of this second class, Tirabassi et al. (2014) observed that assortativity and clustering distributions (but not the degree distribution) approach a Gaussian shape when R approaches R_c. A way to use this observation in an early warning indicator is to define the Kullback–Leibler divergence (KLD), also called relative entropy, between the actual distribution and a Gaussian. Given a one-dimensional distribution $P(x)$ and a Gaussian with the same mean and variance $G(x)$, their relative entropy is defined as

$$KLD \equiv \int_{-\infty}^{\infty} \ln\left(\frac{P(x)}{G(x)}\right) P(x)\, dx. \tag{8.11}$$

KLD values close to zero indicate that $P(x)$ is indeed close to a Gaussian. Tirabassi et al. (2014) used KLD when $P(x)$ is either the assortativity or the clustering distributions (Fig. 8.17). Obviously, KLD quickly drops to zero when approaching the transition point for both the assortativity and clustering distributions. Tirabassi et al. (2014), however, note that the Gaussianization might be a model-specific feature and further analysis is required to assess the generality of this result.

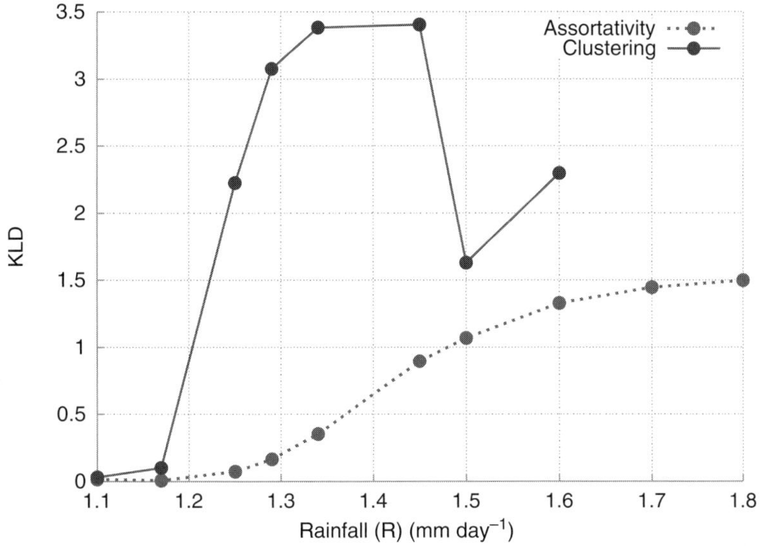

Figure 8.17 KLD values for the distribution of clustering (continuous line) and assortativity (dotted) with respect to Gaussian distributions with same mean and variance. Source: Tirabassi et al. (2014).

The scalar measures are numbers that ideally change monotonically when the transition point is approached. For these indicators it is generally important to monitor their derivative with respect to the bifurcation parameter R in order to sign proximity of the system to the transition point. The absolute value of the indicator itself gives less information than its abrupt change close to the transition (Scheffer et al., 2009; Van Nes and Scheffer, 2007). The classical indicators directly based on CSD are also included in this scalar-based class. For the distribution-based indicators, however, the absolute value carries information in itself. For both classes of indicators quality measures can be defined. First, an ϵ-environment around the bifurcation point is defined by all the R values for which $(R - R_c)/R_c < 0.1$. For a scalar-based indicator J, one then defines the normalized quality measure Q_s^J as

$$Q_s^J \equiv \frac{\left\langle \frac{\partial J}{\partial R} \right\rangle_{R<\epsilon} - \left\langle \frac{\partial J}{\partial R} \right\rangle_{R>\epsilon}}{\left\langle \frac{\partial J}{\partial R} \right\rangle_{R<\epsilon} + \left\langle \frac{\partial J}{\partial R} \right\rangle_{R>\epsilon}}, \tag{8.12}$$

where the brackets indicate the mean over the interval indicated. In this way, if J shows an abrupt change in its derivative close to the transition then we have $Q_s^J \approx 1$. In contrast, if the change of J is merely linear when approaching the tipping point, $Q_s^J \approx 0$.

Table 8.3 *Quality values of different early warning indicators*

Indicator J	**Class**	**Type**	$\mathbf{Q}^J$
Average degree	Scalar	Network	0.963
Variance of degree	Scalar	Network	0.996
Variance of sssortativity	Scalar	Network	0.971
Average clustering	Scalar	Network	0.904
Assortativity Gaussianity	Distribution	Network	0.973
Clustering Gaussianity	Distribution	Network	0.951
Lag-1 auto-correlation	Scalar	Classical	0.468
Spatial correlation	Scalar	Classical	0.730

In the case of distribution-based indicators, an analogous quality measure Q^J_d can be defined by taking into account J itself instead of its derivative, that is:

$$Q^J_\mathrm{d} \equiv \frac{\langle J\rangle_{R>\epsilon} - \langle J\rangle_{R<\epsilon}}{\langle J\rangle_{R<\epsilon} + \langle J\rangle_{R>\epsilon}} \,. \tag{8.13}$$

In Table 8.3 the quality values for both network-based and classical indicators are shown. Obviously, at least for this LPF desertification model, the network-based measures have significantly higher early warning quality than the classical measures.

8.5 Percolation-Based Techniques

The previous sections explored how the variations in correlations, occurring when approaching a transition point, change the topological properties of functional networks constructed from them, and then provide warning signals of the proximity of a bifurcation or regime shift in a climatic or geophysical system.

Here we present an important additional property of these functional networks: As the system approaches a bifurcation and internal correlations increase, networks evolve from a low to a high connectivity state and, *before* reaching its maximum link density at the bifurcation point, a percolation-like transition occurs in the network topology (Meng et al., 2017; Rodríguez-Méndez et al., 2016).

In short, percolation (Newman, 2010; Stauffer and Aharony, 1994) is the phenomenon that occurs in a network when adding links to it starting from the unconnected nodes. For a small number of links the network is composed of many disconnected small components. Under quite general circumstances and in large enough networks, there is a critical number of links at which one of these components becomes suddenly of large size and contains a significant proportion of the nodes. Concepts from percolation theory (Newman, 2010; Stauffer and Aharony, 1994) can then be considered and related to the dynamic transition occurring at the bifurcation. Importantly, metrics can be defined that act as early warnings for this percolation transition, which itself is a precursor of the dynamic transition.

The validity of these general ideas has been tested (Rodríguez-Méndez et al., 2016) by analyzing model systems displaying different types of bifurcations: steady and oscillatory, continuous and discontinuous. In all cases, a percolation transition occurred before the dynamical one, and was characterized with quantities that can be used as early warning signals. In the following we describe the main methodology. In the next sections we will show applications to a simple model system undergoing different types of bifurcations and to El Niño index data.

As usual, one first constructs a functional network from time series of some climate variable at different spatial locations. In the applications of the following sections, the method of construction is that of zero-lag Pearson correlation (see Section 4.2), leading to PCCNs.

By changing a parameter p in the model that has generated the time series or by considering moving time windows centered at successive times (which in this case become analogous to the parameter p), if one deals with observational data, we may be approaching a bifurcation or regime shift in the underlying climatic system and we want to characterize the changing topology of the resulting sequence of networks. In particular, one can study whether the network sequence undergoes a percolation transition as the control parameter p approaches transition points p_d. To do this several metrics are measured: (1) S_1, the relative size of the largest connected component, i.e., the fraction of nodes that are in the largest cluster. It abruptly changes from a value close to 0 to a value close to 1 at the percolation point. (2) The average size of the clusters excluding the largest one. This quantity is maximal at the percolation point (Stauffer and Aharony, 1994). It can be calculated as

$$\sum_s s\, c_s = \frac{1}{N} \sum_s s^2\, n_s \equiv \langle s^2 \rangle, \tag{8.14}$$

where the sum runs over all cluster sizes s excluding the largest one, and c_s is the fraction of nodes belonging to clusters of size s, $c_s = s\, n_s/N$ (n_s is the number of clusters of size s present in the system; Newman, 2010). c_s gives also the probability that a randomly chosen node pertains to a cluster of size s.

The standard percolation indicators, S_1 and $\langle s^2 \rangle$ are not, however, the best precursors. Rodríguez-Méndez et al. (2016) showed that better anticipation can be obtained by exploiting the interplay between the probabilities c_s and the coming transition. In random Erdös–Rényi graphs (see Section 4.1.1) it is possible to perform analytic calculations on the behavior of c_s. As new links are added at random, the percolation occurs when the mean degree $\langle k \rangle$, which acts as the control parameter p, equals 1. Before this, the probabilities c_s that a randomly chosen node belongs to a component of size s are

$$c_s = e^{-s\langle k\rangle}(s\langle k\rangle)^{s-1}/s!. \tag{8.15}$$

These probabilities have a maximum when the mean degree, which in this case is the parameter p, is $\langle k\rangle = p_s = \frac{s-1}{s}$. The succession $\{p_s\}$ of location of maxima of c_s converges to the percolation point $p_\infty = 1$ for increasing component size s, but for low s these maxima could be quite far from the percolation point, and in fact they anticipate it. Rodríguez-Méndez et al. (2016) showed that the early warning character of the maxima of c_s holds also true in non-random functional networks obtained from systems undergoing very different dynamic transitions. This is due to the generic increase in the dynamical correlations of the elementary units of the system that the approach of a global tipping point brings. The sequence of peaks in c_s offers thus a general and versatile tool to predict potential changes in the dynamics at a global scale.

Rodríguez-Méndez et al. (2016) tested the ideas above by analyzing several model systems undergoing different types of bifurcations, and also El Niño index time series. We present in the next section one of the model examples, the Lorenz'96 system, and in Section 8.5.2 we reproduce the analysis of El Niño events.

8.5.1 Percolation in the Lorenz'96 Model Network

The Lorenz'96 model was proposed by Lorenz (1996) as a simplified framework to investigate atmospheric predictability. It reads:

$$\frac{d\psi_k(t)}{dt} = \left[\psi_{k+1}(t) - \psi_{k-2}(t)\right]\psi_{k-1}(t) - \psi_k(t) + \eta_k(t) + p, \tag{8.16}$$

$\psi_k(t)$ is meant to represent the values of some atmospheric variable at different locations k, $k = 1 \ldots N$, arranged in a one-dimensional ring around the globe (and thus having periodic boundary conditions). The structure of (8.16) contains some of the main elements of fluid dynamics, namely dissipation, external forcing, and quadratic nonlinearity through an advection-like term. In addition to the constant forcing p, which will be the control parameter, an additive stochastic perturbation $\eta_k(t)$, uncorrelated in space and time, is added. It is implemented by Rodríguez-Méndez et al. (2016) by adding independent random numbers uniform in $[-a, a]$ ($a = 0.1$) after each time step ($dt = 1/64$) of a fourth-order Runge–Kutta method which is used to integrate the rest of the terms, and N is taken to be $N = 2500$ elements or oscillators. In the absence of the random forcing, three dynamical regimes are easily identified (Karimi and Paul, 2010): (1) For small p, $p < p_1 = 8/9$, the system stabilizes in the homogeneous fixed point $\psi_k(t) = p, \forall\, k,\ t$. (2) A

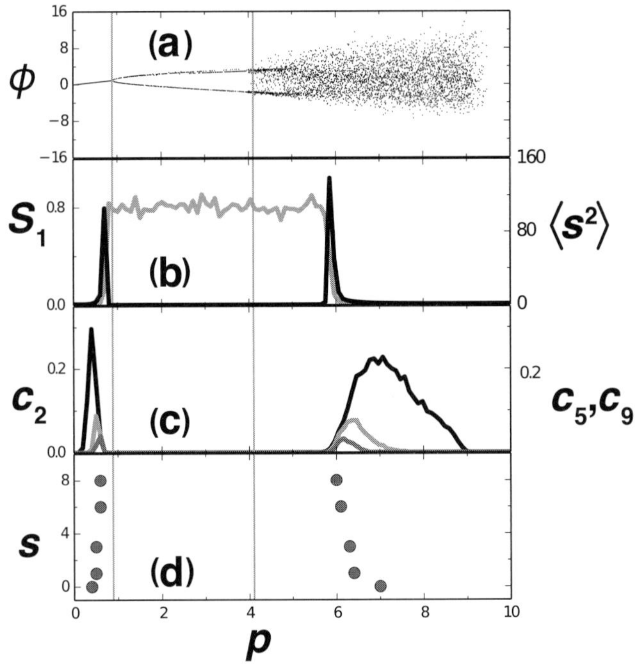

Figure 8.18 Panel (a) shows the bifurcation diagram of the Lorenz'96 model, (8.16), constructed from a two-oscillator Poincaré section (see text). The transition to traveling waves at $p_1 = 8/9$ and to spatiotemporal chaos ($p_2 \approx 4.1$) are shown as vertical lines. Functional networks were built using a threshold of $\gamma = 0.16$. The percolation indicators S_1 (gray) and $\langle s^2 \rangle$ (black) are displayed in panel (b). Panel (c) shows c_2 (upper black curve), c_5 (middle gray curve), and c_9 (lower gray curve). In (d), the circles indicate the value of p, p_s with $s = 2, 5, 9, 16$, and 21, for which the c_s curves attain their respective maxima. These indicators reveal a phase of percolated correlations in a parameter region which includes the interval $[p_1, p_2]$, flanked by two percolation transitions. The curves have been further averaged over 100 realizations of the random noise and initial condition. Source: Rodríguez-Méndez et al. (2016).

Hopf bifurcation occurs at $p = p_1$ so that for intermediate values of p, $8/9 < p < 4.1$ the system is in a traveling-wave state. (3) Beyond $p > p_2 \approx 4.1$ the system becomes spatiotemporally chaotic.

To display the bifurcations observed when integrating (8.16) for $N = 2500$ elements, Rodríguez-Méndez et al. (2016) calculated a Poincaré transversal section in the subspace of two contiguous oscillators: one of the oscillators, say ψ_1, is monitored and when it crosses the value $\psi_1 = 1$ in the increasing direction the value of the contiguous oscillator, say $\phi = \psi_2$, is recorded and displayed. The bifurcation diagram showing the values of these sections ϕ is plotted in Fig. 8.18a

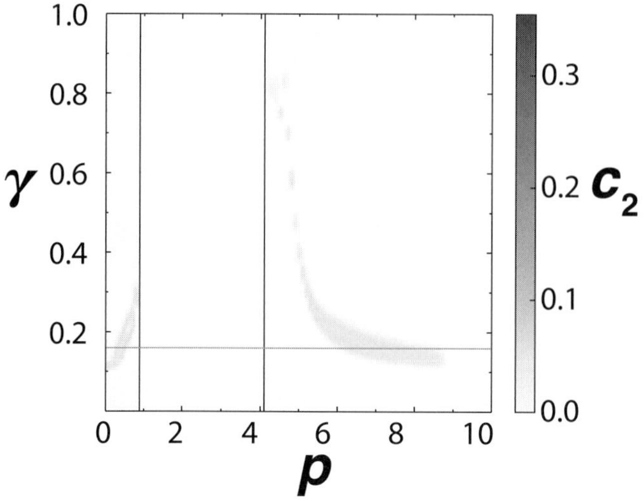

Figure 8.19 The c_2 values, as given by the gray-levels bar, as a function of the control parameter p and the threshold γ used to build the functional network for the Lorenz'96 system. The dynamical transitions to traveling waves and to chaos are indicated by the vertical black lines. The horizontal line identifies the value $\gamma = 0.16$ for which Fig. 8.18 was constructed. Source: Rodríguez-Méndez et al. (2016).

as a function of p. The three regimes described above for the deterministic system are readily identified here also.

Functional networks were constructed from $R = 1000$ snapshots by subtracting the temporal mean to the ψ_k variables to obtain the p_k, interpreting the locations k as nodes and assigning links between pairs of nodes when the zero-lag Pearson correlation is larger than a threshold $\gamma = 0.16$. Figure 8.18b shows the quantities S_1, the size of the largest cluster, and $\langle s^2 \rangle$, the mean cluster size excluding the largest one, for such a network. Panels c and d display the properties of several c_s, the probabilities of randomly chosen nodes to pertain to clusters of size s. These figures clearly identify the presence of a percolated phase at intermediate values of p, started and ended by two percolation transitions. Figure 8.19 shows the quantity c_2 in the (γ, p) parameter plane for this model. We see that for increasing threshold the maxima of c_2 approach the locations p_1 and p_2 at which traveling waves are born via a Hopf bifurcation and at which they destabilize into chaotic behavior, respectively. Thus, the percolating phase is a manifestation of the long-range coherence of the traveling wave state, whereas correlation length remains small in the homogeneous and in the chaotic regime. The quantity c_2 (and indeed the other c_s) clearly anticipates the first bifurcation when increasing p. It also largely anticipates the occurrence of a chaos–order transition when decreasing p from large values.

8.5.2 Percolation in Sea Temperature Networks during El Niño Events

To test the behavior of the precursors proposed before in observed real situations, Rodríguez-Méndez et al. (2016) analyzed SST data from the region of the Pacific used to compute the NINO3.4 index. ENSO (see Section 2.4) is characterized by rather irregular (with average period of about four years) warm (El Niño) and cold (La Niña) episodes departing from the long-term mean temperature in the equatorial Pacific. These oscillations are related to the presence of a Hopf bifurcation in the coupled atmosphere–ocean system (Dijkstra, 2006; Sarachik and Cane, 2010). The bifurcation can be crossed or just approached, being then the oscillation excited by noise. In both cases there should be a build-up of correlations that would become visible in functional networks constructed from temperature time series. In this case there is no control parameter to fix, but rather the equatorial Pacific evolves in time, coupled to the seasonal cycle, leading to changing spatial correlations. We will see that, despite this lack of control, and without using any information on the underlying dynamics, the percolation approach is able to find precursors of the relevant El Niño–La Niña events.

We follow Rodríguez-Méndez et al. (2016). A percolation quantity which is essentially the temporal derivative of the quantity S_1 was also used to characterize El Niño events by Meng et al. (2017). SSTs were obtained from the ERA-Interim reanalysis of the European Centre for Medium-Range Weather Forecasts (ECMWF, 2009), with daily temporal resolution and a spatial resolution of $\Delta x = 0.125^\circ$, in the range of years 1979–2014 (Fig. 8.20). Daily functional networks at day t were built from these time series, computing the zero-lag Pearson correlation with a time window of $R = 200$ days (100 days before and 100 days after time t). The quantities plotted in Fig. 8.20 are further averaged over five days.

Figure 8.20, focuses on three different periods: 1987–1989, during which a strong La Niña occurred, 1996–1998, featuring one El Niño–La Niña pair, and a recent El Niño in 2009. In contrast to the Lorenz'96 example, this system is empirical and the contribution of the noise is more difficult to assess than in a model equation. Therefore, a systematic search for an optimal γ has not been performed. Nevertheless, some exploration of values of γ and how they affect c_2 has been done (Rodríguez-Méndez et al., 2016). Interestingly, it seems that the range of values of γ necessary to observe peaks in c_2 have moved toward lower γ in the early 2000s. Two values of the threshold have been used to produce Figure 8.20: $\gamma = 0.99992$ for the events of 1987–1989 and 1996–1998, and $\gamma = 0.9986$ for the event of 2009. In the two cases, this is where a nice compromise between signal-and-noise is found. In a practical situation, the selection of γ is not a post-hoc process: One can have a clear idea of the range of values to use from the previous events. Once γ is set at a fixed value in this range, if c_2 shows a peak, followed by a sequence of peaks of c_3, c_5, etc., the system is very likely going toward a new El Niño/Niña

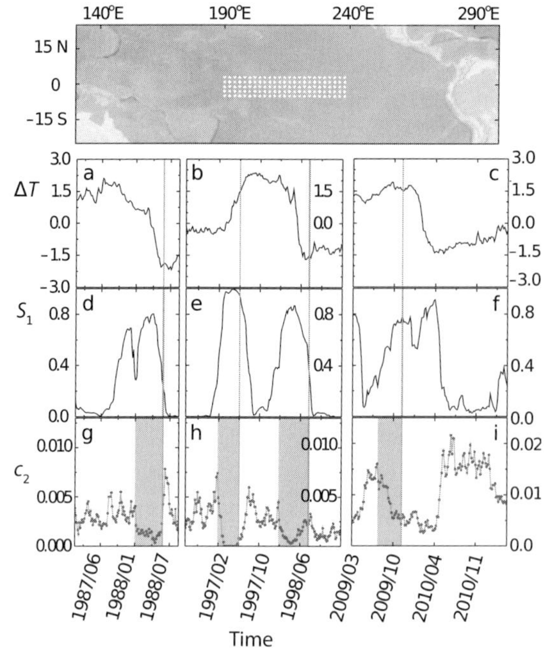

Figure 8.20 Application to El Niño phenomenon. The upper panel shows a map of the area over which the mean SST T is monitored in the NINO3.4. Points denote locations used here as nodes in a functional network. Four events, two La Niña (cold) and two El Niño (warm), are shown in the time axis of panels (a)–(i). Conventional starting dates of the events are marked by vertical lines. In (a)–(c), the SST T is shown as a function of time. A functional network is constructed from correlations at $\gamma = 0.99992$ for 1987–1989 and 1996–1998, and $\gamma = 0.9986$ for 2009. The size of the giant component S_1 is shown in panels (d)–(f), showing percolating phases at a plateau, flanked by two percolation transitions, which occurs before each of the events. Panels (g)–(i) show c_2 in the same time frame. Peaks in c_2 flank both sides of the percolation plateaux. The time by which the peaks of c_2 anticipate the conventional starting date of the event is marked in gray. Source: Rodríguez-Méndez et al. (2016).

event. Figures 8.20a–c display the variations of the ocean superficial temperature and also the moments at which an El Niño or La Niña event are officially declared are marked with a vertical line. Panels d–f, on the other hand, depict the time evolution of the size of the largest connected component S_1, which peaks before or on the arrival of the event. c_2, represented in g–i, also shows maxima way before the corresponding peak of S_1. The anticipatory period since the c_2 peaks to El Niño (La Niña) event is marked in gray in every plot. It corresponds to 240 days in 1988 (Fig. 8.20g), 125 days in 1997 (Fig. 8.20h), 175 days in 1998 (Fig. 8.20h), and 115 days in 2009 (Fig. 8.20i).

In summary, several climate tipping elements have been analyzed with network tools. The same feature of the dynamics close to bifurcation points, critical slowing down, which is captured by classical early warning indicators of an approaching transition, is also responsible for an enhancement of spatial correlations. Network methods are an efficient way to deal with these spatial correlations, filtering much of the noise and providing a clear early warning signal. Network methods, however, require a large set of time series at different spatial locations. The challenge will be to optimize the methodology to reduce the data requirements, as well as the simultaneous improvement of observation networks so that, eventually, operational monitoring of tipping elements such as the MOC could be achieved.

9

Network-Based Prediction

In this chapter, the use of network techniques in climate prediction is described. In Section 9.1, general concepts of predictability are briefly presented, followed by a section on modern machine learning tools (Section 9.2). We next discuss how these tools, combined with climate network methods, can yield new insights into climate prediction studies. We discuss, as particularly relevant examples, the prediction of the onset and withdrawal of the Indian Summer Monsoon (Section 9.3), and the prediction of El Niño events (Section 9.4).

9.1 Concepts of Predictability

The new knowledge of the climate system that is gained through different approaches is eventually incorporated into more advanced models (dynamical or statistical) aimed at making more skillful predictions of future conditions. This process has matured most in weather prediction, where the skill has steadily increased over the last decade (Bauer et al., 2015). Ensemble forecasting, where a suite of initial conditions is propagated forward using an atmospheric model, is the most usual practice. Due to the nonlinear and stochastic behavior of the atmosphere, there is an intrinsic limitation on the prediction skill with increasing lead time, which can be seen from the spread of the ensemble. In the midlatitude atmosphere, the prediction skill becomes low for a lead time of about ten days (Bauer et al., 2015).

Predictability studies of the first kind address how uncertainties in an initial state affect the prediction at a later stage (Lorenz, 1969). Indeed, initial uncertainties can amplify as the prediction lead time increases, thus limiting the skill. Several measures have been developed for the predictability of the first kind in nonlinear systems, such as predictive power (Schneider and Griffies, 1999) and prediction utility (Kleeman, 2002). These are mainly based on the behavior of an initial probability density function (PDF) in time compared to the equilibrium (climatological) PDF. As the equilibrium is approached, these measures approach zero and prediction skill is low.

Predictability studies of the second kind address the predictability of the response of the system to changes in boundary conditions. Examples are the response of the atmospheric circulation due to changes in SST and the climate response to changes in orbital insolation variations. For these studies, a parameter ensemble is well suited as it shows how trajectories are sensitive to parameter variations. Examples are seasonal forecasting efforts and simulations to determine equilibrium climate sensitivity (Stainforth et al., 2005).

Apart from these two types of predictability problems, there are also so-called classification problems which are of more discrete (often yes/no) natures. An example is whether an El Niño will develop next year or not, or whether there will be a strong or weak summer monsoon over India. In these cases, the aim is to develop precursors from which it can be deduced whether events will occur or not. Here, methods based in complex networks have been very useful to infer precursors, as was shown in the previous chapter.

For precursor-type prediction studies, the usual methodology is to develop so-called receiver operating characteristic (ROC) curves. If the PDFs of both true detection and false alarm are known, the ROC curve is generated by plotting the cumulative distribution function of the detection probability on the y-axis (the true positive rate, TPR) versus the cumulative distribution function of the false-alarm probability on the x-axis (the false positive rate, FPR). The development of precursors is closely related to the general technique of machine learning to which we provide an introduction in the context of climate prediction in the next section.

9.2 Machine Learning

Machine learning techniques are nowadays ubiquitous in many fields of science, engineering, and computer science, and they are especially useful for climate prediction (Mitchell, 1997). One prominent engineering example of prediction using machine learning is energy prediction for solar power plants (Sharma et al., 2011). This prediction task can be reduced to learning how the solar plant responds to the climatic conditions, and predicting the future response of the plant using weather forecasts. When machine learning techniques are applied to climate prediction, the task is to "learn" on the dynamics of the climate system from past states and then be able to predict its future states (Bishop, 2006). The originality and advantage of using network approaches is that the temporal information may be already contained in the measures of climate networks so the machine learning techniques will, by default, take the temporal information into account in making predictions of the future states of the system.

The algorithms for machine learning can be divided into three main categories (Russell and Norvig, 2003): supervised learning, unsupervised learning, and reinforcement learning. Supervised learning aims at building a statistical or deterministic model from labeled instances, thus it needs a certain proportion of the available data as the *training set*; then it applies this model to find the unlabeled instances in the rest of the available data, namely the *test set*. In contrast, there are no labeled instances in unsupervised learning, and the goal is to find hidden patterns and structures (e.g., clustering) in the available data on its own. In reinforcement learning, a certain goal is pursued in a dynamic environment without knowing explicitly whether the goal is coming close or not, and the learning process is driven by the feedbacks from the environment (Bishop, 2006). The standard procedure of supervised learning is as follows: The predictor model that generates a reasonable prediction is constructed by performing an appropriate method (e.g., artificial neural network [ANNs], or genetic programming [GP]) on a training set, and then validated on a test set. When the predictor model is validated, the last part of the test set can be used for prediction.

ANNs are a powerful tool for finding boundaries in classification problems, or functions in regression problems (Zou et al., 2009). It is derived by the working principle of a real neural network, and thus has a multilayer structure – an input layer, an output layer, and few (or zero) hidden layers, in which each neuron is connected to all neurons in the previous and following layer. In each layer a number of *perceptrons* (i.e., logistic or other function units which locally discriminate different inputs) can be inserted. Because these units are connected in such a network structure with different layers, when discrimination becomes deeper and deeper, complicated classifications can be computed, far beyond the abilities of other, simpler methods, such as logistic regression or support vector machines. Of course, the computational complexity of an ANN increases dramatically with the number of layers and the mean number of units (neurons) per layer.

In supervised learning, the ANN is initially trained using a training set in order to find the optimal coefficients to connect the different perceptron units. The most common method used for training is called "back-propagation" (Rummelhart, 1986), aimed at minimizing the mean squared error between prediction and actual values in both classification and regression problems. Once the training is complete, the predictor model, i.e., the ANN structure plus the coefficients of the units, can be tested and its performance can be analyzed in the test set.

GP is introduced as a symbolic regression method to find the functional form that fits the available data, and by optimizing the combination and composition of functions in a predetermined function set (Affenzeller et al., 2018). It is an evolutionary-based technique inspired by biological evolution, where each individual is a function (Banzhaf et al., 1998). Once the function base is chosen, a set of initial individuals is generated and evaluated on a training set using a fitness function. Depending on the results of the evaluation, the best individuals are kept, and new ones are generated by applying operations like crossover or mutation, then a new generation is obtained. This procedure is performed using several generations until a set of "best fit" individuals is obtained and then applied to the test set. As an example of application, a model-free feedback-control strategy for flows based on genetic programming was demonstrated by Gautier et al. (2015).

An alternative machine learning approach is known as "reservoir computing" (Lukoševičius and Jaeger, 2009). This technique uses a limited time series of measurements as input to a high-dimensional dynamical system called "reservoir." After the reservoir's response to the data is recorded, linear regression is used to learn a large set of parameters, called the "output weights." The learned output weights are then used to form a modified autonomous reservoir designed to be capable of producing an arbitrarily long time series whose ergodic properties closely resemble those of the input signal. This technique has allowed, for example, accurate calculation of Lyapunov exponents from data, even for large dimensional systems (Pathak et al., 2017, 2018).

9.3 Prediction of the Indian Summer Monsoon

The prediction of the Indian Summer Monsoon (ISM; see Section 2.5) timing is a vital issue for the Indian subcontinent and significantly affects agricultural planning and the gross domestic product of the country, up to 22% of which is determined by agriculture (Subash and Gangwar, 2014). A slight deviation of the monsoon timing manifested as delay (or early arrival) may lead to drastic droughts (floods), causing damage to infrastructure and loss of crops and livelihoods of the population. The onset of the ISM takes place abruptly, and skillful prediction with a lead time of at least two weeks, as well as that of its withdrawal, are important open problems.

9.3.1 The Problem

Various prediction attempts for monsoon onset date (OD) and withdrawal date (WD) have been made for different timescales from short-range (up to 4 days) and medium-range (4–10 days) forecasts using numerical prediction models (Das et al., 2002; Durai and Roy Bhowmik, 2014) to extended-range (10–30 days) and long-range (subseasonal, 30+ days) forecasts (Alessandri et al., 2014; Goswami and

Gouda, 2010). However, most forecasting methods have so far been based on statistical model approaches, which include averaged values of zonal asymmetric temperature anomaly (Prasad, 2005), SST, mean sea level pressure (Wang et al., 2009), tropospheric moisture build-up over areas south of the Indian peninsula (Taniguchi and Koike, 2006), vertically integrated moisture budget (hydrologic onset and withdrawal index) (Fasullo and Webster, 2003), moist static energy (Rajagopalan and Molnar, 2014), outgoing longwave radiation (OLR), and wind fields (Puranik et al., 2013).

The Indian Meteorological Department (IMD) provides a forecast of monsoon OD (on average) 21 days in advance, with an accuracy of ± 4 days with 10% deviation. However, there are certain difficulties in the existing forecasting methods, in particular, of false monsoon onsets mostly related to non-monsoonal atmospheric circulation systems (Flatau et al., 2001). In addition, the methodology used for forecasting of the WD, which determines the length of the monsoon season, has certain issues and limitations, in particular predicting a WD earlier than September 1 or forecasting of a late WD. These limitations require the improvement of OD and WD prediction skills (Webster, 2013).

The ISM is driven by several factors (see Section 2.5), including temperature and pressure gradients between land and ocean (Krishnamurti and Ramanathan, 1982), latent heat release during the monsoon season (Choudhury and Krishnan, 2011), and associated migration of the Intertropical Convergence Zone (ITCZ) (Gadgil, 2003). The high elevation of the Himalayan mountain peaks results in orographic shielding (Boos and Kuang, 2010), which establishes an area of deep convection (Figs. 9.1a, b) over the Indian subcontinent (Joseph et al., 1994). An abrupt and large-scale shift of the regional circulation pattern over the Indian peninsula (Goswami et al., 2010) occurs. In particular, a few days before monsoon onset, outgoing longwave radiation (OLR) shows deep convection over the Bay of Bengal and Arabian Sea, relative humidity increases abruptly in the vertical direction before onset of monsoon (Soman and Krishna Kumar, 1993), and vertically integrated zonal moisture transport increases (Rao et al., 2005). The withdrawal of the monsoon occurs gradually, and is caused by a southward movement of the ITCZ, resulting in an anticyclonic flow over northern and central India, a displacement of the moist marine air with dry continental air, followed by a reduction of rainfall over the Indian subcontinent. The movement of the northern limit of monsoon during the onset and withdrawal of the ISM and its variability are shown in Figs. 9.1c–f.

The IMD considers the onset of the monsoon date as the onset date over the southern tip of the Indian subcontinent (Kerala) region (Fig. 9.1c), black dashed curve). In order to avoid false monsoon onsets, the onset date over the *Eastern Ghats* region was used by Stolbova et al. (2016) as the OD (Fig. 9.1c, red

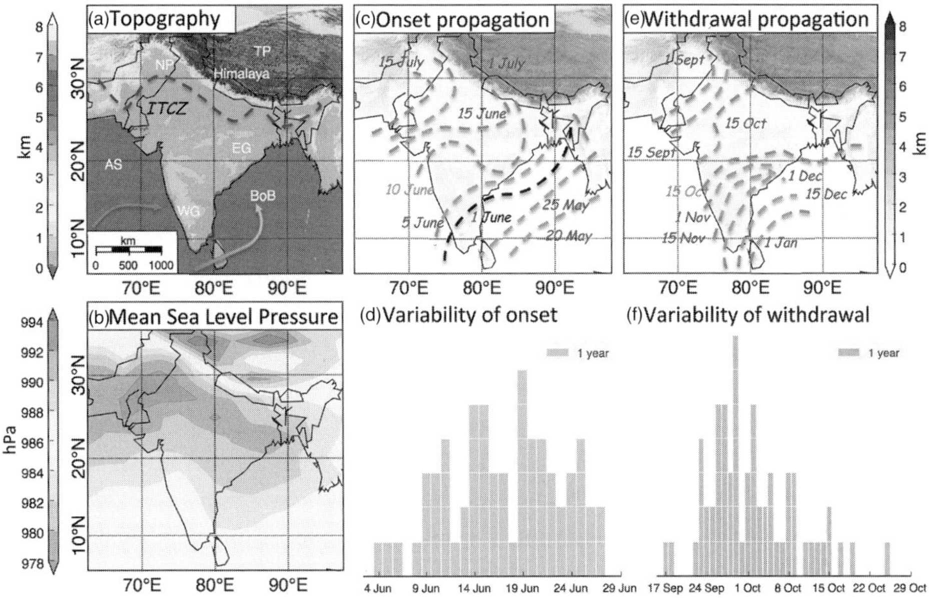

Figure 9.1 (a) Topography of the Indian subcontinent with key features of the Indian Summer Monsoon: Himalaya, Tibetan Plateau (TP), North Pakistan (NP), Eastern Ghats (EG), Western Ghats (WG), Arabian Sea (AS), Bay of Bengal (BoB), Intertropical Convergence Zone (ITCZ). Blue arrows indicate near-sea-level wind direction. (b) Composites of mean sea-level pressure (June to September, 1951–2014 based on NCEP/NCAR data). (c) and (e) Schematic representation of the long-term average propagation (since 1951) of the advance and withdrawal of monsoon over the Indian subcontinent (northern limit of monsoon). Dashed black line shows averaged monsoon onset for the Kerala region forecasted by the IMD, and dashed red line for the Eastern Ghats (the region of main interest in this study). (d) and (f) Histograms of onset and withdrawal dates for the Eastern Ghats region (1951–2014). Source: Stolbova et al. (2016). (A black and white version of this figure appears in some formats. For the color version, please refer to the plate section.)

dashed curve). Most importantly, arrival of the monsoon to the Eastern Ghats marks the arrival of the monsoon onto the Indian subcontinent, as the northern position of the ITCZ during the monsoon is located near the Eastern Ghats (23.5°N).

9.3.2 Climate Networks and the Prediction of the Monsoon

In Stolbova et al. (2016), a novel approach for prediction of monsoon OD and WD was presented. This approach is based on the results of climate network analysis of extreme precipitation over the Indian subcontinent during the pre-monsoon,

monsoon, and post-monsoon season. In particular, using the event synchronization technique and centrality measures, it was discovered that there are two geographic regions in the Indian subcontinent that experience extreme rainfall synchronization only during the monsoon season: the Eastern Ghats and North Pakistan (Stolbova et al., 2014). Later, it was found (Stolbova et al., 2016) that this is caused by the establishment of the monsoon trough, which connects these regions during the monsoon season. As a portion of the ITCZ, the monsoon trough is a convergence zone between the wind patterns of the Southern and Northern Hemispheres.

The maximum northern position of the ITCZ runs through the Eastern Ghats, and therefore it coincides with the hottest place on the Indian subcontinent (around 23.4°N). The topography of the Eastern Ghats creates a favorable condition for the formation of a low-pressure system, and when the ITCZ reaches the Eastern Ghats, two branches of monsoon (Arabian Sea and Bay of Bengal) merge. At the same time, high pressure in North Pakistan and its intricate topography bounded by the Himalaya, create favorable conditions for an anticyclone. Altogether, this creates a general wind pattern in which the center of the monsoon cyclone is near the Eastern Ghats, while the center of the anticyclone is in the North Pakistan region. As a consequence, a significant growth of fluctuations in the centers of the cyclone and the anticyclone systems occurs on the eve of monsoon onset in the Eastern Ghats (Fig. 9.2). Finally, it was shown that the collision of the anticyclone located in North Pakistan and the developing low-pressure system associated with the monsoon marks the onset of the monsoon in the Eastern Ghats. Therefore, fluctuations in North Pakistan and the Eastern Ghats were found to be indicators of this collision and can serve as early warnings for the onset of monsoon.

Based on these factors, the Eastern Ghats and North Pakistan were chosen as optimal observation locations or reference points (RPs) for comparative analysis aiming to predict monsoon onset and withdrawal. Time series of the near-surface air temperature and relative humidity were analyzed at these RPs and an important relation between these two regions was found. This allowed deriving a prediction scheme for forecasting the onset and withdrawal dates of the monsoon as described below.

9.3.3 Prediction of Monsoon Onset and Withdrawal

To develop and test the prediction scheme, near-surface air temperature (T) at 1000 hPa, relative humidity (rh) at 1000 hPa, and wind fields at 700 hPa were used (Stolbova et al., 2016) from two reanalysis gridded daily data sets: ERA40, provided by the ECMWF (Uppala et al., 2005) for the period 1971–2002; and

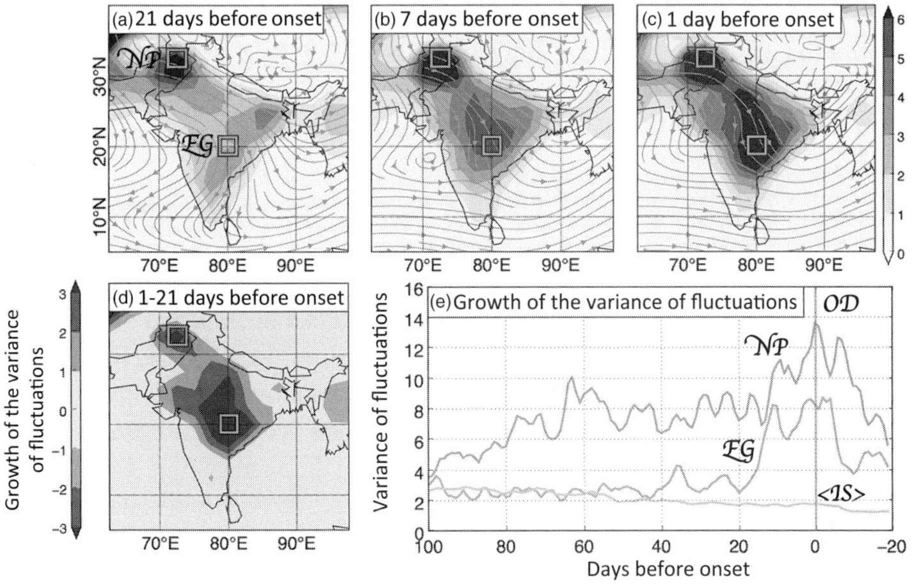

Figure 9.2 Pre-monsoon growth of the variance of fluctuations ($\bar{\sigma}^2$) of the weekly mean values of near-surface air temperature (T) 21 days, 7 days, and 1 day before the monsoon onset at the Eastern Ghats (EG): (a)–(c) (d) = (c) – (a). Composites are for the period 1971–2001 and were calculated from the ERA40 reanalysis data set, 700 hPa winds are indicated by the blue lines. The two boxes refer to RPs: North Pakistan (NP, blue) and EG (pink). E, growth of the variance of fluctuations in NP (blue), EG (pink), and averaged over the Indian subcontinent (<IS>) at approaching OD of the Indian monsoon. OD and WD of the monsoon are taken from Singh and Ranade (2010) and data from the IMD. Source: Stolbova et al. (2016). (A black and white version of this figure appears in some formats. For the color version, please refer to the plate section.)

NCEP/NCAR (Kalnay et al., 1996) for the period 1951–2014. The spatial resolution of both data sets is 2.5 degrees. Data were extracted for the monsoon region (62.5–97.5°E, 5.0–40.0°N; see Fig. 9.1), which results in $15 \times 15 = 225$ grid points.

In the first step of the prediction scheme (Stolbova et al., 2016), the variance σ^2 of temperature (T) and relative humidity (rh) for each grid point are calculated as

$$\sigma^2(x,d,w,y) = < [x(t^*(y)-d-k) - \bar{x}(t^*(y)-d-k)]^2 >_w =$$

$$= \sum_{k=1}^{w} \frac{1}{w}[x(t^*(y)-d-k) - \sum_{i=1}^{w} \frac{x(t^*(y)-d-i)}{w}]^2, \qquad (9.1)$$

where $x(t)$ is a time series, w is the length of the time window, d is the number of days prior to OD, y is a given year, and $t^*(y)$ is the OD in the given year y. The time mean value of σ^2, i.e.,

$$\bar{\sigma}^2(x,d,w) = <\sigma^2(x,d,w,y)>_y = \sum_{y=1}^{Y} \frac{\sigma^2(x,d,w,y)}{Y}, \tag{9.2}$$

where Y is the total number of years, is shown in Fig. 9.2. The results confirm that the Eastern Ghats and the North Pakistan regions (marked by pink and blue boxes) experience the highest growth of $\bar{\sigma}^2$ for temperature (similar results hold for relative humidity), while approaching the transition to monsoon onset.

Next, it was realized that there is an important relation between the Eastern Ghats and the North Pakistan time series. The intersection of the average time series in the RPs for the period from 1951 to 2014 takes place twice. At both times, it coincides with the mean values of the OD and WD as determined by the IMD within the Eastern Ghats with a standard deviation of ± 5 days (Fig. 9.3). Yearly variations in the intersection also occur within a few days of monsoon onset at the Eastern Ghats. This allows one to equalize temperatures in the RPs, and derive a prediction scheme for forecasting the OD and WD of the monsoon in the Eastern Ghats.

This prediction scheme for forecasting the OD and WD was used with the reanalysis data for the period from 1951 to 2014. As a training period, 14 years of data were used prior to the year when a prediction is made. The slopes of the trends for the RPs provided an estimation of an early, normal, or late monsoon arrival: a greater than average slope of T will lead to an earlier than usual OD, and vice versa. Trends of rh in the RPs in comparison with the average trends for the training period add up to the predictability of the OD: higher than average values of relative humidity lead to a late OD, and vice versa. In addition, the tendency of expected early (late) intersection of the time series of rh from the RPs usually leads to an earlier (later) than normal OD.

The analysis of the mean time series from the RPs shows that the OD coincides with the date when T in the Eastern Ghats and in North Pakistan become equal (see Fig. 9.3). Therefore, for the forecasting of the OD, it is necessary to predict when T for the Eastern Ghats will drop abruptly and intersect T for North Pakistan. However, during the pre-monsoon period T for the Eastern Ghats is in a nonlinear saturation state (when T reaches a maximum on the 145th day of the year) and it is a challenge to predict how close the system is to a threshold and when the abrupt transition will occur. During the same time period, T in North Pakistan gradually increases and can be approximated by a linear trend. Hence, using the linear trend one can predict when T in North Pakistan will reach a certain value, which for the Eastern Ghats is a critical threshold for the onset of the monsoon. The value of

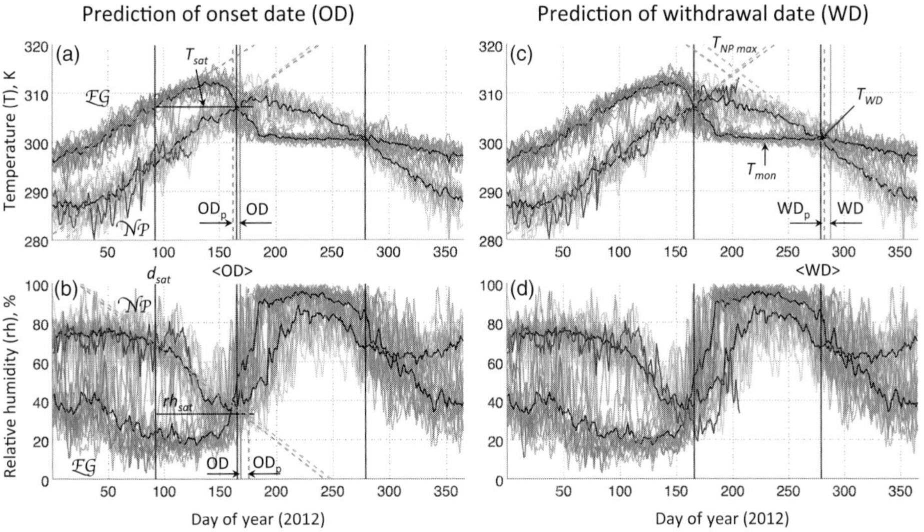

Figure 9.3 Prediction of OD and WD of Indian Monsoon: case study 2012. Left column: prediction of the OD; right column: WD in the Eastern Ghats (EG). (a), (c): air temperature at 1000 hPa; (b), (d) relative humidity at 1000 hPa. Time series from reference points: 14-year mean (black) and 2012 values for North Pakistan (NP) (blue) and the EG (red). Gray lines show time series from the NP and EG for the training period of 14 years. Saturation temperature T_{sat} (a) and saturation humidity rh_{sat} (c) are marked by horizontal black solid lines ($T_{sat} = T_{onset}$, T_{onset} and rh_{sat} calculated as intersection of mean time series for the training period from the EG and NP), and day of the saturation (d_{sat}) (when temperature in the EG in 2012 reaches T_{sat}) with dark blue. The orange line indicate trends to the mean time series in the NP and EG for the training period, light blue is trends for 2012. Black solid lines indicate mean values of the OD (<OD>) and WD (<WD>) for the training period. Dotted gray lines correspond to the predicted onset (OD_p) and withdrawal dates (WD_p), while solid gray lines are actual onset and withdrawal dates for 2012. Source: Stolbova et al. (2016). (A black and white version of this figure appears in some formats. For the color version, please refer to the plate section.)

the critical threshold might be estimated using the average critical threshold from the training period. The OD is determined by identifying the time when the linear trend of T in North Pakistan reaches T_{sat}, which is characterized by monsoon onset temperature in the Eastern Ghats from the training period. When T in North Pakistan fluctuates highly during a year and causes difficulties correctly determining the trend, the trend from the training period can be used for OD forecasting.

Finally, the prediction scheme of the WD is based on the symmetry of T changes in North Pakistan during the year. Knowing the T in the Eastern Ghats from the training period, the trend of North Pakistan in the pre-monsoon period, and the maximum T in North Pakistan, we can estimate the trend of the T decrease in the

North Pakistan region. The WD is then estimated as the intersection of the projected T decrease in North Pakistan and the T in the Eastern Ghats during the monsoon season (see Fig. 9.3c). Variations of rh are too high for WD prediction and the intersection of the rh time series usually takes place one month later than the actual WD (see Fig. 9.3d).

9.3.4 Prediction Skill

The performance of the prediction scheme from the previous subsection is shown in Fig. 9.4. The prediction is regarded as successful if the time difference between the predicted OD and the real one is ≤ 7 days for the OD and ≤ 10 days for WD. The proposed scheme using T results in 74% successful predictions of the OD when made on day 125 of the year (April 10). Predictability of the OD using relative humidity rh is lower than using temperature T (70%), mostly because of the high variability of rh and the associated difficulties in approximating a linear trend. Still, in some cases, the OD prediction based on rh may be useful, if the prediction based on T fails.

While other existing methods forecast the monsoon onset date over Kerala ranging from one month to two weeks prior to the onset date (at the end of April and on May 15), the approach proposed by Stolbova et al. (2016) allowed forecasting the arrival of monsoon onto the Indian subcontinent from 30 to 50 days in advance, and on average two to three weeks earlier than existing methods. The main advantage of the method is that it not only allows long-range forecasting of the monsoon arrival onto the Indian subcontinent, but it also enables estimating the exact onset date, while other methods of long-range forecasting only give a qualitative forecast of early, normal, or late monsoon arrival.

The prediction scheme for WD succeeds in 84% of years when made on day 205 (July 25). Reasons for the limited performance in several of the years are discussed by Stolbova et al. (2016). An important addition of this prediction scheme is the fact that it allows including information about ENSO years in the forecasting of onset and withdrawal dates of monsoon. The results of such inclusion are shown in Figs. 9.4b–d (for a detailed explanation of ENSO inclusion in the forecasting scheme, see Stolbova et al., 2016). The prediction of OD with a separate analysis of El Niño, La Niña, and neutral ENSO years increases the forecast accuracy of the withdrawal date up to 89% for El Niño (nine years) and does not change the accuracy for La Niña years.

9.4 El Niño Prediction

The occurrence of an El Niño event has huge impacts on the weather around the Pacific, causing many climate-related disasters, such as droughts in Australia and

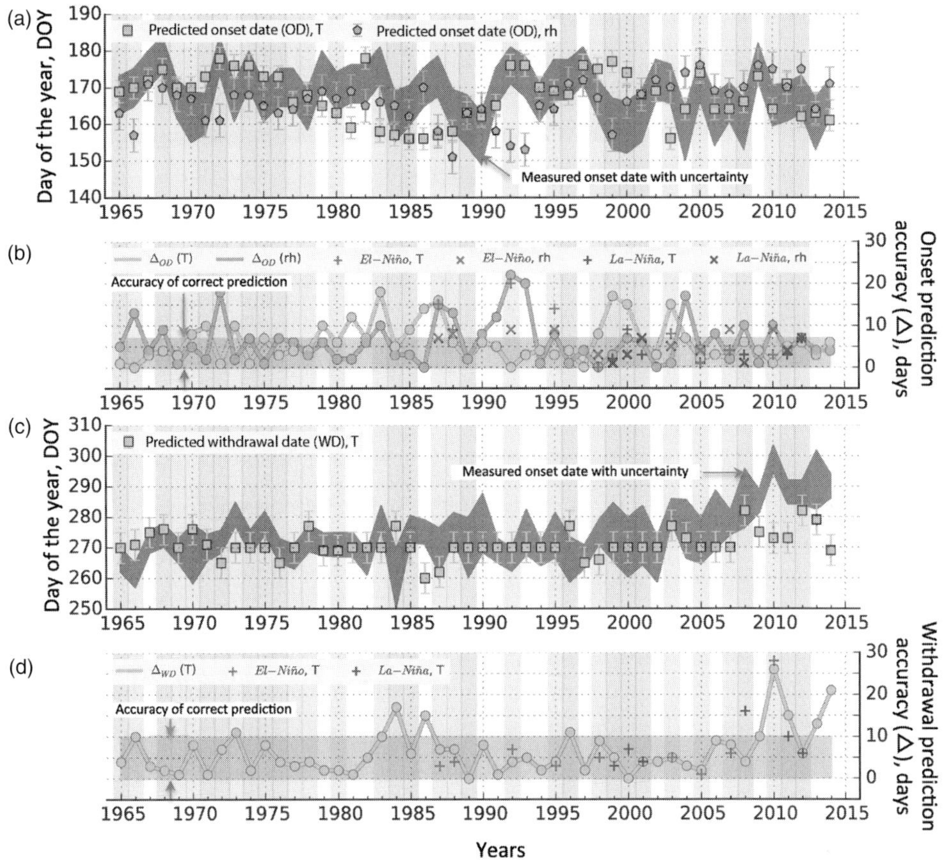

Figure 9.4 Monsoon OD and WD prediction based on *T* (green) and *rh* (orange) and measured (dark blue) OD (a) and WD (c). Red and light-blue shading indicate El Niño and La Niña years. Also shown is the difference (Δ) between the real onset or withdrawal and predicted dates in days ($\Delta_{OD(T)}$: prediction based on temperature, $\Delta_{OD(rh)}$: on relative humidity). Gray shading indicates a range of seven days, within the prediction this is considered accurate (b). The accuracy of prediction of the WD has a range of ten days (gray shadow) (d). Markers "+" (*T*) and "x" (*rh*) show improved prediction based on the training period of 14 years only from preceding El Niño (red) and La Niña years (b and d). Source: Stolbova et al. (2016). (A black and white version of this figure appears in some formats. For the color version, please refer to the plate section.)

torrential rains in South America (Reilly, 2009). For an adequate preparation to such events, it is crucial to develop skillful predictions for whether such an event will occur (say, one year ahead) and if so, how the event will develop in time.

9.4.1 The Problem

Since the 1990s, both dynamical and statistical models have been used to try to predict El Niño events (Chen et al., 2004; Fedorov et al., 2003; Latif and

Barnett, 1994; Yeh et al., 2009). Although about 20 models currently provide ENSO forecasts routinely, good prediction skill is generally limited to about six months ahead. The reason is the presence of the so-called "spring predictability barrier," where errors are greatly amplified due to the coupled feedbacks in the equatorial ocean–atmosphere system (Duan and Wei, 2013; Goddard et al., 2001).

For example, during March–May 2014 most of the models predicted strong El Niño conditions by the end of 2014. However, equatorial Pacific SST anomalies remained relatively small and the peak of the NINO3.4 index (the area-averaged SST anomalies over the region [120°W–170°W ×5°S–5°N]) in late November 2014 did not even exceed 1.0°C. Only by early 2015 did warming conditions appear near the dateline, and subsequently developed into a strongly El Niño event later that year.

The application of climate networks to anomalies in surface air temperature observations in the tropical Pacific has shown that the emergence of El Niño acts as an autonomous component in the climate network (Gozolchiani et al., 2011). A large-scale cooperative mode, linking the El Niño region and the rest of the ocean, builds up one calendar year before the warming event (Ludescher et al., 2013). Based on such findings on the temporal evolution of the climate network, Ludescher et al. (2014) developed a forecasting scheme for El Niño events. A threshold on the average link weight in the reconstructed climate network (Fig. 9.5) can skillfully predict an El Niño event one year ahead. It indeed successfully predicted the onset of the 2014–2016 El Niño event.

9.4.2 Machine Learning Prediction Using Network Measures

In Feng et al. (2016), machine learning techniques incorporating measures of climate networks were used in El Niño predictions, especially for the occurrence of El Niño events and the development of NINO3.4 index over time. The machine learning toolbox employed (ClimateLearn) makes full use of the open-source software packages written in Java, namely Weka (www.cs.waikato.ac.nz/ml/weka) and ECJ (https://cs.gmu.edu/ eclab/projects/ecj). This toolbox allows for basic operations of data mining, i.e., reading, merging, and cleaning data, as well as more sophisticated processes of building time-ordered training and test sets, and calling the scripts for the machine learning methods, such as decision trees, ANNs, and GP.

Most of the supervised learning algorithms used in the context of prediction have a multivariate data set as input. The data set can be represented by a $T \times N$ matrix, ordered in such a way that at time t it has a set of N attributes $x_1(t) \ldots x_N(t)$, $t = 1, \ldots, T$. The goal is to construct a predictor model that generates a prediction at time $t + \tau$ on the quantity y, and τ is the *prediction time*. The training set is built by choosing an interval of time $[t_i^{train}, t_f^{train}]$ $(t_i^{train}, t_f^{train} \in [1, T])$, and the test set

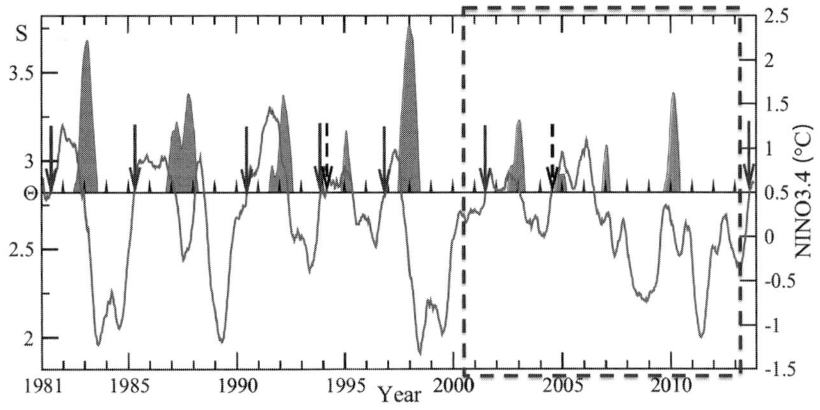

Figure 9.5 Climate network-based El Niño forecasting scheme proposed by Ludescher et al. (2013). The gray curve is the average link weight S of the network, the horizontal line indicates the decision threshold $\theta = 2.82$, and the solid areas show the El Niño events between January 1981 and November 2013. When the link weight crosses the threshold from below, an alarm is given that an El Niño event will start in the following calendar year. Correct predictions are marked by solid arrows and false alarms by dashed arrows. The dashed rectangle indicates the period of the test set we used for prediction by implementing the machine learning toolbox. Source: Ludescher et al. (2014).

is chosen from $[t_i^{test}, t_f^{test}]$ ($t_i^{test}, t_f^{test} \in [1, T]$). Note that for El Niño predictions, the condition $t_i^{test} > t_f^{train}$ has to be satisfied, which means that the instances in the test set happen after the one in the training set, since we are only interested in a chronological prediction for the El Niño event. However, for the cross-validation of the prediction by regression, no such condition applies.

For El Niño predictions, the instance quantity y is the NINO3.4 index, by which the occurrence of an El Niño event is defined as a five consecutive three-month running mean of NINO3.4 index above the threshold of $+0.5°$C (see www.cpc.ncep.noaa.gov). Since the prediction for the occurrence of El Niño events is a classification task, Feng et al. (2016) applied a pure classifier as the predictor model. Therefore the NINO3.4 index needs to be transformed from a numeric quantity to a nominal quantity belonging to two possible classes, i.e., 1 stands for the occurrence of an El Niño event and 0 for its absence. In Feng et al. (2016), an El Niño event was flagged when NINO3.4 values were continuously above the threshold of $+0.4°$C for five months.

9.4.3 Prediction Skill

The time t was chosen from the period of May 1949 to March 2014 with a time interval of 10 days, hence $T = 2365$ (Feng et al., 2016). Eight measures of the same

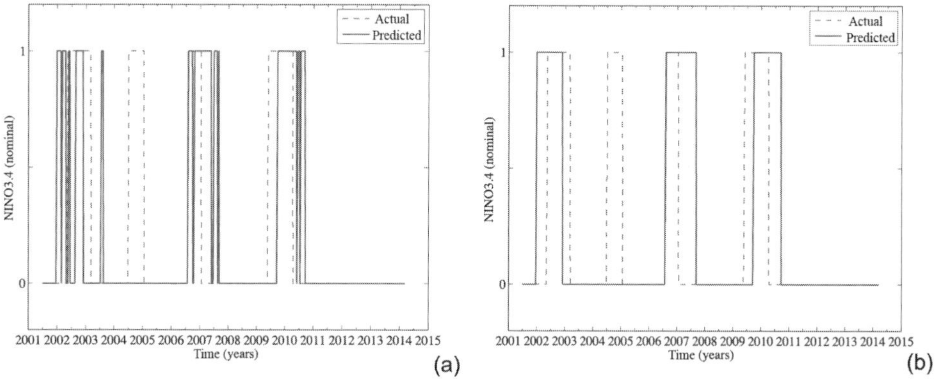

Figure 9.6 Prediction results on the test set from June 2001 to March 2014 (a) without filtering and (b) with filtering, using an ANN with a 3×3 layer structure (three neurons per layer) for the 12-month lead time prediction for the occurrence of El Niño events. Dashed lines are the actual nominal quantity of the NINO3.4 index (1 stands for the occurrence of an El Niño event where NINO3.4 values are continuously above the threshold of $+0.4°$C for five months, and 0 for the absence of such an event), and the solid lines indicate the predicted one.

weighted and directed climate network as in Ludescher et al. (2014) were used as the attributes (with time t as an additional attribute, thus $N = 9$): the averaged values of the maximum correlation $\mathrm{MAX}(C_{ij})$, the minimum correlation $\mathrm{MIN}(C_{ij})$, the maximum delay $\mathrm{MAX}(\Delta t^*)$, the minimum delay $\mathrm{MIN}(\Delta t^*)$, the maximum link weight $\mathrm{MAX}(W_{ij})$, the minimum link weight $\mathrm{MIN}(W_{ij})$, the standard deviation of the correlation $\mathrm{STD}(C_{ij})$, and the mean correlation $\mathrm{MEAN}(C_{ij})$. Supervised learning was used with an ANN having a 3×3 layer structure (three neurons per layer). The training set was from May 1949 to June 2001 (80% of T) and the test set was from June 2001 to March 2014 (20% of T). The prediction time τ is 12 months, similar to that in Ludescher et al. (2014).

Fig. 9.6a shows the classification results on the test set. By applying a specific time series filter, which eliminates the isolated and transient events, and joins the adjacent events, Fig. 9.6b also shows that the forecasting scheme gives accurate alarms 12 months ahead for the El Niño events in 2002, 2006, and 2009, without a false alarm in 2004. Compared with the results in Fig. 9.5 (inside the rectangle), the machine learning toolbox gives an improved prediction for the occurrence of El Niño events when using more measures of the same complex network as in Ludescher et al. (2014).

One advantage of using supervised learning for prediction is that the predictor model is constructed automatically from the training set without subjective decisions like the choice of thresholds. However, the available data for prediction as well as the number of instances is limited. For example, only a few El Niño events occurred between May 1949 and March 2014. The accuracy of the prediction will hence mostly depend on the length of the training set. As a consequence, one needs to choose proper proportions of the available data for the training/test set, to avoid "under-training."

To demonstrate that the proportion used for the test set (20% of T) gives the best performance, ROC-type analysis was performed (Feng et al., 2016) by varying the proportion from 16% to 30% of T as the test set. With a proportion between 16% and 20%, the averaged hit rate $D = 0.90$ and the averaged false-alarm rate $\alpha = 0.10$, for 21–25%, $D = 0.71$ and $\alpha = 0.29$, and from 26% to 30%, $D = 0.21$ and $\alpha = 0.79$. Thus, to have a high hit rate together with a low false-alarm rate, the best proportion for the test set is $\leq$20%. Of course, one should also maximize the length of the test set to incorporate more El Niño events for testing, and this motivated the choice of 20%.

The short-term development of NINO3.4 index is strongly related to the stability of the Pacific background state and the occurrence of westerly wind bursts (WWBs) near the dateline. Using the SST data, Feng and Dijkstra (2017) found that the stability of the Pacific climate can be measured by the skewness of the degree distribution, indicated by S_d, of a Pearson correlation climate network (PCCN) reconstructed from these observations (see Section 4.2). As a proxy of WWB activity, the time series of the second principal component (PC2) of the wind stress residual was used.

Next, the machine learning toolbox was used to investigate the importance of S_d and PC2 for the El Niño development by supervised learning regression. The attributes were therefore the background stability index $x_1 = S_d$, the westerly wind burst measure x_2 = PC2, and the time $x_3 = t$ from November 1961 to October 2014 with a time interval of one month; hence $T = 636$ and $N = 3$. The training set was from November 1961 to April 2004 (80% of T) and the test set was from May 2004 to October 2014 (20% of T). The quality of the predicted results in the test set was measured by the normalized root mean squared error (NRMSE) between the actual time series of NINO3.4 index $y^A(t)$ and the predicted one $y^P(t)$, defined as

$$NRMSE(y^A, y^P) = \frac{1}{\max_n(y^A, y^P) - \min_n(y^A, y^P)} \sqrt{\frac{1}{n} \sum_{t_i^{test} \leq t \leq t_f^{test}} (y^A(t) - y^P(t))^2}, \tag{9.3}$$

where n is the number of data points in the time series.

An ANN with a 2×1 layer structure (two neurons in the first layer and one neuron in the second layer) was used to do the regression. Since the optimal prediction time τ that gives a reasonable prediction $y(t+\tau)$ at time $t+\tau$ is unknown, τ was varied from 1 up to 12 months. Figure 9.7 shows the regression results on the test set for

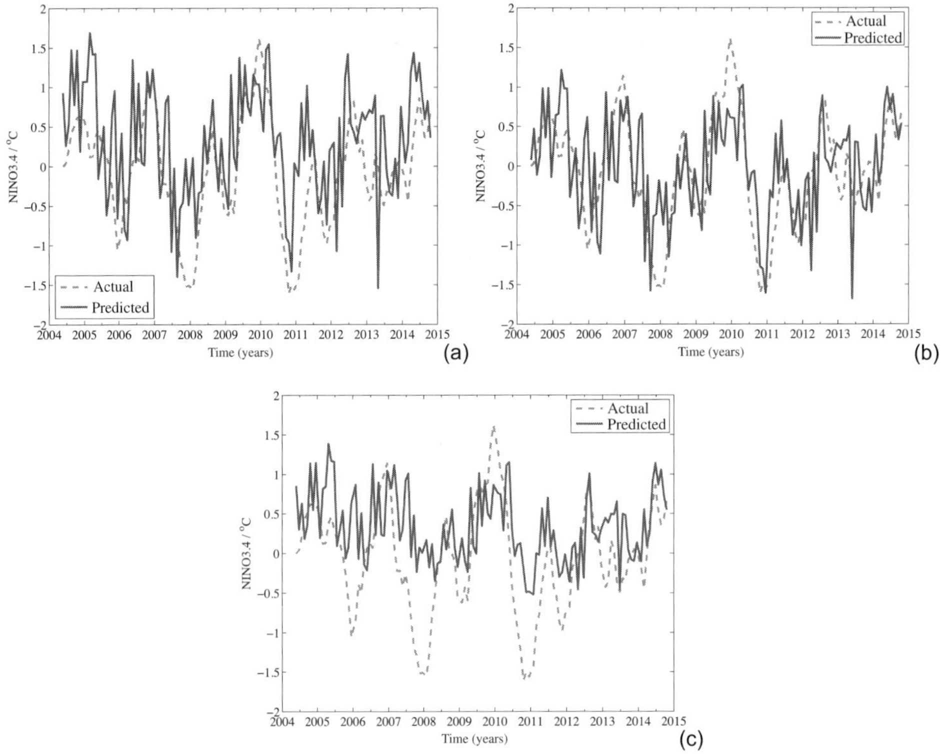

Figure 9.7 Regression results on the test set from May 2004 to October 2014 using an ANN with a 2×1 layer structure (two neurons in the first layer and one neuron in the second) for the prediction of the NINO3.4 index with a lead time of (a) two months (NRMSE = 0.23), (b) three months (NRMSE = 0.18), and (c) four months (NRMSE = 0.22). Dashed lines are the actual values of NINO3.4 index, and solid lines indicate the predicted ones.

the 2–4 months ahead as NINO3.4 forecasts. The best prediction is given at τ = three months, with the smallest value of the NRMSE = 0.18 (Fig. 9.7b). Due to the irregular behavior of the PC2 representing the WWBs, the predicted NINO3.4 indices in Fig. 9.7 show more fluctuation than the actual one. When a three-month running mean is applied to the predicted NINO3.4 index (three months lead time, Fig. 9.7b) as well as the actual one, the forecast has greater accuracy (NRMSE = 0.14).

To test the robustness of the result of the regression of the three-month lead time NINO3.4 index forecast (Fig. 9.7b), a series of cross-validations were performed by Feng et al. (2016) by keeping constant certain percentage splits between the training set and test set (70–30, 75–25, 80–20, and 85–15), and randomly choosing 200 initial times of the test set t_i^{test} from November 1961 to October 2014 for each percentage split. The peak values of the NRMSE remained near 0.17, independent of the choices of the percentage splits and t_i^{test}. Therefore the regression result in Fig. 9.7b was considered robust.

In summary, network-based methods are useful in prediction studies, likely because long-range correlations can be efficiently captured. Network-based prediction schemes have been shown to be quite successful, but it is unclear why this occurs. Building on the success of complex network approaches to investigate various aspects of climate variability, the machine learning techniques provide an innovative and efficient way to novel predictions of the occurrence and development of El Niño events.

References

Achuthavarier, D., Krishnamurthy, V., Kirtman, B. P., and Huang, B. (2012). Role of the Indian Ocean in the ENSO–Indian summer monsoon teleconnection in the NCEP climate forecast system. *Journal of Climate*, **25**(7), 2490–2508.

Affenzeller, M., Wagner, S., Winkler, S., and Beham, A. (2018). *Genetic Algorithms and Genetic Programming: Modern Concepts and Practical Applications*. Chapman and Hall/CRC.

Albert, R. and Barabási, A. (2002). Statistical mechanics of complex networks. *Reviews of Modern Physics*, **74**, 47–96.

Alessandri, A., Borrelli, A., Cherchi, A., et al. (2014). Prediction of Indian summer monsoon onset using dynamical sub-seasonal forecasts: effects of realistic initialization of the atmosphere. *Monthly Weather Review*, **143**, 778–793.

Alley, R. B., Marotzke, J., Nordhaus, W. D., et al. (2003). Abrupt climate change. *Nature*, **299**, 2005–2010.

Almeida, R. A. F. de, Nobre, P., Haarsma, R. J., and Campos, E. J. D. (2007). Negative ocean-atmosphere feedback in the South Atlantic Convergence Zone. *Geophysical Research Letters*, **34**(18), L18809.

Altman, N. and Krzywinski, M. (2017). Interpreting P values. *Nature Methods*, **14**, 213.

Ananthakrishnan, R. and Soman, M. K. (1990). The onset of the southwest monsoon in 1990. *Current Science*, **61**(7), 447–453.

Andersen, T., Carstensen, J., Hernández-García, E., and Duarte, C. M. (2009). Ecological thresholds and regime shifts: approaches to identification. *Trends in Ecology & Evolution*, **24**(1), 49–57.

Andronova, N. and Schlesinger, M. (2000). Causes of global temperature changes during the 19th and 20th centuries. *Geophysical Research Letters*, **27**(14), 2137–2140.

Arizmendi, F. and Barreiro, M. (2017). ENSO teleconnections in the southern hemisphere: a climate network view. *Chaos*, **27**(9), 093109.

Arizmendi, F., Marti, A., and Barreiro, M. (2014). Evolution of atmospheric connectivity in the 20th century. *Nonlinear Processes in Geophysics*, **21**, 825–839.

Arizmendi, F., Barreiro, M. B., and Masoller, C. (2017). Identifying large-scale patterns of unpredictability and response to insolation in atmospheric data. *Scientific Reports*, **7**, 45676.

Ashkenazy, Y., Feliks, Y., Gildor, H., and Tziperman, E. (2008). Asymmetry of daily temperature records. *Journal of the Atmospheric Sciences*, **65**(10), 3327–3336.

AVISO (2013). *SSALTO/DUACS User Handbook: (M)SLA and (M)ADT Near-Real Time and Delayed Time Products*. Centre national d'études spatiales.

Baccala, L. A. and Sameshima, K. (2001). Partial directed coherence: a new concept in neural structure determination. *Biological Cybernetics*, **84**, 463.

Balasis, G., Donner, R. V., Potirakis, S. M., et al. (2013). Statistical mechanics and information-theoretic perspectives on complexity in the earth system. *Entropy*, **15**(11), 4844–4888.

Bandt, C. and Pompe, B. (2002). Permutation entropy: a natural complexity measure for time series. *Physical Review Letters*, **88**(17), 174102.

Banzhaf, W., Nordin, P., Keller, R. E., and Francone, F. D. (1998). *Genetic Programming: An Introduction*, volume 1. Morgan Kaufmann.

Barabási, A.-L. (2009). Scale-free networks: a decade and beyond. *Science*, **325**(5939), 412–413.

Barabási, A.-L. and Albert, R. (1999). Emergence of scaling in random networks. *Science*, **286**(5439), 509–512.

Barnosky, A. D., Hadly, E. A., Bascompte, J., et al. (2012). Approaching a state shift in Earth's biosphere. *Nature*, **486**(7401), 52–58.

Barreiro, M., Chang, P., and Saravanan, R. (2002). Variability of the South Atlantic Convergence Zone as simulated by an atmospheric general circulation model. *Journal of Climate*, **15**, 745.

Barreiro, M., Chang, P., and Saravanan, R. (2005). Simulated precipitation response to SST forcing and potential predictability in the region of the South Atlantic Convergence Zone. *Climate Dynamics*, **24**, 105–114.

Barreiro, M., Marti, A. C., and Masoller, C. (2011). Inferring long memory processes in the climate network via ordinal pattern analysis. *Chaos*, **21**(1), 013101.

Barros, V., Doyle, M., González, M., et al. (2002). Climate variability over subtropical South America and the South American monsoon: a review. *Meteorologica*, **27**(1–2), 33–57.

Barsugli, J. J. and Battisti, D. S. (1998). The basic effects of atmosphere–ocean thermal coupling on midlatitude variability. *Journal of the Atmospheric Sciences*, **55**, 477–493.

Barthélemy, M. (2011). Spatial networks. *Physics Reports*, **499**(1–3), 1–101.

Bathiany, S., Dijkstra, H., Crucifix, M., et al. (2016). Beyond bifurcation: using complex models to understand and predict abrupt climate change. *Dynamics and Statistics of the Climate System*, **1**(1), dzw004.

Bauer, P., Thorpe, A., and Brunet, G. (2015). The quiet revolution of numerical weather prediction. *Nature*, **525**(7567), 47–55.

Benestad, R. E., Sutton, R. T., and Anderson, D. L. T. (2002). The effect of El Niño on intraseasonal Kelvin waves. *Quarterly Journal of the Royal Meteorological Society*, **128**(582), 1277–1291.

Berbery, E. H. and Vera, C. S. (1996). Characteristics of the Southern Hemisphere winter storm track with filtered and unfiltered data. *Journal of the Atmospheric Sciences*, **53**(3), 468–481.

Berezin, Y., Gozolchiani, A., Guez, O., and Havlin, S. (2012). Stability of climate networks with time. *Scientific Reports*, **2**, 1–8.

Bishop, C. M. (2006). *Pattern Recognition and Machine Learning*. Springer.

Black, E., Blackburn, M., Harrison, G., Hoskins, B., and Methven, J. (2004). Factors contributing to the summer 2003 European heatwave. *Weather*, **59**, 217–223.

Blayo, É., Bocquet, M., Cosme, E., and Cugliandolo, L. F. (2014). *Advanced Data Assimilation for Geosciences: Lecture Notes of the Les Houches School of Physics: Special Issue, June 2012*. Oxford University Press.

Boers, N., Bookhagen, B., Barbosa, H. M. J., et al. (2014). Prediction of extreme floods in the eastern Central Andes based on a complex networks approach. *Nature Communications*, **5**, 5199.

Boffetta, G., Cencini, M., Falcioni, M., and Vulpiani, A. (2002). Predictability: a way to characterize complexity. *Physics Reports*, **356**(6), 367–474.

Bombardi, R. J., Carvalho, L. M. V., Jones, C., and Reibota, M. S. (2013). Precipitation over eastern South America and the South Atlantic Sea surface temperature during neutral ENSO periods. *Climate Dynamics*, **42**, 1553–1568.

Boos, W. R. and Kuang, Z. (2010). Dominant control of the South Asian monsoon by orographic insulation versus plateau heating. *Nature*, **463**(7278), 218–222.

Booth, B. B., Dunstone, N. J., Halloran, P. R., Andrews, T., and Bellouin, N. (2012). Aerosols implicated as a prime driver of twentieth-century North Atlantic climate variability. *Nature*, **484**(7393), 228–232.

Bracco, A., Kucharski, F., Kallummal, R., and Molteni, F. (2004). Internal variability, external forcing and climate trends in multi-decadal AGCM ensembles. *Climate Dynamics*, **23**(6), 659–678.

Bradley, E. and Kantz, H. (2015). Nonlinear time-series analysis revisited. *Chaos*, **25**, 097610.

Bretherton, F. (1988). *Earth System Science: A Closer View*. NASA.

Broecker, W. S. (2006). Abrupt climate change revisited. *Global and Planetary Change*, **54**(3), 211–215.

Brovkin, V. and Claussen, M. (1998). On the stability of the atmosphere–vegetation system in the Sahara/Sahel region. *Journal of Climate*, **103**(D24), 31613–32624.

Bryan, F. O. (1986). High-latitude salinity effects and interhemispheric thermohaline circulations. *Nature*, **323**, 301–304.

Budyko, M. I. (1969). The effect of solar radiations on the climate on the Earth. *Tellus*, **21**, 611–619.

Caesar, L., Rahmstorf, S., Robinson, A., Feulner, G., and Saba, V. (2018). Observed fingerprint of a weakening Atlantic Ocean overturning circulation. *Nature*, **556**, 191–196.

Cai, W. and Cowan, T. (2009). La Niña Modoki impacts Australia autumn rainfall variability. *Geophysical Research Letters*, **36**(12), L12805.

Caldarelli, G. (2007). *Scale-Free Networks: Complex Webs in Nature and Technology*. Oxford University Press.

Carvalho, L. M. V., Jones, C., and Liebmann, B. (2002). Extreme precipitation events in southeastern South America and large-scale convective patterns in the South Atlantic convergence zone. *Journal of Climate*, **15**, 2377–2394.

Carvalho, L. M. V., Jones, C., and Liebmann, B. (2004). The South Atlantic Convergence Zone: intensity, form, persistence, and relationships with intraseasonal to interannual activity and extreme rainfall. *Journal of Climate*, **17**, 88–107.

Castiglione, P., Falcioni, M., Lesne, A., and Vulpiani, A. (2010). *Chaos and Coarse Graining in Statistical Mechanics*. Cambridge University Press.

Cellucci, C. J., Albano, A. M., and Rapp, P. E. (2005). Statistical validation of mutual information calculations: comparison of alternative numerical algorithms. *Physical Review E*, **71**, 066208.

Cencini, M., Cecconi, F., and Vulpiani, A. (2010). *Chaos: From Simple Models to Complex Systems*. World Scientific.

Chaikin, P. and Lubensly, T. (1995). *Principles of Condensed Matter Physics*. Cambridge University Press.

Chan, A., Dehne, F., and Taylor, R. (2005). CGMgraph/CGMlib: implementing and testing CGM graph algorithms on PC clusters and shared memory machines. *International Journal of High Performance Computing Applications*, **19**, 81–97.

Chang, E. (1993). Downstream development of baroclinic waves as inferred from regression analysis. *Journal of the Atmospheric Sciences*, **50**, 2038–2053.

Chang, E. K. M. (1999). Characteristics of wave packets in the upper troposphere. Part II: seasonal and hemispheric variations. *Journal of the Atmospheric Sciences*, **56**(11), 1729–1747.

Chang, E. K. M. and Yu, D. B. (1999). Characteristics of wave packets in the upper troposphere. Part I: Northern Hemisphere winter. *Journal of the Atmospheric Sciences*, **56**(11), 1708–1728.

Chaves, R. R. and Nobre, P. (2004). Interactions between sea surface temperature over the South Atlantic Ocean and the South Atlantic Convergence Zone. *Geophysical Research Letters*, **31**, L03204-1.

Chelton, D. B. and Schlax, M. (1996). Global observations of oceanic Rossby waves. *Science*, **272**, 234–238.

Chelton, D. B., Schlax, M., Lyman, J., and Johnson, G. (2003). Equatorially trapped Rossby waves in the presence of meridionally sheared baroclinic flow in the Pacific Ocean. *Progress in Oceanography*, **56**, 323–380.

Chen, D., Cane, M. A., Kaplan, A., Zebiak, S. E., and Huang, D. (2004). Predictability of El Niño over the past 148 years. *Nature*, **428**(6984), 733–736.

Choudhury, A. D. and Krishnan, R. (2011). Dynamical response of the South Asian monsoon trough to latent heating from stratiform and convective precipitation. *Journal of the Atmospheric Sciences*, **68**(6), 1347–1363.

Cimponeriu, L., Rosenblum, M., and Pikovsky, A. (2004). Estimation of delay in coupling from time series. *Physical Review E*, **70**, 046213.

Clarke, A. J. (2008). *An Introduction to the Dynamics of El Niño the Southern Oscillation*. Elsevier.

Claussen, M., Mysak, L., Weaver, A., et al. (2002). Earth system models of intermediate complexity: closing the gap in the spectrum of climate system models. *Climate Dynamics*, **18**, 579–586.

Clement, A. C. and Peterson, L. C. (2008). Mechanisms of abrupt climate change of the last glacial period. *Reviews of Geophysics*, **46**, RG4002.

Collins, M., Knutti, R., Arblaster, J., et al. (2013). Long-term climate change: projections, commitments and irreversibility. In T. Stocker, D. Qin, G.-K. Plattner, editors, *Climate Change 2013: The Physical Science Basis. Contribution of Working Group I to the Fifth Assessment Report of the Intergovernmental Panel on Climate Change*, pages 1029–1136. University of Cambridge Press.

Cover, T. M. and Thomas, J. A. (2006). *Elements of Information Theory*, 2nd ed. Wiley.

Csardi, G. and Nepusz, T. (2006). The igraph software package for complex network research. *InterJournal*, **1695**.

Cunningham, S. A., Kanzow, T., Rayner, D., et al. (2007). Temporal variability of the Atlantic Meridional Overturning Circulation at 26.5 N. *Science*, **317**(5840), 935–938.

Da Costa, E. D. and Colin de Verdiere, A. C. (2004). The 7.7 year North Atlantic oscillation. *Quarterly Journal of the Royal Meteorological Society*, **128**, 797–817.

Dakos, V., Scheffer, M., van Nes, E. H., et al. (2008). Slowing down as an early warning signal for abrupt climate change. *PNAS*, **105**(38), 14308–14312.

Dakos, V., van Nes, E. H., Donangelo, R., Fort, H., and Scheffer, M. (2010). Spatial correlation as leading indicator of catastrophic shifts. *Theoretical Ecology*, **3**(3), 163–174.

Dakos, V., Kéfi, S., Rietkerk, M., van Nes, E. H., and Scheffer, M. (2011). Slowing down in spatially patterned ecosystems at the brink of collapse. *The American Naturalist*, **177**(6), 154–166.

Dakos, V., Carpenter, S. R., van Nes, E. H., and Scheffer, M. (2015). Resilience indicators: prospects and limitations for early warnings of regime shifts. *Philosophical Transactions of the Royal Society of London B: Biological Sciences*, **370**(1659), 20130263.

Dansgaard, W. (1993). Evidence for general instability of past climate from a 250-kyr ice-core record. *Nature*, **364**, 218–220.

Das, S., Mitra, A. K., Iyengar, G. R., and Singh, J. (2002). Skill of medium-range forecasts over the Indian Monsoon region using different parameterizations of deep convection. *Weather and Forecasting*, **17**(1992), 1194–1210.

Daubechies, I. (1992). *Ten Lectures on Wavelets*. SIAM.

De Niet, A., Wubs, F., van Scheltinga, A. T., and Dijkstra, H. A. (2007). A tailored solver for bifurcation analysis of ocean–climate models. *Journal of Computational Physics*, **227**(1), 654–679.

DeAngelis, D. L., Post, W. M., and Travis, C. C. (1986). *Positive Feedback in Natural Systems*. Springer-Verlag.

Dee, D., Uppala, S., Simmons, et al. (2011). The ERA-Interim reanalysis: configuration and performance of the data assimilation system. *Quarterly Journal of the Royal Meteorological Society*, **137**(656), 553–597.

Delcroix, T., Boulanger, J.-P., Masia, F., and Menkes, C. (1994). Geosat-derived sea level and surface current anomalies in the equatorial Pacific during the 1986-1989 El Niño and La Niña. *Journal of Geophysical Research*, **99**, 25093–25107.

Dellnitz, M., Hessel-von Molo, M., Metzner, P., Preis, R., and Schütte, C. (2006). Graph algorithms for dynamical systems. In A. Mielke, editor, *Analysis, Modeling and Simulation of Multiscale Problems*, pages 619–645. Springer Verlag.

Dellnitz, M., Froyland, G., Horenkamp, C., Padberg-Gehle, K., and Sen Gupta, A. (2009). Seasonal variability of the subpolar gyres in the Southern Ocean: a numerical investigation based on transfer operators. *Nonlinear Processes in Geophysics*, **16**(6), 655–663.

Dellnitz, M., Froyland, G., Horenkamp, C., Padberg-Gehle, K., and Sen Gupta, A. (2009b). Seasonal variability of the subpolar gyres in the southern ocean: a numerical investigation based on transfer operators. *Nonlinear Processes in Geophysics*, **16**(6), 655–663.

Delworth, T. L. and Mann, M. E. (2000). Observed and simulated multidecadal variability in the Northern Hemisphere. *Climate Dynamics*, **16**, 661–676.

Den Toom, M., Dijkstra, H. A., and Wubs, F. W. (2011). Spurious multiple equilibria introduced by convective adjustment. *Ocean Modelling*, **38**(1–2), 126–137.

Deser, C. and Blackmon, M. L. (1993). Surface climate variations over the North Atlantic Ocean during winter: 1900–1989. *Journal of Climate*, **6**, 1743–1753.

Deza, J. I. (2015). Climate networks constructed by using information-theoretic measures and ordinal time-series analysis. PhD thesis, Universitat Politècnica de Catalunya.

Deza, J. I., Barreiro, M., and Masoller, C. (2013). Inferring interdependencies in climate networks constructed at inter-annual, intra-season and longer time scales. *The European Physical Journal Special Topics*, **222**(2), 511–523.

Deza, J., Masoller, C., and Barreiro, M. (2014). Distinguishing the effects of internal and forced atmospheric variability in climate networks. *Nonlinear Processes in Geophysics*, **21**, 617–631.

Deza, J., Barreiro, M., and Masoller, C. (2015). Assessing the direction of climate interactions by means of complex networks and information theoretic tools. *Chaos*, **25**, 033105.

Dijkstra, E. W. (1959). A note on two problems in connexion with graphs. *Numerische Mathematik*, **1**(1), 269–271.

Dijkstra, H. A. (2006). Interaction of SST modes in the North Atlantic Ocean. *Journal of Physical Oceanography*, **36**, 286–299.

Dijkstra, H. A. (2013). *Nonlinear Climate Dynamics*. Cambridge University Press.

Dijkstra, H. A. and Ghil, M. (2005). Low-frequency variability of the large-scale ocean circulation: a dynamical systems approach. *Reviews of Geophysics*, **43**(3), RG3002.

Dijkstra, H. A., Frankcombe, L. M., and Von der Heydt, A. S. (2008). A stochastic dynamical systems view of the Atlantic Multidecadal Oscillation. *Philosophical Transactions of the Royal Society A*, **366**(1875), 2543–2558.

Dobson, I., Carreras, B. A., Lynch, V. E., and Newman, D. E. (2007). Complex systems analysis of series of blackouts: cascading failure, critical points, and self-organization. *Chaos*, **17**(2), 026103.

Donangelo, R., Fort, H., Dakos, V., Scheffer, M., and van Nes, E. H. (2010). Early warnings for catastrophic shifts in ecosystems: comparison between spatial and temporal indicators. *International Journal of Bifurcation and Chaos*, **20**, 315–321.

Donges, J. F., Zou, Y., Marwan, N., and Kurths, J. (2009a). Complex networks in climate dynamics. *The European Physical Journal Special Topics*, **174**(1), 157–179.

Donges, J. F., Zou, Y., and Marwan, N. (2009b). The backbone of the climate network. *EPL*, **87**, 48007.

Donges, J. F., Heitzig, J., Beronov, B., et al. (2015). Unified functional network and nonlinear time series analysis for complex systems science: the pyunicorn package. *Chaos*, **25**, 1–26.

Donner, R. V., Zou, Y., Donges, J. F., et al. (2010). Recurrence networks: a novel paradigm for nonlinear time series analysis. *New Journal of Physics*, **12**(3), 033025.

Donner, R. V., Small, M., Donges, J. F., et al. (2011). Recurrence-based time series analysis by means of complex network methods. *International Journal of Bifurcation and Chaos*, **21**(4), 1019–1046.

d'Ovidio, F., Fernandez, V., Hernandez-García, E., and Lopez, C. (2004). Mixing structures in the Mediterranean Sea from finite-size Lyapunov exponents. *Geophysics Research Letters*, **31**, L17203.

Drumond, A., Nieto, R., Gimeno, L., and Ambrizzi, T. (2008). A Lagrangian identification of major sources of moisture over Central Brazil and La Plata basin. *Journal of Geophysical Research*, **113**, D14128.

Duan, W. and Wei, C. (2013). The "spring predictability barrier" for ENSO predictions and its possible mechanism: results from a fully coupled model. *International Journal of Climatology*, **33**(5), 1280–1292.

Dubois, M., Rossi, V., Ser-Giacomi, E., et al. (2016). Linking basin-scale connectivity, oceanography and population dynamics for the conservation and management of marine ecosystems. *Global Ecology and Biogeography*, **25**(5), 503–515.

Dukowicz, J. K. and Smith, R. D. (1994). Implicit free-surface method for the Bryan–Cox–Semtner ocean model. *Journal of Geophysical Research*, **99**(C4), 7991–8014.

Durai, V. R. and Roy Bhowmik, S. K. (2014). Prediction of Indian summer monsoon in short to medium range time scale with high resolution global forecast system (GFS) T574 and T382. *Climate Dynamics*, **42**, 1527–1551.

ECMWF (2009). European Centre for Medium-Range Weather Forecasts, 2009: ERA-Interim Project. Research data archive at the National Center for Atmospheric Research, Computational and Information Systems Laboratory, Boulder, CO. Available at http://apps.ecmwf.int/datasets/data/interim-full-daily.

Eichler, M., Dahlhaus, R., and Sandkuhler, J. (2003). Partial directed coherence: a new concept in neural structure determination. *Biological Cybernetics*, **89**, 289.

Eisenman, I., Yu, L., and Tziperman, E. (2005). Westerly wind bursts: ENSO's tail rather than the dog? *Journal of Climate*, **18**, 5224–3238.

Enfield, D. B., Mestas-Nunes, A. M., and Trimble, P. (2001). The Atlantic multidecadal oscillation and its relation to rainfall and river flows in the continental US. *Geophysics Research Letters*, **28**, 2077–2080.

Fasullo, J. and Webster, P. J. (2003). A hydrological definition of Indian Monsoon onset and withdrawal. *Journal of Climate*, **16**, 3200–3211.

Fedorov, A., Harper, S., Philander, S., Winter, B., and Wittenberg, A. (2003). How predictable is El Niño? *Bulletin of the American Meteorological Society*, **84**(7), 911–919.

Feldhoff, J. H., Donner, R. V., Donges, J. F., Marwan, N., and Kurths, J. (2012). Geometric detection of coupling directions by means of inter-system recurrence networks. *Physics Letters A*, **376**(46), 3504–3513.

Feng, Q. and Dijkstra, H. (2014). Are North Atlantic multidecadal SST anomalies westward propagating? *Geophysical Research Letters*, **41**, 541–546.

Feng, Q. and Dijkstra, H. A. (2017). Climate network stability measures of El Niño variability. *Chaos*, **27**(3), 035801-15.

Feng, Q. Y., Viebahn, J. P., and Dijkstra, H. A. (2014). Deep ocean early warning signals of an Atlantic MOC collapse. *Geophysical Research Letters*, **41**, 6008–6014.

Feng, Q., Vasile, R., Segond, M., et al. (2016). Climatelearn: a machine-learning approach for climate prediction using network measures. *Geoscientific Model Development Discussion*. DOI: 10.5194/gmd-2015-273

Feng, S. and Hu, Q. (2008). How the North Atlantic Multidecadal Oscillation may have influenced the Indian summer monsoon during the past two millennia. *Geophysical Research Letters*, **35**, L01707.

Flatau, M. K., Flatau, P. J., and Rudnick, D. (2001). The dynamics of double monsoon onsets. *Journal of Climate*, **14**, 4130–4146.

Fortunato, S. (2010). Community detection in graphs. *Physics Reports*, **486**, 75–174.

Fortunato, S. and Barthélemy, M. (2007). Resolution limit in community detection. *PNAS*, **104**(1), 36–41.

Fortunato, S. and Hric, D. (2016). Community detection in networks: a user guide. *Physics Reports*, **659**, 1–44.

Frankcombe, L., Dijkstra, H., and von der Heydt, A. (2008). Sub-surface signatures of the Atlantic Multidecadal Oscillation. *Geophysical Research Letters*, **35**, L19602.

Frankcombe, L. M. and Dijkstra, H. A. (2009). Coherent multidecadal variability in North Atlantic sea level. *Geophysical Research Letters*, **36**, L15604.

Frankcombe, L. M., von der Heydt, A. S., and Dijkstra, H. A. (2010). North Atlantic multidecadal climate variability: an investigation of dominant time scales and processes. *Journal of Climate*, **23**(13), 3626–3638.

Frankignoul, C. and Hasselmann, K. (1977). Stochastic climate models. II: application to sea-surface temperature anomalies and thermocline variability. *Tellus*, **29**, 289–305.

Froyland, G. (2005). Statistically optimal almost-invariant sets. *Physica D: Nonlinear Phenomena*, **200**(3–4), 205–219.

Froyland, G. and Dellnitz, M. (2003). Detecting and locating near-optimal almost-invariant sets and cycles. *SIAM Journal on Scientific Computing*, **24**(6), 1839–1863.

Froyland, G. and Padberg-Gehle, K. (2012). Finite-time entropy: a probabilistic approach for measuring nonlinear stretching. *Physica D: Nonlinear Phenomena*, **241**(19), 1612–1628.

Froyland, G., Padberg, K., England, M. H., and Treguier, A. M. (2007). Detection of coherent oceanic structures via transfer operators. *Physical Review Letters*, **98**(22), 224503.

Froyland, G., Santitissadeekorn, N., and Monahan, A. (2010). Transport in time-dependent dynamical systems: finite-time coherent sets. *Chaos*, **20**(4), 043116.

Froyland, G., Horenkamp, C., Rossi, V., Santitissadeekorn, N., and Gupta, A. S. (2012). Three-dimensional characterization and tracking of an Agulhas ring. *Ocean Modelling*, **52**, 69–75.

Froyland, G., Stuart, R. M., and van Sebille, E. (2014). How well-connected is the surface of the global ocean? *Chaos*, **24**(3), 033126.

Gadgil, S. (2003). The Indian monsoon and its variability. *Annual Review of Earth and Planetary Sciences*, **31**, 429–467.

Gadgil, S. (2004). Extremes of the Indian summer monsoon rainfall, ENSO and equatorial Indian Ocean oscillation. *Geophysical Research Letters*, **31**(12), L12213.

Ganachaud, A. and Wunsch, C. (2000). Improved estimates of global ocean circulation, heat transport and mixing from hydrographic data. *Nature*, **408**, 453–457.

Gautier, N., Aider, J., Duriez, T., et al. (2015). Closed-loop separation control using machine learning. *Journal of Fluid Mechanics*, **770**, 442–457.

Gill, A. E. (1982). *Atmosphere–Ocean Dynamics*. Academic Press.

Gladwell, M. (2000). *The Tipping Point*. Little Brown.

Glantz, M. H., Katz, R. W., and Nicholls, N. (1991). *Teleconnections Linking Worldwide Climate Anomalies*. Information Systems Division, National Agricultural Library.

Goddard, L., Mason, S. J., Zebiak, S. E., et al. (2001). Current approaches to seasonal to interannual climate predictions. *International Journal of Climatology*, **21**(9), 1111–1152.

Goswami, B. N., Kulkarni, J. R., Mujumdar, V. R., and Chattopadhyay, R. (2010). On factors responsible for recent secular trend in the onset phase of monsoon intraseasonal oscillations. *International Journal of Climatology*, **30**(14), 2240–2246.

Goswami, P. and Gouda, K. C. (2010). Evaluation of a dynamical basis for advance forecasting of the date of onset of monsoon rainfall over India. *Monthly Weather Review*, **138**.

Gozolchiani, A., Havlin, S., and Yamasaki, K. (2011). Emergence of El Niño as an autonomous component in the climate network. *Physical Review Letters*, **107**(14), 148501.

Granger, C. W. J. (1969). Investigating causal relations by econometric and cross-spectral methods. *Econometrica*, **37**, 424–438.

Gray, W. M., Sheaffer, J. D., and Landsea, C. W. (1997). Variability of Atlantic hurricane activity. In H. Diaz and D. Pulwarthy, editors, *Hurricanes*, pages 15–53. Springer.

Gregor, D. and Lumsdaine, A. (2005). The parallel bgl: a generic library for distributed graph computations. In O. I. U. Publications, editor, *In Parallel Object-Oriented Scientific Computing (POOSC)*. OSL Indiana University Publications.

Guckenheimer, J. and Holmes, P. (1990). *Nonlinear Oscillations, Dynamical Systems and Bifurcations of Vector Fields, 2nd edition*. Springer-Verlag.

Guez, O. C., Gozolchiani, A., and Havlin, S. (2014). Influence of autocorrelation on the topology of the climate network. *Physical Review E*, **90**(6), 062814.

Guttal, V. and Jayaprakash, C. (2007). Impact of noise in bistable ecological systems. *Ecological Modelling*, **201**, 420–428.

Guttal, V. and Jayaprakash, C. (2009). Spatial variance and spatial skewness: leading indicators of regime shifts in spatial ecological systems. *Theoretical Ecology*, **2**(1), 3–12.

Hagberg, A. A., Schult, D. A., and Swart, P. J. (2008). Exploring network structure, dynamics, and function using NetworkX. In G. Varoquaux, T. Vaught, and J. Millman, editors, *SciPy 2008: Proceedings of the 7th Python in Science Conference*, pages 11–15.

Haken, H. (1977). *Synergetics: An Introduction. Nonequilibrium Phase Transitions and Self-Organization in Physics, Chemistry and Biology*. Springer-Verlag.

Haller, G. (2015). Lagrangian coherent structures. *Annual Review of Fluid Mechanics*, **47**(1), 137–162.

Halpin-Healy, T. and Zhang, Y.-C. (1995). Kinetic roughening phenomena, stochastic growth, directed polymers and all that: aspects of multidisciplinary statistical mechanics. *Physics Reports*, **254**(4–6), 215–414.

Hamed, K. H. and Rao, A. R. (1998). A modified Mann–Kendall trend test for autocorrelated data. *Journal of Hydrology*, **204**(1), 182–196.

Hamming, R. W. (1950). Error detecting and error correcting codes. *Bell System Technical Journal*, **26**, 147–157.

Hartmann, D. L. (1994). *Global Physical Climatology*. Academic Press.

Hasselmann, K. (1976). Stochastic climate models. I: theory. *Tellus*, **28**, 473–485.

Hawkins, E., Smith, R. S., Allison, L. C., et al. (2011). Bistability of the Atlantic overturning circulation in a global climate model and links to ocean freshwater transport. *Geophysical Research Letters*, **38**(10), L10605.

Hegger, R., Kantz, H., and Schreiber, T. (1999). Practical implementation of nonlinear time series methods: the TISEAN package. *Chaos*, **9**(2), 413–435.

Held, H. and Kleinen, T. (2004). Detection of climate system bifurcations by degenerate fingerprinting. *Geophysical Research Letters*, **31**, L23207.

Held, I. M., Ting, M., and Wang, H. (2002). Northern winter stationary waves: theory and modeling. *Journal of Climate*, **15**, 2125–2144.

Hernández-Carrasco, I., López, C., Hernández-García, E., and Turiel, A. (2012). Seasonal and regional characterization of horizontal stirring in the global ocean. *Journal of Geophysical Research*, **117**, C10007.

Hlavackovaschindler, K., Paluš, M., Vejmelka, M., and Bhattacharya, J. (2007). Causality detection based on information-theoretic approaches in time series analysis. *Physics Reports*, **441**(1), 1–46.

Hlinka, J., Hartman, D., Vejmelka, M., et al. (2013). Reliability of inference of directed climate networks using conditional mutual information. *Entropy*, **15**, 2023–2045.

Holme, P. and Saramäki, J. (2012). Temporal networks. *Physics Reports*, **519**(3), 97–125.

Huisman, S. E., den Toom, M., Dijkstra, H. A., and Drijfhout, S. (2010). An indicator of the multiple equilibria regime of the Atlantic Meridional Overturning Circulation. *Journal of Physical Oceanography*, **40**(3), 551–567.

Ihshaish, H., Tantet, A., Dijkzeul, J. C. M., and Dijkstra, H. A. (2015). ParaGraph: a parallel toolbox for the construction and analysis of large complex climate networks. *Geoscientific Model Development*, **8**(10), 3321–3331.

Imkeller, P. and Von Storch, J.-S. (2001). *Stochastic Climate Models*, volume 49. Springer Science & Business Media.

James, I. (1994). *Introduction to Circulating Atmospheres*. Cambridge University Press.

Johns, W., Baringer, M., and Beal, L. (2011). Continuous, array-based estimates of Atlantic Ocean heat transport at 26.5 N. *Journal of Climate*, **24**, 2429–2449.

Johnsen, S. J., Clausen, H. B., Dansgaard, W., et al. (1992). Irregular glacial interstadials recorded in a new Greenland ice core. *Nature*, **359**, 311–313.

Joseph, P. V., Eischeid, J. K., and Pyle, R. J. (1994). Interannual variability of the onset of the Indian Summer Monsoon and its association with atmospheric features, El Niño, and sea surface temperature anomalies. *Journal of Climate*, **7**, 81–105.

Junquas, C., Vera, C., Li, L., and Treut, H. L. (2012). Summer precipitation variability over Southeastern South America in a global warming scenario. *Climate Dynamics*, **38**, 1867–1883.

Kalnay, E., Kanamitsu, M., Kistler, R., et al. (1996). The NCEP/NCAR reanalysis 40-year project. *Bulletin of the American Meteorological Society*, **77**(3), 437–471.

Kantz, H. and Schreiber, T. (2003). *Nonlinear Time Series Analysis*, volume 7. Cambridge University Press.

Kao, H.-Y. and Yu, J.-Y. (2009). Contrasting Eastern-Pacific and Central-Pacific types of ENSO. *Journal of Climate*, **22**(3), 615–632.

Karimi, A. and Paul, M. R. (2010). Extensive chaos in the Lorenz-96 model. *Chaos*, **20**(4), 043105.

Keller, H. B. (1977). Numerical solution of bifurcation and nonlinear eigenvalue problems. In P. H. Rabinowitz, editor, *Applications of Bifurcation Theory*. Academic Press.

Kempe, D., Kleinberg, J., and Kumar, A. (2000). Connectivity and inference problems for temporal networks. In *Proceedings of the Thirty-Second Annual ACM Symposium on Theory of Computing*, pages 504–513. ACM.

Kerr, R. A. (2000). A North Atlantic climate pacemaker for the centuries. *Science*, **288**, 1984–1986.

Kessler, W. S. (1990). Observations of long Rossby waves in the northern tropical Pacific. *Journal of Geophysical Research*, **95**, 5183–5217.

Kim, H. and Anderson, R. (2012). Temporal node centrality in complex networks. *Physical Review E*, **85**(2), 026107.

Kim, H.-M., Webster, P. J., and Curry, J. A. (2009). Impact of shifting patterns of Pacific Ocean warming on North Atlantic tropical cyclones. *Science*, **325**(5936), 77–80.

Kistler, R., Collins, W., Saha, S., et al. (2001). The NCEP-NCAR 50-year reanalysis: monthly means CD-ROM and documentation. *Bulletin of the American Meteorological Society*, **82**(2), 247–267.

Kleeman, R. (2002). Measuring dynamical prediction utility using relative entropy. *Journal of the Atmospheric Sciences*, **59**, 2057–2072.

Knight, J. R., Allan, R. J., Folland, C. K., Vellinga, M., and Mann, M. E. (2005). A signature of persistent natural thermohaline circulation cycles in observed climate. *Geophysical Research Letters*, **32**, L20708.

Knight, J. R., Folland, C. K., and Scaife, A. A. (2006). Climate impacts of the Atlantic Multidecadal Oscillation. *Geophysical Research Letters*, **33**, L17706.

Kochen, M. (1989). *The Small World*. Ablex.

Kodama, Y. M. (1992). Large-scale common features of subtropical precipitation zones (the Baiu frontal zone, the SPCZ, and the SACZ) Part I: characteristics of subtropical frontal zones. *Journal of the Meteorological Society of Japan*, **70**, 813–836.

Kostakos, V. (2009). Temporal graphs. *Physica A*, **388**(6), 1007–1023.

Krishnamurti, T. N. and Ramanathan, Y. (1982). Sensitivity of the Monsoon onset to differential heating. *Journal of the Atmospheric Sciences*, **39**, 1290–1306.

Kuehn, C. (2011). A mathematical framework for critical transitions: bifurcations, fast–slow systems and stochastic dynamics. *Physica D*, **240**(12), 1020–1035.

Kuehn, C. (2013). A mathematical framework for critical transitions: normal forms, variance and applications. *Journal of Nonlinear Science*, **23**(23), 457–510.

T. Kuhlbrodt, A. Griesel, M. Montoya, A. Levermann, M. Hofmann, and S. Rahmstorf (2007). On the driving processes of the atlantic meridional overturning circulation, Rev. *Geophysics*, **45**, 1–32.

Kushnir, Y. (1994). Interdecadal variations in North Atlantic sea surface temperature and associated atmospheric conditions. *Journal of Physical Oceanography*, **7**, 141–157.

G. Lancaster et al. (2018). Surrogate data for hypothesis testing of physical systems. *Physics Reports*, **748**, 1–60

Large, W. G. and Yeager, S. (2004). Diurnal to decadal global forcing for ocean and sea-ice models: the data sets and flux climatologies. Technical report, National Center for Atmospheric Research.

Larkin, N. K. and Harrison, D. (2005). On the definition of El Niño and associated seasonal average U.S. weather anomalies. *Geophysical Research Letters*, **32**(13), L13705.

Latif, M. (1998). Dynamics of interdecadal variability in coupled ocean–atmosphere models. *Journal of Climate*, **11**, 602–624.

Latif, M. and Barnett, T. P. (1994). Causes of decadal climate variability over the North Pacific and North America. *Science*, **266**, 634–637.

Legler, D. and O'Brien, J. (1988). Tropical Pacific wind stress analysis for TOGA. Intergovernmental Oceanographic Commission.

Lenton, T. M. (2011). Early warning of climate tipping points. *Nature Climate Change*, **1**(4), 201–209.

Lenton, T. M., Held, H., Kriegler, E., et al. (2008). Tipping elements in the Earth's climate system. *PNAS*, **105**(6), 1786–1793.

Lenton, T. M., Livina, V. N., Dakos, V., van Nes, E. H., and Scheffer, M. (2012). Early warning of climate tipping points from critical slowing down: comparing methods to improve robustness. *Philosophical Transactions. Series A, Mathematical, Physical, and Engineering Sciences*, **370**(1962), 1185–1204.

Lentz, H. H., Selhorst, T., and Sokolov, I. M. (2013). Unfolding accessibility provides a macroscopic approach to temporal networks. *Physical Review Letters*, **110**(11), 118701.

Levermann, A. (2011). When glacial giants roll over. *Nature*, **472**, 43–44.

Levermann, A., Schewe, J., Petoukhov, V., and Held, H. (2009). Basic mechanism for abrupt monsoon transitions. *PNAS*, **106**(49), 20572–20577.

Levnajić, Z. and Mezić, I. (2010). Ergodic theory and visualization: I. Mesochronic plots for visualization of ergodic partition and invariant sets. *Chaos*, **20**(3), 033114.

Lian, T., Chen, D., Tang, Y., and Wu, Q. (2014). Effects of westerly wind bursts on El Niño: a new perspective. *Geophysical Research Letters*, **41**(10), 3522–3527.

Lifland, J. (2003). The North Atlantic Oscillation: climatic significance and environmental impact. *Eos, Transactions of the American Geophysical Union*, **84**(8), 73–73.

Livina, V. N. and Lenton, T. M. (2007). A modified method for detecting incipient bifurcations in a dynamical system. *Geophysical Research Letters*, **34**(3). DOI: 10.1029/2006GL028672

Livina, V. N., Kwasniok, F., and Lenton, T. M. (2010). Potential analysis reveals changing number of climate states during the last 60 kyr. *Climate of the Past*, **6**(1), 77–82.

Livina, V. N., Kwasniok, F., Lohmann, G., Kantelhardt, J. W., and Lenton, T. M. (2011). Changing climate states and stability: from Pliocene to present. *Climate Dynamics*, **37**(11–12), 2437–2453.

Lorenz, E. N. (1969). The predictability of a flow which possesses many scales of motion. *Tellus*, **21**(3), 289–307.

Lorenz, E. N. (1996). Predictability: a problem partly solved. In T. Palmer and R. Hagedorn, editors, *Proceedings of the Seminar on Predictability, Vol. I, ECMWF Seminar*, pages 40–58. ECMWF.

Lovejoy, S. (2014). Scaling fluctuation analysis and statistical hypothesis testing of anthropogenic warming. *Climate Dynamics*, **42**(9–10), 2339–2351.

Ludescher, J., Gozolchiani, A., Bogachev, M. I., Bunde, A., Havlin, S., and Schellnhuber, H. J. (2013). Improved El Niño forecasting by cooperativity detection. *PNAS*, **110**(29), 11742–11745.

Ludescher, J., Gozolchiani, A., Bogachev, M. I., et al. (2014). Very early warning of next El Niño. *PNAS*, **111**(6), 2064–2066.

Lukoševičius, M. and Jaeger, H. (2009). Reservoir computing approaches to recurrent neural network training. *Computer Science Review*, **3**(3), 127–149.

Lumsdaine, A., Gregor, D., Hendrickson, B., and Berry, J. W. (2007). Challenges in parallel graph processing. *Parallel Processing Letters*, **17**(1), 5–20.

Lyman, J. M., Chelton, D. B., Deszoeke, R. A., and Samelson, R. M. (2005). Tropical instability waves as a resonance between equatorial Rossby waves. *Journal of Physical Oceanography*, **35**, 232–254.

Malik, N., Marwan, N., and Kurths, J. (2010). Spatial structures and directionalities in Monsoonal precipitation over South Asia. *Nonlinear Processes in Geophysics*, **17**(5), 371–381.

Malik, N., Bookhagen, B., Marwan, N., and Kurths, J. (2012). Analysis of spatial and temporal extreme monsoonal rainfall over South Asia using complex networks. *Climate Dynamics*, **39**(3–4), 971–987.

Mancho, A. M., Small, D., and Wiggins, S. (2006). A tutorial on dynamical systems concepts applied to Lagrangian transport in oceanic flows defined as finite time data sets: theoretical and computational issues. *Physics Reports*, **437**(3–4), 55–124.

Mancho, A. M., Wiggins, S., Curbelo, J., and Mendoza, C. (2013). Lagrangian descriptors: a method for revealing phase space structures of general time dependent dynamical systems. *Communications in Nonlinear Science and Numerical Simulation*, **18**, 3530–3557.

Mantua, N. J. and Hare, S. R. (2002). The Pacific decadal oscillation. *Journal of Oceanography*, **58**(1), 35–44.

Marotzke, J. (2000). Abrupt climate change and thermohaline circulation: mechanisms and predictability. *PNAS*, **97**, 1347–1350.

Marshall, J. and Plumb, R. A. (2008). *Atmosphere, Ocean and Climate Dynamics: An Introductory Text*. Academic Press.

Martin, E., Paczuski, M., and Davidsen, J. (2013). Interpretation of link fluctuations in climate networks during El Niño periods. *EPL*, **102**(4), 48003.

Martin-Gomez, V. and Barreiro, M. (2016). Complex network analysis of ocean's influence on spring time rainfall variability over southeastern South America during the 20th century. *International Journal of Climatology*, **36**, 1344–1358.

Martin-Gomez, V. and Barreiro, M. (2017). Effect of future climate change on the coupling between the tropical oceans and precipitation over southeastern South America. *Climatic Change*, **141**, 315–329.

Martin-Gomez, V., Hernandez-Garcia, E., Barreiro, M., and Lopez, C. (2016). Interdecadal variability of southeastern South America rainfall and moisture sources during the austral summertime. *Journal of Climate*, **29**, 6751–6763.

Marwan, N., Donges, J. F., Zou, Y., Donner, R. V., and Kurths, J. (2009). Complex network approach for recurrence analysis of time series. *Physics Letters A*, **373**(46), 4246–4254.

Masuda, N., Klemm, K., and Eguíluz, V. M. (2013). Temporal networks: slowing down diffusion by long lasting interactions. *Physical Review Letters*, **111**, 188701.

Matsueda, M. (2011). Predictability of Euro-Russian blocking in summer 2010. *Geophysical Research Letters*, **38**, L06801.

McNeall, D., Halloran, P. R., Good, P., and Betts, R. A. (2011). Analyzing abrupt and nonlinear climate changes and their impacts. *Wiley Interdisciplinary Reviews: Climate Change*, **2**(5), 663–686.

McPhaden, M. J. (1999). Genesis and evolution of the 1997–98 El Niño. *Science*, **283**, 950–954.

Mehlhorn, K. and Näher, S. (1995). Leda: a platform for combinatorial and geometric computing. *Communications of the ACM*, **38**(1), 96–102.

Meng, J., Fan, J., Ashkenazy, Y., and Havlin, S. (2017). Percolation framework to describe El Niño conditions. *Chaos*, **27**(3), 035807.

Menkes, C. E., Lengaigne, M., Vialard, J., et al. (2014). About the role of Westerly wind events in the possible development of an El Niño in 2014. *Geophysical Research Letters*, **41**(18), 6476–6483.

Milgram, S. (1967). The small world problem. *Psychology Today*, **1**(1), 61–67.

Millot, C. and Taupier-Letage, I. (2005). Circulation in the Mediterranean sea. In S. Goffredo and Z. Dubinsky, editors, *The Mediterranean Sea*, pages 29–66. Springer.

Mitchell, J.M. (1976). An overview of climate variability and its causal mechanisms, *Quaternary Research*, **6**, 481–493.

Mitchell, T. (1997). *Machine Learning*. McGraw-Hill.

Mokhov, I., Smirnov, D., Nakonechny, P., et al. (2011). Alternating mutual influence of El-Niño/Southern Oscillation and Indian monsoon. *Geophysical Research Letters*, **38**(8).

Molkenthin, N., Rehfeld, K., Marwan, N., and Kurths, J. (2014). Networks from flows: from dynamics to topology. *Scientific Reports*, **4**. DOI: 10.1038/srep04119

Molteni, F. (2003). Atmospheric simulations using a GCM with simplified physical parameterizations: I. Model climatology and variability in multi-decadal experiments. *Climate Dynamics*, **20**, 175–191.

Moron, V., Vautard, R., and Ghil, M. (1998). Trends, interdecadal and interannual oscillations in global sea-surface temperature. *Climate Dynamics*, **14**, 545–569.

Mosquera-Vásquez, K., Dewitte, B., and Illig, S. (2014). The Central Pacific El Niño intraseasonal Kelvin wave. *Journal of Geophysical Research: Oceans*, **119**(10), 6605–6621.

Mudelsee, M. (2014). *Climate Time Series Analysis* Springer.

Mudelsee, M. and Bermejo, M. A. (2017). Optimal heavy tail estimation: part 1. Order selection. *Nonlinear Processes in Geophysics*, **24**(4), 737–744.

Mukhin, D., Gavrilov, A., Feigin, A., Loskutov, E., and Kurths, J. (2015). Principal nonlinear dynamical modes of climate variability. *Scientific Reports*, **5**, 15510.

Muller, R. A., Curry, J., Groom, D., et al. (2013). Decadal variations in the global atmospheric land temperatures. *Journal of Geophysical Research: Atmospheres*, **118**(11), 5280–5286.

Navarra, A. and Simoncini, V. A. (2010). *Guide to Empirical Orthogonal Functions for Climate Data Analysis*. Springer-Verlag.

Newman, M. (2010). *Networks: An introduction*. Oxford University Press.

Newman, M. (2004). Analysis of weighted networks. *Physical Review E*, **70**(5), 056131.

Newman, M. and Girvan, M. (2004). Finding and evaluating community structure in networks. *Physical Review E*, **69**(2), 026113.

Newman, M., Barabási, A.-L., and Watts, D. J. (2006). *The Structure and Dynamics of Networks*. Princeton University Press.

Nilsson-Jacobi, M., André, C., Doos, K., and Jonsson, P. R. (2012). Identification of subpopulations from connectivity matrices. *Ecography*, **35**, 1004–1016.

Nogués-Paegle, J. and Mo, K. C. (1997). Alternating wet and dry conditions over South America during summer. *Monthly Weather Review*, **125**, 279–291.

Oddo, P., Adani, M., Pinardi, N., et al. (2009). A nested Atlantic–Mediterranean sea general circulation model for operational forecasting. *Ocean Science*, **5**(4), 461.

Overpeck, J. T. and Cole, J. E. (2006). Abrupt change in Earth's climate system. *Annual Review of Environment and Resources*, **31**(1), 1–31.

Paladin, G. and Vulpiani, A. (1987). Anomalous scaling laws in multifractal objects. *Physics Reports*, **156**(4), 147–225.

Palus, M. (2007). From nonlinearity to causality: statistical testing and inference of physical mechanisms underlying complex dynamics. *Contemporary Physics*, **48**, 307.

Palus, M. and Stefanovska, A. (2003). Direction of coupling from phases of interacting oscillators: an information-theoretic approach. *Physical Review E*, **67**, 055201.

Pathak, J., Lu, Z., Hunt, B., Girvan, M., and Ott, E. (2017). Using machine learning to replicate chaotic attractors and calculate Lyapunov exponents from data. *Chaos*, **27**, 121102.

Pathak, J., Hunt, B., Girvan, M., Lu, Z., and Ott, E. (2018). Model-free prediction of large spatiotemporally chaotic systems from data: a reservoir computing approach. *Physical Review Letters*, **120**, 024102.

Peacock, T. and Dabiri, J. (2010). Introduction to focus issue: Lagrangian coherent structures. *Chaos*, **20**(1), 017501.

Philander, S. G. H. (1990). *El Niño and the Southern Oscillation.* Academic Press.

Prasad, V. S. (2005). Onset and withdrawal of Indian summer monsoon. *Geophysical Research Letters*, **32**(20), L20715.

Preis, R., Dellnitz, M., Hessel, M., Schütte, C., and Meerbach, E. (2004). Dominant paths between almost invariant sets of dynamical systems. Preprint 154 of the DFG Schwerpunktprogramm 1095, available from www2.math.uni-paderborn.de/ags/ag-dellnitz.

Puranik, S. S., Ray, K. C. S., Sen, P. N., and Kumar, P. P. (2013). An index for predicting the onset of monsoon over Kerala. *Current Science*, **105**(7).

Quail, T., Shrier, A., and Glass, L. (2015). Predicting the onset of period-doubling bifurcations in noisy cardiac systems. *PNAS*, **112**(30), 9358–9363.

Quian Quiroga, R., Kreuz, T., and Grassberger, P. (2002). Event synchronization: a simple and fast method to measure synchronicity and time delay patterns. *Physical Review E*, **66**(4), 041904.

Rahmstorf, S. (2000). The thermohaline circulation: a system with dangerous thresholds? *Climatic Change*, **46**, 247–256.

Rahmstorf, S., Crucifix, M., Ganopolski, A., et al. (2005). Thermohaline circulation hysteresis: a model intercomparison. *Geophysical Research Letters*, **L23605**, 1–5.

Rajagopalan, B. and Molnar, P. (2014). Combining regional moist static energy and ENSO for forecasting of early and late season Indian monsoon rainfall and its extremes. *Geophysical Research Letters*, **41**, 4323–4331.

Rao, P. L. S., Mohanty, U. C., and Ramesh, K. J. (2005). The evolution and retreat features of the summer monsoon over India. *Meteorological Applications*, **12**, 241.

Rayner, N., Parker, D., Horton, E., et al. (2003). Global analyses of sea surface temperature, sea ice, and night marine air temperature since the late nineteenth century. *Journal of Geophysical Research*, **108**(D14), 4407.

Rehfeld, K., Marwan, N., Heitzig, J., and Kurths, J. (2011). Comparison of correlation analysis techniques for irregularly sampled time series. *Nonlinear Processes in Geophysics*, **18**(3), 389–404.

Rehfeld, K., Marwan, N., Breitenbach, S. F. M., and Kurths, J. (2012). Late Holocene Asian Summer Monsoon dynamics from small but complex networks of palaeoclimate data. *Climate Dynamics*, **41**, 3–19.

Reilly, B. (2009). *Disaster and Human History: Case Studies in Nature, Society and Catastrophe*. McFarland.

Rényi, A. (1970). *Probability Theory*. North-Holland.

Ribeiro, B., Perra, N., and Baronchelli, A. (2013). Quantifying the effect of temporal resolution on time-varying networks. *Scientific Reports*, **3**, 3006.

Robertson, A. W. and Mechoso, C. R. (2000). Interannual and interdecadal variability of the South Atlantic Convergence Zone. *Monthly Weather Review*, **128**(8), 2947–2957.

Rodríguez-Méndez, V., Eguíluz, V. M., Hernández-García, E., and Ramasco, J. J. (2016). Percolation-based precursors of transitions in extended systems. *Scientific Reports*, **6**, 295525.

Rosenblum, M. and Pikovsky, A. (2001). Detecting direction of coupling in interacting oscillators. *Physical Review E*, **64**, 045202.

Rossi, V., López, C., Hernández-García, E., et al. (2009). Surface mixing and biological activity in the four Eastern Boundary upwelling systems. *Nonlinear Processes in Geophysics*, **16**, 557–568.

Rossi, V., Ser-Giacomi, E., López, C., and Hernández-García, E. (2014). Hydrodynamic provinces and oceanic connectivity from a transport network help designing marine reserves. *Geophysical Research Letters*, **41**(8), 2883–2891.

Rosvall, M. and Bergstrom, C. T. (2008). Maps of random walks on complex networks reveal community structure. *PNAS*, **105**(4), 1118–1123.

Roulston, M. and Neelin, J. (2000). The response of an ENSO model to climate noise, weather noise and intraseasonal forcing. *Geophysical Research Letters*, **27**, 3723–3726.

Roundy, P. E. and Kiladis, G. N. (2006). Observed relationships between oceanic kelvin waves and atmospheric forcing. *Journal of Climate*, **19**(20), 5253–5272.

Rummelhart, D. (1986). Learning representations by back-propagation errors. *Nature*, **323**, 533–536.

Runge, J., Petoukhov, V., Donges, J. F., et al. (2015). Identifying causal gateways and mediators in complex spatio-temporal systems. *Nature Communications*, **6**, 8502.

Russell, S. J. and Norvig, P. (2003). *Artificial Intelligence: A Modern Approach*, 2nd edition. Pearson Education.

Sabeerali, C. T., Rao, S. A., Ajayamohan, R. S., and Murtugudde, R. (2011). On the relationship between Indian summer monsoon withdrawal and Indo-Pacific SST anomalies before and after 1976/1977 climate shift. *Climate Dynamics*, **39**(3–4), 841–859.

Saha, S. and Saha, K. (1980). A hypothesis on onset, advance and withdrawal of the Indian summer monsoon. *Pure and Applied Geophysics*, **118**(2), 1066–1075.

Saha, S., Moorthi, S., and Pan, H.-L., et al. (2010). The NCEP climate forecast system reanalysis. *Bulletin of the American Meteorological Society*, **91**, 1015–1057.

Saltzmann, B. (2001). *Dynamical Paleoclimatology*. Academic Press.

Samelson, R. and Wiggins, S. (2006). *Lagrangian Transport in Geophysical Jets and Waves*. Springer.

Sameshima, K. and Baccala, L. A. (1999). Using partial directed coherence to describe neuronal ensemble interactions. *Journal of Neuroscience Methods*, **94**, 93.

Sankar, S., Kumar, M. R. R., Reason, C., and Paula, D. (2011). On the relative roles of El Niño and Indian Ocean dipole events on the Monsoon onset over Kerala. *Theoretical and Applied Climatology*, **103**, 359–374.

Santitissadeekorn, N. and Bollt, E. (2007). Identifying stochastic basin hopping by partitioning with graph modularity. *Physica D: Nonlinear Phenomena*, **231**(2), 95–107.

Santitissadeekorn, N., Froyland, G., and Monahan, A. (2010). Optimally coherent sets in geophysical flows: a transfer-operator approach to delimiting the stratospheric polar vortex. *Physical Review E*, **82**(5), 056311.

Sarachik, E. S. and Cane, M. A. (2010). *The El Niño-Southern Oscillation Phenomenon*. Cambridge University Press.

Schaub, M., Lambiotte, R., and Barahona, M. (2012). Encoding dynamics for multiscale community detection: Markov time sweeping for the map equation. *Physical Review E*, **86**, 026112.

Scheffer, M., Carpenter, S., Foley, J. A., Folke, C., and Walker, B. (2001). Catastrophic shifts in ecosystems. *Nature*, **413**(6856), 591–596.

Scheffer, M., Bascompte, J., Brock, W. A., and Brovkin, V. (2009). Early-warning signals for critical transitions. *Nature*, **461**(7260), 53–59.

Scheffer, M., Carpenter, S. R., Lenton, T. M., et al. (2012). Anticipating critical transitions. *Science*, **338**(6105), 344–348.

Schlesinger, M. E. and Ramankutty, N. (1994). An oscillation in the global climate system of period 65–70 years. *Nature*, **367**, 723–726.

Schneider, T. and Griffies, S. (1999). A conceptual framework for predictability studies. *Journal of Climate*, **12**(10), 3133–3155.

Schneider, U., Becker, A., Finger, P., et al. (2011). GPCC full data reanalysis version 6.0 at 1.0: monthly land-surface precipitation from rain-gauges built on GTS-based and historic data. Data set.

Schwarz, G. (1978). Estimating the dimension of a model. *Annals of Statistics*, **6**(2), 461–464.

Seber, G. A. and Lee, A. J. (2012). *Linear Regression Analysis* Wiley.

Sellers, W. D. (1969). A global climate model based on the energy balance of the earth–atmosphere system. *Journal of Applied Meteorology*, **8**, 392–400.

Ser-Giacomi, E., Vasile, R., Recuerda, I., Hernández-García, E., and López, C. (2015a). Dominant transport pathways in an atmospheric blocking event. *Chaos*, **25**, 087413.

Ser-Giacomi, E., Rossi, V., López, C., and Hernández-García, E. (2015b). Flow networks: a characterization of geophysical fluid transport. *Chaos*, **25**(3), 036404.

Ser-Giacomi, E., Vasile, R., Hernández-García, E., and López, C. (2015c). Most probable paths in temporal weighted networks: an application to ocean transport. *Physical review E*, **92**, 036404.

Shadden, S. C., Lekien, F., and Marsden, J. E. (2005). Definition and properties of Lagrangian coherent structures from finite-time Lyapunov exponents in two-dimensional aperiodic flows. *Physica D*, **212**(3–4), 271–304.

Shannon, C. E. (1948). A mathematical theory of communication. *Bell System Technical Journal*, **27**, 623–656.

Sharma, N., Sharma, P., Irwin, D., and Shenoy, P. (2011). Predicting solar generation from weather forecasts using machine learning. In *2011 IEEE International Conference on Smart Grid Communications (SmartGridComm)*, pages 528–533. IEEE.

Shnerb, N. M., Sarah, P., Lavee, H., and Solomon, S. (2003). Reactive grass and vegetation patterns. *Physical Review Letters*, **90**.

Shukla, J. (1998). Predictability in the midst of chaos: a scientific basis for climate forecasting. *Science*, **282**, 728–731.

Siek, J., Lee, L.-Q., and Lumsdaine, A. (2002). *The Boost Graph Library: User Guide and Reference Manual*. Addison-Wesley Longman Publishing Co.

Simmons, A., Jones, P., Bechtold, V., et al. (2005). Comparison of trends and low-frequency variability in CRU, ERA-40, and NCEP/NCAR analyses of surface air temperature. *Journal of Geophysical Research-Atmospheres*, **109**, D24115.

Simmons, A. J., Willett, K. M., Jones, P. D., Thorne, P. W., and Dee, D. P. (2010). Comparison of trends and low-frequency variability in CRU, ERA-40, and NCEP/NCAR analyses of surface air temperature. *Journal of Geophysical Research*, **115**, D01110.

Singh, N. and Ranade, A. A. (2010). *Determination of Onset and Withdrawal Dates of Summer Monsoon across India using NCEP/NCAR Re-analysis Data Set*. Indian Institute of Tropical Meteorology.

Smeed, D. A., McCarthy, G., Cunningham, S. A., et al. (2013). Observed decline of the Atlantic meridional overturning circulation 2004 to 2012. *Ocean Science Discussions*, **10**(5), 1619–1645.

Smith, T. M. and Reynolds, R. W. (2003). Extended reconstruction of global sea surface temperatures based on COADS data (1854–1997). *Journal of Climate*, **16**(10), 1495–1510.

Smith, T. M. and Reynolds, R. W. (2004). Improved extended reconstruction of SST (1854–1997). *Journal of Climate*, **17**, 2466–2477.

Smith, T. M., Reynolds, R., Peterson, T., and Lawrimore, J. (2008). Improvements to NOAA's historical merged land–ocean surface temperature analysis (1880–2006). *Journal of Climate*, **21**, 2283–2296.

Soman, M. K. and Krishna Kumar, K. (1993). Space–time evolution of the meteorological features associated with the onset of the Indian summer monsoon. *Monthly Weather Review*, **121**, 1177–1194.

Speetjens, M., Lauret, M., Nijmeijer, H., and Anderson, P. (2013). Footprints of Lagrangian flow structures in Eulerian concentration distributions in periodic mixing flows. *Physica D: Nonlinear Phenomena*, **250**, 20–33.

Stainforth, D. A., Aina, T., Christensen, C., and Collins, M. (2005). Uncertainty in predictions of the climate response to rising levels of greenhouse gases. *Nature*, **433**(7024), 403–406.

Starnini, M., Baronchelli, A., Barrat, A., and Pastor-Satorras, R. (2012). Random walks on temporal networks. *Physical Review E*, **85**(5), 056115.

Stauffer, D. and Aharony, A. (1994). *Introduction to Percolation Theory*, 2nd edition. Taylor & Francis.

Steuer, R., Kurths, J., Daub, C. O., Weise, J., and Selbig, J. (2002). The mutual information: detecting and evaluating dependencies between variables. *Bioinformatics*, **18**(2), S231–S240.

Stohl, A., Forster, C., Frank, A., Seibert, P., and Wotawa, G. (2005). Technical note: the Lagrangian particle dispersion model FLEXPART version 6.2. *Atmospheric Chemistry and Physics*, **5**, 2461–2474.

Stohl, A., Sodemann, H., Eckhardt, S., et al. (2011). The Lagrangian particle dispersion model FLEXPART version 8.2. FLEXPART User Guide.

Stolbova, V., Martin, P., Bookhagen, B., Marwan, N., and Kurths, J. (2014). Topology and seasonal evolution of the network of extreme precipitation over the Indian subcontinent and Sri Lanka. *Nonlinear Processes in Geophysics*, **21**, 901–917.

Stolbova, V., Surovyatkina, E., Bookhagen, B., and Kurths, J. (2016). Tipping elements of the Indian monsoon: prediction of onset and withdrawal. *Geophysical Research Letters*, **43**, 3982–3990.

Stommel, H. (1961). Thermohaline convection with two stable regimes of flow. *Tellus*, **13**, 244–230.

Straus, D. M. and Shukla, J. (2010). Distinguishing between the SST-forced variability and internal variability in mid latitudes: analysis of observations and GCM simulations. *Quarterly Journal of the Royal Meteorological Society*, **126**(567), 2323–2350.

Strogatz, S. H. (2001). Exploring complex networks. *Nature*, **410**(6825), 268–276.

Subash, N. and Gangwar, B. (2014). Statistical analysis of Indian rainfall and rice productivity anomalies over the last decades. *International Journal of Climatology*, **34**(7), 2378–2392.

Sugihara, G., May, R., Ye, H., et al. (2012). Detecting causality in complex ecosystems. *Science*, **338**, 496–500.

Sutton, R. T. and Hodson, D. L. (2005). Atlantic Ocean forcing of North American and European summer climate. *Science*, **309**, 115–118.

Tallapragada, P. and Ross, S. D. (2013). A set oriented definition of finite-time Lyapunov exponents and coherent sets. *Communications in Nonlinear Science and Numerical Simulation*, **18**(5), 1106–1126.

Tang, J., Musolesi, M., Mascolo, C., Latora, V., and Nicosia, V. (2010a). Analysing information flows and key mediators through temporal centrality metrics. In *Proceedings of the 3rd Workshop on Social Network Systems*, page 3. ACM.

Tang, J., Scellato, S., Musolesi, M., Mascolo, C., and Latora, V. (2010b). Small-world behavior in time-varying graphs. *Physical Review E*, **81**(5), 055101.

Taniguchi, K. and Koike, T. (2006). Comparison of definitions of Indian summer monsoon onset: better representation of rapid transitions of atmospheric conditions. *Geophysical Research Letters*, **33**(2), L02709.

Tantet, A. and Dijkstra, H. A. (2014). An interaction network perspective on the relation between patterns of sea surface temperature variability and global mean surface temperature. *Earth System Dynamics*, **5**(1), 1–14.

Te Raa, L. A. and Dijkstra, H. A. (2002). Instability of the thermohaline ocean circulation on interdecadal timescales. *Journal of Physical Oceanography*, **32**(1), 138–160.

Tél, T. and Gruiz, M. (2006). *Chaotic Dynamics: An Introduction Based on Classical Mechanics*. Cambridge University Press.

Thomas, C. J., Lambrechts, J., Wolanski, E., et al. (2014). Numerical modelling and graph theory tools to study ecological connectivity in the Great Barrier Reef. *Ecological Modelling*, **272**, 160–174.

Thompson, J. M. T. and Sieber, J. (2011). Predicting climate tipping as a noisy bifurcation: a review. *International Journal of Bifurcation and Chaos*, **21**(2), 399–423.

Thornalley, D. J. R., Oppo, D. W., Ortega, P., et al. (2018). Anomalously weak Labrador Sea convection and Atlantic overturning during the past 150 years. *Nature*, **556**, 227–230.

Tietsche, S., Notz, D., Jungclaus, J., and Marotzke, J. (2011). Recovery mechanisms of Arctic summer sea ice. *Geophysical Research Letters*, **38**(2). DOI: 10.1029/2010GL045698

Ting, M., Kushnir, Y., Seager, R., and Li, C. (2009). Forced and internal twentieth-century SST trends in the North Atlantic. *Journal of Climate*, **22**(6), 1469–1481.

Tirabassi, G. (2015). *Disentangling Climatic Interactions and Detecting Tipping Points by Means of Complex Networks*. Universitat Politecnica de Catalunya.

Tirabassi, G. and Masoller, C. (2013). On the effects of lag-times in networks constructed from similarities of monthly fluctuations of climate fields. *EPL*, **102**, 59003.

Tirabassi, G. and Masoller, C. (2016). Unravelling the community structure of the climate system by using lags and symbolic time-series analysis. *Scientific Reports*, **6**, 29804.

Tirabassi, G., Viebahn, J., Dakos, V., et al. (2014). Interaction network based early-warning indicators of vegetation transitions. *Ecological Complex.*, **19**, 148–157.

Tirabassi, G., Masoller, C., and Barreiro, M. (2015a). A study of the air–sea interaction in the South Atlantic Convergence Zone through Granger causality. *International Journal of Climatology*, **35**, 3440–3453.

Tirabassi, G., Sevilla-Escoboza, R., Buldu, J. M., and Masoller, C. (2015b). Inferring the connectivity of coupled oscillators from time-series statistical similarity analysis. *Scientific Reports*, **5**, 10829.

Tirabassi, G., Sommerlade, L., and Masoller, C. (2017). Inferring directed climatic interactions with renormalized partial directed coherence and directed partial correlation. *Chaos*, **27**, 035815.

Toshiaki, S., Roundy, P., and Kiladis, G. (2008). Variability of intraseasonal Kelvin waves in the equatorial Pacific Ocean. *Journal of Physical Oceanography*, **38**, 921–944.

Tourre, Y. M., Rajagopalan, B., and Kushnir, Y. (1999). Dominant patterns of climate variability in the Atlantic Ocean during the last 136 Years. *Journal of Climate*, **12**, 2285–2299.

Trenberth, K. E., Branstator, G. W., Karoly, D., et al. (1998). Progress during TOGA in understanding and modeling global teleconnections associated with tropical sea surface temperatures. *Journal of Geophysical Research*, **103**(C7), 14291.

Trenberth, K. E., Fasullo, J., and Smith, L. (2005). Trends and variability in column-integrated atmospheric water vapor. *Climate Dynamics*, **24**, 741–758.

Tsonis, A. A. and Roebber, P. J. (2004). The architecture of the climate network. *Physica A: Statistical Mechanics and its Applications*, **333**, 497–504.

Tsonis, A. A. and Swanson, K. L. (2006). What do networks have to do with climate? *Bulletin of the American Meteorological Society*, **87**(5), 585–595.

Tsonis, A. A. and Swanson, K. L. (2008). Topology and predictability of El Niño and La Niña networks. *Physical Review Letters*, **100**(22), 228502.

Tupikina, L., Molkenthin, N., Lopez, C., et al. (2016). Correlation networks from flows: the case of forced and time-dependent advection-diffusion dynamics. *PLoS ONE*, **11**(4), e0153703.

Turner, J. (2004). The El Niño-southern oscillation and Antarctica. *International Journal of Climatology*, **24**, 1–31.

Uppala, S. M., Kallberg, P. W., Simmons, A. J., et al. (2005). The ERA-40 re-analysis. *Quarterly Journal of the Royal Meteorological Society*, **131**(612), 2961–3012.

van de Leemput, I. A., Wichers, M., Cramer, A. O. J., et al. (2014). Critical slowing down as early warning for the onset and termination of depression. *PNAS*, **111**(1), 87–92.

van der Mheen, M., Dijkstra, H. A., Gozolchiani, A., et al. (2013). Interaction network based early warning indicators for the Atlantic MOC collapse. *Geophysical Research Letters*, **40**(11), 2714–2719.

Van Nes, E. H. and Scheffer, M. (2007). Slow recovery from perturbations as a generic indicator of a nearby catastrophic shift. *The American Naturalist*, **169**(6), 738–747.

van Westen, R. M. and Dijkstra, H. A. (2017). Southern Ocean origin of multidecadal variability in the North Brazil current. *Geophysical Research Letters*, **44**(20), 10540–10548.

Van Wijk, B., Stam, C., and Daffertshofer, A. (2010). Comparing brain networks of different size and connectivity density using graph theory. *PLoS ONE*, **5**, e13701.

Veraart, A. J., Faassen, E. J., Dakos, V., et al. (2012). Recovery rates reflect distance to a tipping point in a living system. *Nature*, **481**(7381), 357–359.

Vianna, M. L. and Menezes, V. V. (2013). Bidecadal sea level modes in the North and Ssouth Atlantic Oceans. *Geophysical Research Letters*, **40**(22), 5926–5931.

Wakata, Y. (2007). Frequency-wavenumber spectra of equatorial waves detected from satellite altimeter data. *Journal of Oceanography*, **63**, 483–490.

Walin, G. (1985). The thermohaline circulation and the control of ice ages. *Palaeogeography, Palaeoclimatology, Palaeoecology*, **50**, 323–332.

Wang, B., Ding, Q., and Joseph, P. V. (2009). Objective definition of the Indian summer monsoon onset. *Journal of Climate*, **22**(12), 3303–3316.

Wang, C. and Picaut, J. (2004). *Understanding ENSO Physics: A Review*. Wiley.

Wang, R., Dearing, J. A., Langdon, P. G., et al. (2012). Flickering gives early warning signals of a critical transition to a eutrophic lake state. *Nature*, **492**(7429), 419–422.

Wang, Y., Gozolchiani, A., Ashkenazy, et al. (2013). Dominant imprint of Rossby waves in the climate network. *Physical Review Letters*, **111**, 138501.

Wang, Y., Gozolchiani, A., Ashkenazy, Y., and Havlin, S. (2016a). Oceanic El-Niño wave dynamics and climate networks. *New Journal of Physics*, **18**, 033021.

Wang, Y., Zhou, D., Bunde, A., and Havlin, S. (2016b). Testing reanalysis data sets in Antarctica: trends, persistence properties, and trend significance. *Journal of Geophysical Research-Atmospheres*, **121**(21), 12839–12855.

Wasserstein, R. and Lazar, N. A. (2016). The ASA's statement on p-values: context, process, and purpose. *American Statistics*, **70**, 129–133.

Watts, D. J. and Strogatz, S. H. (1998). Collective dynamics of 'small-world' networks. *Nature*, **393**(6684), 440–442.

Waugh, D. W. and Abraham, E. R. (2008). Stirring in the global surface ocean. *Geophysical Research Letters*, **35**, L20605.

Webster, P. (1998). Monsoon: processes, predictability, and the prospects for prediction. *Journal of Geophysical Research*, **103**(14), 14451–14510.

Webster, P. J. (2013). Improve weather forecasts for the developing world. *Nature*, **493**, 17–19.

Weijer, W. and Dijkstra, H. A. (2001). A bifurcation study of the three-dimensional thermohaline ocean circulation: the double hemispheric case. *Journal of Marine Research*, **59**, 599–631.

Weijer, W., Maltrud, M. E., Hecht, M. W., Dijkstra, H. A., and Kliphuis, M. A. (2012). Response of the Atlantic Ocean circulation to Greenland Ice Sheet melting in a strongly-eddying ocean model. *Geophysical Research Letters*, **39**(9), L09606.

Wiggins, S. (2005). The dynamical systems approach to Lagrangian transport in oceanic flows. *Annual Review of Fluid Mechanics*, **37**, 295–328.

Willett, C. S., Leben, R. R., and Lavín, M. F. (2006). Eddies and tropical instability waves in the eastern tropical Pacific: a review. *Progress in Oceanography*, **69**, 218–238.

Wu, L., Liu, Z., Gallimore, R., et al. (2003). Pacific decadal variability: the tropical Pacific mode and the North Pacific mode. *Journal of Climate*, **16**(8), 1101–1120.

Wu, R., Chen, J., and Chen, W. (2012). Different types of ENSO influences on the Indian summer monsoon variability. *Journal of Climate*, **25**(3), 903–920.

Yamasaki, K., Gozolchiani, A., and Havlin, S. (2008). Climate networks around the globe are significantly affected by El Niño. *Physical Review Letters*, **100**(22), 228501.

Yan, W., Woodard, R., and Sornette, D. (2010). Diagnosis and prediction of tipping points in financial markets: crashes and rebounds. *Physics Procedia*, **3**(5), 1641–1657.

Yeh, S.-W., Kug, J.-S., Dewitte, B., et al. (2009). El Niño in a changing climate. *Nature*, **461**(7263), 511–514.

Yuan, Y. and Yan, H. (2013). Different types of La Niña events and different responses of the tropical atmosphere. *Chinese Science Bulletin*, **58**(3), 406–415.

Zappala, D., Barreiro, M., and Masoller, C. (2016). Global atmospheric dynamics investigated by using Hilbert frequency analysis. *Entropy*, **18**(11), 408.

Zappala, D., Barreiro, M., and Masoller, C. (2018). Quantifying changes in spatial patterns of surface air temperature dynamics over several decades. *Earth System Dynamics*, **9**, 383–391.

Zebiak, S. E. and Cane, M. A. (1987). A model El Niño-Southern Oscillation. *Monthly Weather Review*, **115**, 2262–2278.

Zhang, C. (2005). Madden-Julian Oscillation. *Reviews in Geopyhsics*, **43**(2004), 1–36.

Zhang, R. and Delworth, T. L. (2006). Impact of Atlantic multidecadal oscillations on India/Sahel rainfall and Atlantic hurricanes. *Geophysical Research Letters*, **33**, L17712.

Zhou, D., Gozolchiani, A., Ashkenazy, Y., and Havlin, S. (2015). Teleconnection paths via climate network direct link detection. *Physical Review Letters*, **115**, 268501.

Zou, J., Han, Y., and So, S.-S. (2009). Overview of artificial neural networks. In D. Livingstone, editor, *Artificial Neural Networks*, pages 14–22. Springer.

Copyright Acknowledgments

We gratefully acknowledge the following copyright holders who have kindly provided permission to reproduce the figures indicated. Sources of all figures are also referenced in each figure caption.

Chapter 3: Fig. 3.7 by AIP (Arizmendi et al., 2017).

Chapter 4: Fig. 4.1 by AIP (Donges et al., 2015). Fig. 4.2 by AIP (Barreiro et al., 2011). Figs. 4.4 and 4.5 by AIP (Deza et al., 2015). Figs. 4.6–4.8 by Nature Publishing (Tirabassi and Masoller, 2016). Fig. 4.10 by AIP (Froyland et al., 2014). Fig. 4.11 by AGU (Rossi et al., 2014).

Chapter 5: Fig. 5.1 by AIP (Donges et al., 2015). Figs. 5.2–5.5 by AIP (Feng and Dijkstra, 2017). Figs. 5.6–5.9 by EGU (Ihshaish et al., 2015).

Chapter 6: Figs. 6.1 and 6.2 by AIP (Arizmendi and Barreiro, 2017). Figs. 6.3–6.6 by EGU (Arizmendi et al., 2014). Figs. 6.7–6.9 by AIP (Wang et al., 2013). Figs. 6.13 and 6.14 by Springer (Rehfeld et al., 2012). Figs. 6.15 and 6.16 by Springer (Malik et al., 2010). Figs. 6.17 and 6.18 by EGU (Stolbova et al., 2014). Fig. 6.19 and 6.20 by Wiley (Tirabassi et al., 2015a). Figs. 6.21 and 6.22 by AMS (Martin-Gomez et al., 2016).

Chapter 7: Figs. 7.1–7.3, and 7.4 by IOP (Wang et al., 2016a). Figs. 7.6–7.8 by AGU (Feng and Dijkstra, 2014). Figs. 7.9–7.11, 7.12–7.15 by AIP (Ser-Giacomi et al., 2015b). Figs. 7.16–7.19 by AIP (Ser-Giacomi et al., 2015c).

Chapter 8: Figs. 8.1 and 8.2 by OUP (Bathiany et al., 2016). Figs. 8.3–8.7 by AGU (van der Mheen et al., 2013). Figs. 8.4, 8.8, 8.9, and 8.10 by AGU

(Feng et al., 2014). Figs. 8.11–8.17 by Elsevier (Tirabassi et al., 2014). Figs. 8.18–8.20 by Nature Publishing (Rodríguez-Méndez et al., 2016).

Figs. 9.1–9.4 by AGU (Stolbova et al., 2016). Fig. 9.5 by US National Academy of Sciences (Ludescher et al., 2014).

Index